METHODS in MICROBIOLOGY

METHODS in MICROBIOLOGY

Edited by

T. BERGAN

Department of Microbiology,
Institute of Pharmacy and Department of Microbiology,
Aker Hospital, University of Oslo,
Oslo, Norway

J. R. NORRIS

Agricultural Research Council,
Meat Research Institute,
Bristol, England

Volume 13

 1979

ACADEMIC PRESS

A Subsidiary of Harcourt Brace Jovanovich, Publishers
London · New York · Toronto · Sydney · San Francisco

ACADEMIC PRESS INC. (LONDON) LTD
24–28 Oval Road
London NW1

U.S. Edition published by
ACADEMIC PRESS INC.
111 Fifth Avenue
New York, New York 10003

British Library Cataloguing in Publication Data

Methods in Microbiology.
Vol. 13
1. Microbiology–Technique
I. Bergan, T
II. Norris, John Robert
576′.028   QR65   68–57745

ISBN: 0–12–521513–4

PRINTED IN GREAT BRITAIN BY
ADLARD AND SON LIMITED
DORKING, SURREY

LIST OF CONTRIBUTORS

J. H. Bates, *Veterans Administration Hospital and University of Arkansas School of Medicine, Little Rock, Arkansas 72201, U.S.A.*

T. Bergan, *Department of Microbiology, Institute of Pharmacy and Department of Microbiology, Aker Hospital, University of Oslo, Oslo, Norway.*

H. W. B. Engel, *Rijks Instituut Voor de Volksgezondheid, Bilthoven, Netherlands.*

H. Ernø, *FAO/WHO Collaborating Centre for Animal Mycoplasmas, Institute of Medical Microbiology, University of Aarhus, Denmark*

E. A. Freundt, *FAO/WHO Collaborating Centre for Animal Mycoplasmas, Institute of Medical Microbiology, University of Aarhus, Denmark*

K. Höhne, *University of Würzburg, Federal Republic of Germany*

Mary Ann Gerencser, *Department of Microbiology, Medical Centre, West Virginia University, Morgantown, W.V. 26506, U.S.A.*

Ruth M. Lemcke, *The Lister Institute of Preventive Medicine, University of London, England*

D. E. Mahony, *Department of Microbiology, Faculty of Medicine, Dalhousie University, Halifax, Nova Scotia, Canada*

Paula Maximescu, *The Cantacuzino Institute, Bucharest, Romania*

Eugenia Meitert, *The Cantacuzino Institute, Bucharest, Romania*

W. B. Redmond, *Veterans Administration Hospital and Emory University School of Medicine, Atlanta, Georgia 30333, U.S.A.*

Alice Saragea, *The Cantacuzino Institute, Bucharest, Romania*

W. B. Schaefer, *National Jewish Hospital and Research Center, Denver, Colorado 80206, U.S.A.*

H. P. R. Seeliger, *University of Würzburg, Federal Republic of Germany*

Iv. Stoev, *Institute Against Swine Diseases, Vratza, Bulgaria*

H. Ullmann, *Department of Microbiology, Institute of Hygiene, University of Tübingen, Silcherstr. 7, 74 Tübingen, West Germany*

PREFACE

Volume 13 of "Methods in Microbiology" is a continuation of the series of Volumes describing the methods available for typing major pathogens by serology, bacteriophage typing, and bacteriocin typing.

This Volume has a Chapter dealing with bacteriophage typing and bacteriocin typing of clostridia. Chapters II and III detail the typing of *Listeria monocytogenes* and *Erysipelothrix insidiosa*. In Chapter IV, the bacteriology and epidemiology of diphtheria bacilli are discussed. Chapter V presents bacteriophage typing of the *Shigella* species. Then follow serology and phage typing of *Actinomyces*, *Arachnia*, and *Mycobacterium*. Chapter IX describes the characterisation of *Mycoplasma* and Chapter X deals with *Campylobacter*.

We have been fortunate in having had the good collaboration of internationally recognised authorities. The Chapter on phage typing of *Shigella* originally had a contributor who unfortunately had to withdraw because of other commitments. Since this happened two years after the deadline and we could not proceed without this important topic, we prepared it ourselves in consultation with experts in the field. We are grateful for this support.

We hope that Volumes 10–13 will serve as useful references for microbiologists who need to know most methods available to type bacteria in the context of their epidemiological relationships. The Volumes were originally planned five years ago. Since then there have been significant developments in the characterisation of several groups of bacteria. When the time is ripe for it, we hope to collect the methodology for typing further microbes in a later volume of "Methods in Microbiology".

T. BERGAN
J. R. NORRIS

August, 1979

CONTENTS

CONTENTS OF PUBLISHED VOLUMES

Bacteriocin, Bacteriophage and other Epidemiological Typing Methods for the Genus *Clostridium*

D. E. MAHONY

Department of Microbiology, Faculty of Medicine, Dalhousie University, Halifax, Nova Scotia, Canada

I. INTRODUCTION

Many of the bacteria classified as clostridia are harmless. Their habitat is in the soil and intestinal tract of man and animals; yet some of these same species, under appropriate conditions, are capable of causing destructive and sometimes fatal disease such as botulism, gas gangrene, clostridial cellulitis, tetanus and the non-fatal but frequently encountered *Clostridium perfringens* food poisoning. Because such organisms are ubiquitous, the possible typing of these bacteria presents an intriguing problem. Just what is the origin of a hospital acquired case of gas gangrene—spore-laden dust in an operating theatre, inadequately sterilised instruments or materials, or the patient's own flora? What is the source of an enterotoxin-producing strain of *C. perfringens* in an outbreak of food poisoning? Similar questions might be asked of botulism and tetanus. Means are available for answering a few of these questions.

Historically, typing of only pathogenic bacteria has been considered and this Chapter will be restricted to the pathogenic species of *Clostridium* associated with gas gangrene, botulism, tetanus and food poisoning. Unfortunately, little work has been done on bacteriocin or bacteriophage typing of clostridia and perhaps one of the major objectives of this contribution is to review much of our knowledge on clostridial bacteriocins and bacteriophages with a view to suggesting potential typing schemes for various species. An appreciation of some of the difficulties which might impede the development of typing schemes should also be considered. A bacteriocin typing scheme for *C. perfringens* developed in our laboratory is presented as one possible model for typing the clostridia.

II. APPROACHES TO BACTERIOCIN TYPING

There are two approaches to bacteriocin typing. The first involves examining the ability of isolates to produce bacteriocins active against a standard set of indicator bacteria. The host range of any bacteriocin produced by the test strains then defines the typing pattern. One method of performing this test is to inoculate a plate of an appropriate medium with a single wide diametrical streak of the test organism. After growth (18 or more hours) the bacteria are scraped off the plate and the plate exposed to chloroform vapours. Subsequently the plate is aired to remove the chloroform before the standard indicator bacteria are streaked across the plate at right angles to the original streak. The plates are incubated to allow growth of the indicator bacteria. If bacteriocin has been produced by the test strain, the growth of one or more of the indicator strains should be inhibited where the streakings intersect.

The second approach tests the susceptibility of unknowns to a standard set of bacteriocins. This approach is less time consuming than the former since only one incubation period is required. The test organism is simply inoculated confluently over the surface of an agar plate as is done for antibiotic sensitivity testing, and defined volumes of the different bacteriocins are dropped on to sectors of the seeded plate. The pattern of inhibition which develops after subsequent growth of the lawn defines the bacteriocin type of the organism.

There are factors in favour of each method and further detail may be found in the literature (Shannon, 1957; MacPherson and Gillies, 1969).

III. GROWTH AND IDENTIFICATION OF CLOSTRIDIA

Many excellent references may be found on the growth and identification of the genus *Clostridium* (Dowell and Hawkins, 1974; Holdeman and Moore, 1972; Smith, 1975; Smith and Holdeman, 1968; Willis, 1969). In most laboratories, cooked meat or thioglycollate-containing broths and the Gaspak anaerobic jar (Bioquest, B.B.L.) are used in cultivation. It is beyond the scope of this Chapter to discuss the many variations in growth techniques and identification procedures.

The clostridia are usually Gram-positive, spore-forming, catalase-negative anaerobic bacteria. Some species are much more aerotolerant than others, but anaerobic techniques must be applied to assure the isolation of any member of the genus. *C. perfringens* is often reluctant to produce spores and is also non-motile. In addition, a double zone of haemolysis surrounds the smooth, entire-edged colonies of this organism on human or sheep blood agar plates; the inner zone of beta haemolysis is produced by the theta toxin while the outer weaker zone is caused by the alpha toxin. Some isolates, however, may lack the theta toxin or produce rough, irregular colonies which will confuse identification. The organism can be specifically identified by its saccharolytic activity on various sugars and neutralisation of its toxins with specific antisera. Some clostridia may only be specifically identified by toxin neutralisation tests, e.g. *C. botulism* and *C. tetani*. Newer analytical tools such as gas–liquid chromatography have enhanced the identification of many anaerobic bacteria by detecting specific organic acids produced by the fermentation of defined carbohydrates. This technology is described in detail by Holdeman and Moore (1972).

IV. NATURE AND MODE OF ACTION OF CLOSTRIDIAL BACTERIOCINS

A. Bacteriocins of *C. perfringens*

1. *Incidence of bacteriocinogenicity in* C. perfringens

Although the first observation of bacteriocin-like activity in this species was made by Smith in 1959 while studying lysogeny, the first major screening of large numbers of *C. perfringens* strains for bacteriocins was carried out by Sasarman and Antohi (1963). From 24 strains tested against one another, four bacteriocin-producing strains were identified: strain 1241 (toxin type E), 2077 (type F) and P24 and 353 (non-typed strains). Undoubtedly more bacteriocins would have been detected in supernatant fluids of cultures had these investigators not heated the supernatant fluid at 70°C for 1 h before spotting on to the plates. Using these bacteriocins, 237 strains of *C. perfringens* were tested for susceptibility. Sasarman and Antohi concluded that 81·4% of their strains were typable and 12 different typing patterns were obtained. Sixty-seven per cent of the strains fell within the first five types while the remaining seven types represented small numbers of strains. There was no apparent association between either the production of, or susceptibility to, bacteriocins and the classical toxin typing of these organisms.

This latter fact was also confirmed by Uchiyama (1966a) when 74 strains of *C. perfringens* were divided into four groups based upon their ability to produce bacteriocins. Twelve bacteriocin-producing strains were detected. In testing 14 strains of toxin types B, C, D, E and F, there was no type specificity in the action of bacteriocin. Only *C. perfringens* was sensitive to these bacteriocins during a survey of a number of Gram-positive and Gram-negative organisms. Also Tubylewicz (1966a) reported that 5 of 35 strains of *C. perfringens* Type A were bacteriocinogenic. A further study by Sasarman and Antohi (1968) showed that of 251 strains of *C. perfringens* examined, 12 elaborated bacteriocins (no heat treatment of the supernatant fluids). Hirano *et al.* (1972) showed that 14 of 176 strains produced bacteriocin while Mahony and Butler (1971) detected four bacteriocinogenic strains amongst 33 tested. In 1974, Mahony used ten bacteriocins to type 274 cultures of *C. perfringens* and observed 50 different bacteriocin typing patterns. Currently, 65 typing patterns have been observed (Mahony and Swantee, 1978).

The naming of bacteriocins of *C. perfringens* is still unresolved. Some authors call such bacteriocins "perfringocins", while others have chosen the term "welchicins".

2. *Production and detection of bacteriocins*

The methods used to produce and detect bacteriocins of *C. perfringens* are similar to those employed for bacteriocins of other bacteria with the exception that growth of the organisms is carried out under anaerobic conditions.

Spontaneous production of bacteriocins has been reported by Uchiyama (1966a), Sasarman and Antohi (1968), Mahony and Butler (1971), and Hirano and Imamura (1971). Tubylewicz (1966a) used ultraviolet light induction to get an increased production of the active principle. Mahony and Butler (1971) were unable to induce one strain (strain 28), although one other strain was later shown to produce a higher titre after treatment with UV light (Mahony, 1973). In 1977, Mahony reported the induction of six of ten bacteriocinogenic strains of *C. perfringens* using Mitomycin C. The response to this antibiotic was peculiar in that induction occurred only after removal of Mitomycin C from the treated culture. Cell death was associated with induction of bacteriocin.

Sasarman and Antohi (1971) showed that the majority of their welchicin-producing strains gave the highest yield of bacteriocin when grown at the organism's optimum growth temperature (37°C). At 50°C bacteriocin was not produced, and at room temperature production was reduced. Other investigators have reported a temperature of 37°C for bacteriocin production. Mahony and Butler (1971) studied the production of bacteriocin 28, and found that optimum production was obtained in the late log phase between 2 and 3 h after initiation of growth. Uchiyama (1966b) found that maximum titre was achieved after 3 h of growth. In contrast, a bacteriocin of *C. perfringens* type A described by Clarke *et al.* (1975) was released into the culture fluid in the stationary phase of bacterial growth. With the method used by Tubylewicz (1966a) bacteriocin production was assayed after overnight incubation.

Ionesco *et al.* (1974) have described independent synthesis of bacteriocin and bacteriophage in one strain of *C. perfringens*. Both factors could be induced by UV light. Bacteriocin production demonstrated a lag period of 70 min while the maximum titre was achieved by 180 min after induction.

3. *Host range*

All the bacteriocins obtained by Tubylewicz (1966a) were active against *C. perfringens* and, under aerobic conditions, did not inhibit the growth of 37 bacterial strains belonging to ten other genera. Similar results have been published by Uchiyama (1966a). Sasarman and Antohi (1968) reported a wider range of activity for a bacteriocin (welchicin A) produced by *C. perfringens* type E (strain 1241). In addition to activity against 100% of *C. perfringens* tested including the producing strain, they found activity against *Clostridium oedematiens*, *Clostridium bifermentans*, and *Clostridium fallax*. Sensitivity to this bacteriocin (welchicin A) was also demonstrated by various species of the genus *Bacillus* and subsequently (1971) these workers also reported activity against *Corynebacterium diphtheriae*, staphylococci and streptococci. Sensitivity to a similar host range was later reported by the same authors (1970), where they showed that welchicin B (produced by *C. perfringens* type D, strain 366) was also active on Gram-positive organisms outside the family *Bacillaceae*. Sasarman and Antohi (1971), described the existence of a new third class of welchicin which was active on *C. perfringens* and *Streptococcus*, but not on many strains of the other genera listed above.

A perfringocin active on *C. pasteurianum* has been described by Clarke *et al.* (1975).

Resistance of the producing strain to its own bacteriocin is a usual characteristic of bacteriocinogenic organisms, although one of the five bacteriocins studied by Tubylewicz (1966a), bacteriocin d, was found to be weakly active against the strain 496 from which it was derived, and one of the strains reported by Sasarman and Antohi (1968), strain 1241, type E, produced a welchicin A active against all the strains tested, including the producer organism.

4. *Chemical nature and antigenicity*

Most bacteriocins have been thermolabile (Tubylewicz and Uchiyama, 1966b; Mahony and Butler, 1971). Temperature sensitivity of the welchicins has been variable (Sasarman and Antohi, 1971). The bacteriocins active only against *C. perfringens* and *Streptococcus* showed reduced activity or complete inactivation above 50°C, while the activity of those inhibitory only to *C. perfringens* or to *Clostridium*, *Bacillus*, and *Streptococcus* was not reduced by temperatures up to and including 90°C for a period of 2 h. Hirano and Imamura (1972c) reported two types of *C. perfringens* bacteriocin—one heat stable (type S) and the other thermolabile (type L). The perfringocin described by Clarke *et al.* (1975) is heat stable (100°C, 10 min). We currently have two bacteriocins which are similarly heat stable (Mahony, unpublished data).

The sensitivity of bacteriocin to proteolytic enzymes was tested by Tubylewicz (1966b). Both trypsin and papain completely inactivated perfringocin. Sasarman and Antohi (1971), Mahony and Butler (1971) and Clarke (1975) reported similar results. The L type of Hirano and Imamura (1972a, b, c) was trypsin sensitive, but the S type was not affected by trypsin. Recently, we have observed two trypsin-resistant bacteriocins of *C. perfringens* (Mahony, unpublished data).

Tubylewicz (1966b) studied the influence of UV rays on bacteriocins. He found that, within a range of 537–12,888 erg/mm², UV light did not bring about any detectable change in the activity of bacteriocins *a* and *b*. Similar findings were reported by Hirano and Imamura (1972) for both their types of bacteriocins.

The influence of pH on bacteriocin activity was studied by Tubylewicz (1966b) who noted stability over a pH range of 4–10. Uchiyama (1966b) reported that their bacteriocins were inactivated below pH 3 and over pH 9. The S type bacteriocin of Hirano and Imamura (1972a, b, c) was active at pH values 3 and 9, while the L type displayed a narrower activity range. The bacteriocin of Clarke *et al.* (1975) was stable from pH 2 to 12 and we have observed two bacteriocins which are similarly stable (Mahony, unpublished data). Tubylewicz (1966b) found the bacteriocins to be non-dialysable; activity was able to pass through Seitz E.K. filters. Hirano and Imamura (1972a, b, c) reported that some activity of the S type was lost on dialysis, while Uchiyama (1966b) found that the substance was not filterable through Seitz and Chamberland L3 bacterial filters. Also the bacteriocin was not sensitive to chloroform. Mahony and Butler (1971) found that their bacteriocin 28 was not affected by commercially available antitoxin.

The first attempt to purify the bacteriocins of *C. perfringens* and to determine their chemical composition was carried out by Tubylewicz (1966b). The procedure was modified in 1968. It involved precipitation of bacterial supernatant fluids with ammonium sulphate (60–70% of saturation) followed by ether–ethanol extraction. Active perfringocin was obtained from the supernatant fluids of four bacteriocinogenic cultures following this method. All the preparations of active perfringocins studied by Tubylewicz (1968) were described as protein–saccharide complexes.

Of the two types of bacteriocin studied by Hirano and Imamura (1972c), the L type appeared to be protein in nature—possibly a lytic enzyme involved in bacteriophage replication. The S type seemed to be a more stable toxic substance with a lower molecular weight. We (Mahony, unpublished data) have shown by column chromatography using Sephadex G150 that bacteriocin 28 of *C. perfringens* has a molecular weight of about 100,000. The biological activity of the bacteriocin becomes very unstable as purification progresses. The perfringocin described by Clarke *et al.* (1975) seemed to be a pure protein with a molecular weight of about 76,000 as shown by gel electrophoresis.

Tubylewicz (1970) studied the antigenic properties of four bacteriocins using several serological methods. Immune sera were obtained by immunisation of rabbits with partially purified bacteriocins. The bacteriocin activity was neutralised by such immune sera. The results of double diffusion tests in gels indicated that bacteriocins *a*, *b*, *c* and *d* are related antigenically, although not identical. Immunoelectrophoresis of bacteriocins *a* and *b* caused them to move towards the negative pole. Two components of these bacteriocins had different mobilities. Complement fixation tests were performed, but were not specific in that cross-reactions occurred.

5. *Mode of action*

Studies on the mode of action of the bacteriocins of *C. perfringens* have only recently been reported. Mahony and Butler (1971) followed the growth of a broth culture of a sensitive strain of *C. perfringens* after the addition of bacteriocin produced by strain 28. The degree of bacterial inhibition was dependent on the concentration of the bacteriocin, and the growth phase of the indicator. There was an apparent lag in action when the bacteriocin was added to a young 30 min culture, but when the bacteriocin was added to a culture at 120 min or later, growth was halted much more quickly. Broth cultures treated with bacteriocin continued to produce gas, although there was marked inhibition of growth as measured by viable count and optical density. There was no complete clearing of the treated culture. Bacteriocin 28 appears to act on sensitive indicator strains of *C. perfringens* through inhibition of cell wall synthesis or removal of the pre-existing cell wall. This is suggested by the development of spheroplasts from bacteriocin-treated cells in liquid media, and of L-form-type colonies on solid media (Mahony *et al.*, 1971). Other bacteriocins which we have studied inhibit protein, DNA or RNA synthesis or a combination thereof (Mahony, unpublished data).

In studying the mode of action of their lytic type of bacteriocin, Hirano and Imamura (1972a, c) compared it to that of lysozyme, since the lytic factor caused a loss of the Gram-staining property and disintegration of sensitive cells. However, it differed from lysozyme in being active only on strains of *C. perfringens* and not on any other Gram-positive species tested. Both living and dead cells were affected by the bacteriocin with cells killed by chloroform treatment lysing more rapidly, possibly due to loss of some protective mechanism found on the outside of the cell wall of the living bacteria. Clarke *et al.* (1975) indicated a lytic action of their perfringocin on cells of *C. pasteurianum*.

Ionesco and Bouanchaud (1973) and Ionesco *et al.* (1976) have provided evidence that bacteriocin production in *C. perfringens* type A (strain BP6K-N_5) is controlled by a plasmid with a molecular weight of 5.7×10^6 daltons (Ionesco *et al.*, 1976). Sebald and Ionesco (1974) showed that the initiation of spore germination in this strain was inhibited by the N_5 bacteriocin and subsequently Wolff and Ionesco (1975) purified the bacteriocin by ion exchange and Sephadex chromatography. The protein of 82,000 molecular weight ran as a single band in polyacrylamide gels. Ionesco and Wolff (1975) describe the mode of action of bacteriocin N_5. Simultaneous inhibition of DNA, RNA and protein synthesis occurred in sensitive cells treated with this bacteriocin.

To summarise published results to date on the bacteriocins of *C. perfringens* we may say: some are spontaneously produced while others can be induced by UV light or Mitomycin C. No bacteriophage-like particles have been reported with bacteriocin activity. Bacteriocins showing various host ranges have been demonstrated; some intraspecies specific; some intrageneric; and others intergeneric. Two of 25 bacteriocins described are active against the producing strain. Some are sensitive to proteolytic enzymes, high temperatures and extremes of pH, while others are resistant. One type appears to be dialysable, the other is not. Attempts to purify the bacteriocins of *C. perfringens* define them as proteins or protein–saccharide complexes. The bacteriocins seem to represent a heterogeneous group of molecules and more experimental work is required to adequately describe their chemistry and mode(s) of action.

B. Bacteriocins of *C. botulinum*

Bacteriocins of *C. botulinum* were first reported in 1966 by two different groups of investigators. Kautter *et al.* (1966) described bacteriocins produced by non-toxigenic strains of *C. botulinum* (resembling toxigenic type E organisms) which were bacteriolytic for vegetative cells and prevented germination of spores of *C. botulinum* type E. Of the species examined only *C. botulinum* type E and, to a lesser extent, *C. perfringens* and *C. acetobutylicum* were sensitive to this type of bacteriocin which was named boticin E. Boticin E was heat stable, dialysable, unaffected by chloroform and inactivated by trypsin. Ethyl alcohol and acetone precipitates of the bacteriocin were fully active but trichloroacetic acid precipitates were only partially so.

One boticin E (S5) was composed of two components (Ellison and Kautter, 1970): a lower molecular weight boticin ranging between 5000 and 30,000 daltons and accounting for about 80% of the inhibitory activity and a high molecular weight component in excess of 40×10^6 daltons. Both components were resistant to boiling and sensitive to trypsin. The lower molecular weight fraction was most susceptible to trypsin. The lower molecular weight boticin was stable over a pH range of 1·1–9·5. Both boticins prevented the outgrowth of spores and had bactericidal activity for vegetative cells of the indicator strain of *C. botulinum* type E, 070. The bacteriocin stopped cell growth and caused an extreme drop in viability within 9 min. The action of the small bacteriocin could not be reversed by subsequent addition of trypsin suggesting rapid uptake of the bacteriocin. Methods of purification and handling of the boticins described in this Chapter should be of value in further studies of these molecules.

Electron microscopy of the indicator strain after treatment with the lower molecular weight boticin (now named S5) (Ellison *et al.*, 1971) revealed aggregation of DNA, apparent structural rearrangement of mesosomes and eventual dissolution of cell contents leaving bacterial ghosts with intact cell walls and remnants of the cytoplasmic membrane and internal structures.

Only a few strains of those clostridia which are non-toxigenic but otherwise identical to *C. botulinum* type E produce boticins (Anastasio *et al.*,

1971). Proteolytic strains of types A, B and F were resistant to boticin action. The boticin adsorbed only to sensitive organisms and acted only on growing cells; oxygenation of the culture prevented lysis by bacteriocin. The fact that protoplasts and L-forms of *C. botulinum* had the same susceptibility pattern as the parent suggested that cell wall receptor sites were not critical for bacteriocin action. The bacteriocin action on sensitive spores would seem to result from inhibition of germination rather than of outgrowth.

In contrast to the host range described above, Ueda and Takagi (1972) found that high molecular weight, sedimentable bacteriocins induced from proteolytic type B strains were active against vegetative cells and germination of spores of proteolytic *C. botulinum* strains.

Lau *et al.* (1974) described another bacteriocin of a non-toxigenic type E strain which resembled a phage tail. This boticin "P" could be purified by ammonium sulphate precipitation, Sephadex gel filtration and ultracentrifugation. The bacteriocin exerted a static effect on vegetative growth and outgrowth of spores, but did not inhibit the initial events of germination. Only *C. botulinum* types B and E were susceptible to this bacteriocin although only a limited number of strains of types A, B, C, D, E and F were tested.

The last work to be cited refers to one of the original reports on bacteriocins of *C. botulinum* made by Beerens and Tahon (1967) in Moscow in 1966. Twelve strains of *C. botulinum* (types A, B and C) and 29 strains of *C. sporogenes* were examined. The observations were as follows: 10/12 strains of *C. botulinum* produced substances active against certain other strains of *C. botulinum*; 11/12 strains were sensitive to one or more strains different from themselves. No correlation was found between the sensitivity of a strain and its toxigenic type nor was there any correlation between bacteriocin production and toxigenicity. When the results presented by Beerens and Tahon are examined for bacteriocin typing, amongst the 12 strains tested, 11 different bacteriocin typing patterns are observed. Bacteriocin typing of *C. botulinum* might indeed be feasible.

C. Other bacteriocins

Schallehn (1975) found that two of 17 strains of *Clostridium septicum* were bacteriocinogenic. One bacteriocin was studied and found to be spontaneously produced during the logarithmic growth phase of the bacterium. This bacteriocin was sensitive to heat and proteolytic enzymes, resistant to chloroform and not inducible by Mitomycin C. The molecular weight was determined to be 20,000 by Sephadex filtration. In 1976, Schallehn and Krämer examined the mode of action of the bacteriocin and found a total cessation of RNA and protein synthesis followed by a cessation of DNA synthesis in the bacteriocin-treated indicator strain of *C. septicum*.

A bacteriocin of the non-pathogenic organism, *C. butyricum*, was described by Clarke *et al.* (1975) and Clarke and Morris (1975). Clarke and Morris (1976) showed that this trypsin-sensitive bacteriocin had activity against a number of clostridial species resulting in death but not lysis of indicator cells. Synthesis of RNA, DNA and protein was inhibited and membrane damage was apparent as indicated by ion leakage.

Bacteriocins of *C. sporogenes* (Betz and Anderson, 1964) are briefly discussed in the Section on bacteriophages of *C. sporogenes*.

V. BACTERIOPHAGES OF CLOSTRIDIUM

A. Bacteriophages of *C. perfringens* and other clostridia associated with gas gangrene

In the early 1940s, Russian workers isolated bacteriophages active against *C. perfringens*, *C. oedematiens*, *C. septicum*, *C. histolyticum* and *C. putrificus*. These phages were immunogenic, caused complete lysis of indicator bacteria and lysed 96% of *C. perfringens* strains tested. Such phages were used with apparent success in the phagotherapy of gas gangrene in humans. A review of the Russian literature on this subject may be found in the bibliography of Spencer (1953).

Tsyp *et al.* (1944) isolated *C. perfringens* bacteriophages from sewage waters. These lysed type C organisms, but other types were not easily lysed. In 1964, Bychkov described development of a *C. perfringens* bacteriophage with a head 50–60 nm in diameter and a tail of 15–20 nm, and a 45–50 min latent period.

A second group of papers dealing with bacteriophages of *C. perfringens* originated in France. In 1947, Kréguer *et al.* isolated bacteriophages active against one of four strains of *C. perfringens* type A. Guélin (1949) isolated phage "M" which had a latent period of 15–20 min followed by a rising titre of infectious particles until 60–90 min after infection. Guélin also noted that greater phage yields were obtained by using a low multiplicity of infection. Guélin and Kréguer (1950) could not demonstrate any change in toxin production in *C. perfringens* type A upon infection with a phage designated 80b. Other papers by Guélin (1950a, b, 1953) discuss the isolation of phages from river water and some rather inconclusive results on host cell morphologies. The first electron microscopy on clostridial phages was reported by Elford *et al.* in 1953, when two phages of *C. perfringens* were described. Phage "M" possessed a head diameter of 35 nm while phage "W" had a head diameter of 60 nm. Both had tail structures of 120 × 15 nm.

In 1953, Sames and McLung isolated eight phages active on *C. perfringens*, but one of the larger surveys of such bacteriophages is to be found in the doctoral thesis of Sames (1956). This work described the isolation of 33 bacteriophages belonging to three serological groups and four plaque types. One-step growth curves were described with burst sizes ranging from 150 to 1900 plaque forming units (pfu).

In 1959, Smith studied *C. perfringens* types A, B, C, D, E and F (now considered type C) for lysogeny and found that 12 of 49 strains of type A, ten of 31 type B and ten of 26 type C and none of 38 type D, five type E or three type F were lysogenic. The temperate phages attacked only strains of the same type as that from which they originated. Virulent phages, which lysed strains belonging to types A, B, C, D and F, were also isolated although a high proportion of bacterial strains remained resistant to all the phages isolated. Induction of the lysogenic strains was successful with UV light, nitrogen mustard and thioglycollate.

Gáspár and Tolnai (1959) performed one of the more informative characterisations of a virulent *C. perfringens* phage when they studied one of Guélin's phages designated 808/3a. A device was designed in this experiment for obtaining one-step growth curves of the phage. The phage-bacteria culture was constantly agitated with bubbling nitrogen and samples were moved at various times for plating. A modified Wilson–Blair medium was employed for plaque counts and a thick overlay of agar containing phage and bacteria was used to produce anaerobic conditions without the aid of an anaerobic jar. The latent period of this phage was 45 min and the burst size, 452 pfu. Adsorption and immunological experiments were also conducted. In a second paper (Gáspár, 1960) describing the effect of aeration upon phage production, it was shown that aeration of the culture any time up to 45 min after infection prevented phage multiplication, whereas aeration at 45 min had no marked effect on phage production. Aeration of the culture at the time of infection greatly prolonged the latent period, the lag depending on how long aerobic conditions were maintained. Aeration was shown to have no adverse effect upon phage viability itself.

Vieu *et al.* (1965) described a phage which was 70 nm long and had a polyhedric head 40 nm in diameter and a short tail 30 nm long. The tail portion possessed a contracted plate 8 nm long and 34 nm in diameter containing four to seven distinct structures. The tail core which projected beyond this plate was consistently demonstrated in all preparations.

Ohtomo *et al.* (1966) described a one-step growth experiment on a phage of *C. perfringens* which demonstrated a latent period of 23 min and a burst size of 140–150 pfu. Oxygen inhibited the multiplication of this phage, but the inhibition was reversible with the return of anaerobic conditions.

Imamura (1966a, b) surveyed 140 strains of *C. perfringens* from sewage and faeces for lysogeny. Seventeen per cent of the strains were lysogenic, but the host range of the phages was very narrow. In a later paper Hirano and Imamura (1972b), examining 17 strains of known toxin types A to F, found lysogeny in only types B, C and E. These results differ from Smith's (1959) indicating that lysogeny is not restricted to certain toxin types.

The potential of phage typing *C. perfringens* was shown by Imbert (1968). With a battery of 15 phages isolated from sewer water, 85% of 181 strains of *C. perfringens* could be divided into 51 lysotypes. The most frequent typing pattern represented 13% of the typable cultures, while many of the typing patterns contained only one or two strains.

Hirano and Yonekura (1967) presented electron microscopy data on 15 free phages and 16 temperate phages. All virulent phages had a head diameter of 40 nm and a contractile tail 30–40 nm long, while temperate phages had much longer tail structures of 140–220 nm which were relatively simple in design. A temperate phage of the latter morphology, described by Mahony and Kalz (1968) exhibited a latent period of 45 min and a 17-fold yield. Lysogenisation by this phage resulted in the production of an increased number of heat resistant spores by the lysogenised strain and an altered sporulation time (Stewart and Johnson, 1977). A second phage (Mahony and Easterbrook, 1970) also showed a latent period of 45 min. Head forms were visible intracellularly 30 min post-infection and tails by 35 min. Assembly seemed complete 40 min post-infection. In 1971, Bradley and Hoeniger noted structural changes in *C. perfringens* infected with a short-tailed bacteriophage originally isolated by Vieu *et al.* in 1965. Ionesco *et al.* (1974) demonstrated a lag period of 50 min for phage induced by UV light in a strain which was both lysogenic and bacteriocinogenic. Differences in the production of these two elements were described.

Yonekura *et al.* (1972) described the isolation of 22 virulent bacteriophages of *C. perfringens* from sewage and concluded that the host ranges of nine such phages were quite restricted. Some 50% of the strains were resistant to any phage. The host range was variable partially due to host-induced modification and other unexplained factors. These authors concluded that phage typing of *C. perfringens* was not feasible.

Imamura and Nakama (1974) tried to isolate phages of *C. perfringens* from human intestinal contents and faeces but found phages in only three of 48 samples. These phages possessed the same morphology as that described for temperate phages (Hirano and Yonekura, 1967). Four of 30 isolates of *C. perfringens* were lysogenic and the morphologies of the phages carried by these strains were similar to those of the temperate group. In 1976, Grant and Riemann described four temperate phages isolated from *C. perfringens* type C which belonged to two classes. Three of the phages were homoimmune and serologically related to the other phages. One was inducible with Mitomycin C. The temperate nature of two of the phages was confirmed by serially passing the lysogenic strains through phage-specific antisera. Transduction attempts with these phages were unsuccessful.

Paquette and Fredette (1977) described four temperate phages of *C. perfringens* type A in a study where ten of 23 strains were found to carry phages. One of the phages had complex morphology similar to that described earlier by Vieu *et al.* (1965) and to the free phages of Hirano and Yonekura (1967). The other phages had simple tail structures. The latent periods of three of the phages ranged from 20 to 30 min with bursts of 90–145 pfu. One lysogenic strain was inducible with UV irradiation.

A bacteriophage isolated from one of three lysogenic strains of *C. histolyticum* was described by Guélin *et al.* (1966). Only one of 14 strains was sensitive to this phage. Adsorption of this phage appeared to be slower when heat-killed cells were used. Electron photomicrographs of the phage showed a 80 nm polyhedral head, a 240 nm striated tail, and end plates 30 nm in width.

Popovitch and Sebald (1967) isolated 14 phages active on *C. histolyticum* and used these to type 34 strains of the species. Nine lysotypes were observed; one strain was not typable. Phage stocks had a titre of at least 10^8 pfu. The data presented in this paper have been reconstructed in the light of potential phage typing and are presented in Table I. Some isolated phages are interesting because they have a host range extending to other clostridial species (Sebald and Popovitch, 1967).

Clostridium sporogenes is associated with gas gangrene infections. Frenkel (1940) first described the isolation of a phage of *C. sporogenes* from river water. In 1964, Betz and Anderson isolated four more phages from soil and sewage and studied them with eight phages obtained from Dr L. S. McClung of Indiana University. These 12 phages constituted three groups based upon plaque morphology, host range, receptor sites and serological relationships. None of 25 strains of *C. sporogenes* tested was lysogenic although bacteriocin-like substances were produced by nine and 20 were sensitive to one or more bacteriocins. Six of the above phages were studied by electron microscopy and information related to one-step growth and DNA characterisation was presented (Betz, 1968). Most of these phages had elongated six-sided heads and simple, tubular, non-contractile tails. Latent periods were about 1 h and average yield sizes ranged from 17 to 150 pfu/ml. The guanine plus cytosine ratio of the phage DNA (39·5%) exceeded that of the host bacteria (33·2%). Although these authors were not pursuing a typing scheme, a number of phage and bacteriocin typing patterns might be extractable from their data.

The intracellular development of one of these phages (F1) has been

TABLE I

Lysotypes of 34 strains of *C. histolyticum*[a]

	Phages							
Lysotypes	ω	φ2	6σ	σ2	θtu	θy	πa	Percentage of strains tested in lysotype
I	+	−	−	−	−	−	−	38
II	−	+	+	−	−	−	−	3
III	+	+	−	−	−	−	−	12
IV	+	+	+	−	−	−	−	12
V	+	+	+	+	−	−	−	12
VI	+	+	+	+	+	−	−	12
VII	+	−	−	−	−	+	−	3
VIII	−	−	−	−	−	−	+	3
IX	−	−	−	−	−	−	−	3

+, Inhibition (lysis); −, No inhibition.
[a]Data reconstructed from Popovitch and Sebald (1967).

described by Hoeniger and Bradley (1971) and RNA transcription from the two strands of DNA described by Taylor and Guha (1974, 1975).

Schallehn and Lenz (1975) noted a bacteriophage in the culture fluids of *C. novyi* type A, although they did not find a sensitive indicator strain for this phage. Eklund *et al.* (1974, 1976) also studied bacteriophages of *C. novyi* and Eklund *et al.* (1976) showed that production of alpha toxin of *C. novyi* types A and B was influenced by bacteriophages. When the type A organism lost its phage, the resulting non-toxigenic bacterium more closely resembled *C. botulinum* types C and D than the other types of *C. novyi*.

B. Bacteriophages of other clostridia

1. *Bacteriophages of* C. botulinum

Bacteriophages are commonly associated with *C. botulinum* and phages may, at least in some cases, govern toxin production. In 1968, Vinet *et al.* briefly described a phage of *C. botulinum* type C. Inoue and Iida (1968) described phages induced from strains of *C. botulinum* types A–F by UV light and/or Mitomycin C treatment, grouping them into three morphological categories associated with various toxin types. Eklund *et al.* (1969) found no phages from a non-toxigenic strain resembling type E. Toxigenic strains of types A, B, E and F on the other hand produced such particles.

In 1970, Inoue and Iida showed that if a Mitomycin C-induced toxigenic culture of *C. botulinum* type C was added to a non-toxigenic mutant of the same strain (isolated by treatment of the toxigenic strain with Acridine orange), non-toxigenic bacteria were converted to toxin production. Further evidence supporting the role of phage in toxin production was provided by Inoue and Iida (1971) when they showed that a lysate from a type C strain could convert a non-toxigenic type D strain into a type C toxin-producing strain. Eklund *et al.* (1971) demonstrated the existence of two phages carried by *C. botulinum* type C, only one of which had converting ability. Toxin-producing strains lost toxigenicity upon addition of phage antisera. In 1972, Eklund *et al.* in quantitating *C. botulinum* phages found the composition of the soft agar overlays to be important. This consisted of trypticase, peptone, yeast extract, glucose, 0·7% agar, 2·5% NaCl and 0·5% lactalysate as well as catalase and cysteine hydrochloride. The association of phage and toxigenicity in type D strains was also confirmed. Oguma *et al.* (1973) produced further evidence for phage conversion when phage lysates from already converted strains could convert non-toxic strains of types C and D.

Eklund and Poysky (1974) could interconvert non-toxigenic type C and D strains with phages from toxigenic type C and D and postulated that toxigenic types C and D might lose their prophage in their natural environment and could then be converted to either type C or D toxin-producers by specific phages. However, *C. botulinum* might lose prophages and still produce toxins. In such strains either prophages remain non-induced or not all toxin production in *C. botulinum* is phage-controlled. Of even greater interest, perhaps, was the demonstration of interspecies conversion (Eklund *et al.*, 1974) where a non-toxigenic mutant of *C. botulinum* type C could be converted not only to toxigenic type C or D by respective phages, but also to *C. novyi* type A by a phage from that organism.

Hariharan and Mitchell (1976) observed that all toxigenic strains of *C. botulinum* type C which they examined were infected with one or two types of phages and, that in highly toxigenic strains, the phage–bacterium relationship was characterised by a stable lysogenic type of association. When Oguma (1976) examined the stability of toxigenicity in *C. botulinum* types C and D, he found that most converted strains lost toxigenicity upon serial transfer with or without phage antiserum, suggesting that pseudolysogeny rather than true lysogenic conversion was occurring. He also suggested that non-converting mutants of the phages may have arisen and that this type of phage would render the bacteria phage-resistant but non-toxigenic, thus explaining the loss of toxigenicity commonly observed in some type C and D strains of *C. botulinum*.

Oguma *et al.* (1976, 1977) reported that haemagglutinin production in *C. botulinum* types C and D seemed to be controlled by bacteriophages and that this property could be transmitted separately or concomitantly with toxin production.

Dolman and Chang (1972) described a large number of different phage morphologies which they tried to associate with different toxigenic types of *C. botulinum*. Perhaps a significant look at bacteriophages in relationship to the taxonomy of *C. botulinum* is presented by Sugiyama and King (1972). Rather than looking at lysogenic cultures as a source of phages, these workers isolated from bottom sediments phages which were virulent for wild-type *C. botulinum*. The host ranges indicated seven different phages which fell into three morphological groups. The hosts were restricted to non-proteolytic types B, E and D and to a few non-toxigenic strains resembling type E; none of the phages was active on proteolytic types A, B or F or on cultures producing type C or D toxin. This, in addition to serological evidence (Solomon *et al.*, 1971; Batty and Walker, 1965; Takumi and Kawata, 1976), DNA homologies (Lee and Rieman, 1970a, b) and metabolic properties of types C and D, proteolytic A, D and F and non-proteolytic B, E and F, indicates that serological toxin typing bears little or no relationship to the natural relationship of these organisms.

In 1973, Mitsui *et al.* reported an inducible lysin produced after UV light or Mitomycin C treatment of *C. botulinum* A190. This lysin was active against freeze–thawed cells and cells treated in other manners, but not against viable cells. Kiritani *et al.* (1973) made some interesting observations concerning numerical taxonomy of *C. botulinum* and *C. sporogenes* strains and their susceptibilities to induced lysins and to Mitomycin C. Based upon numerical taxonomy and lysin spectra, toxin activity was the only criterion to differentiate proteolytic *C. botulinum* types A, B and F from *C. sporogenes*. These organisms, all belonging to phenon I, were susceptible to the *C. botulinum* A190 lysin. Phenon II contained non-proteolytic types B, C, D, E and F. Although a number of botulinum strains were more sensitive than *C. sporogenes* to Mitomycin C, the species cannot be reliably differentiated on this basis. It appears that phage host ranges of potential typing schemes of *C. botulinum* have not yet been described. With the soft agar overlay method of Eklund *et al.* (1972), it should be possible to examine strains with a view of typing.

2. *Bacteriophages of* C. tetani

Little has been reported on the phages of *C. tetani*. Cowles (1934) isolated phage from crude sewage by an enrichment technique using five bacterial strains. Two of these were sensitive to phage and lysogenisation of the sensitive bacteria was demonstrated. Cowles examined toxin production in *C. tetani* without demonstrating that bacteriophages were responsible for toxin production.

Prescott and Altenbern (1967a) described lysis of *C. tetani* following treatment with Mitomycin C or UV irradiation. A subsequent paper

(Prescott and Altenbern, 1967b) showed that one phage had a sheathed tail while that of a second was simple in structure. Each was about 100 nm long, the head diameter being about 65 nm. Prescott and Altenbern could not demonstrate plaque formation with their induced cultures, but Roseman and Richardson (1969) successfully showed plaques of phages from 11 of 26 toxin producing strains of *C. tetani*.

3. Other bacteriophages

This review has been restricted to the bacteriophage and bacteriocins of the pathogenic species of *Clostridium*. A number of other papers deal with non-pathogenic species. One might draw attention, in particular, to the work of Hongo and Murata in Japan and their extensive studies on the bacteriophages of *Clostridium saccharoperbutylacetonicum* (Hongo *et al.*, 1969). For a review on the classification of bacteriophages of *Clostridium*, see Ackerman (1974).

VI. PRODUCTION OF BACTERIOCIN FOR TYPING OF *C. PERFRINGENS*

A. Preparation of bacteriocin

The bacteriocin-producing strain of *C. perfringens* is grown for 18 h in cooked meat medium (Difco). At this time, the bacteria are subcultured by adding a 10% inoculum to Brain Heart Infusion Broth (Difco) which has been previously boiled for 10 min and cooled to 37°C. After 2·5 h incubation at 37°C, 30 ml are transferred to 500 ml of freshly boiled and cooled Brain Heart Infusion Broth. After 5 h of incubation, the culture is centrifuged at $6000 \times g$ for 10 min to sediment the bacteria. The supernatant is removed and the bacteriocin is precipitated by the addition of powdered ammonium sulphate to 40% saturation (28 g/100 ml) at 4°C. The fluid is constantly stirred by a magnetic stirrer during this procedure. Good yields of bacteriocin are attainable in 1–2 h, although we usually allow precipitation to continue overnight. The precipitate is collected by centrifugation at $6000 \times g$ for 10 min and resuspended in 5 ml of sterile Brain Heart Infusion Broth. Buffers might be substituted for broth, although we have found broth preferable for preserving bacteriocin activity. The bacteriocin should be stored at 4°C or preferably at -70°C for long-term storage. The method of obtaining higher yields of some bacteriocins by induction with Mitomycin C is described elsewhere (Mahony, 1977).

B. Titration of bacteriocin

We have consistently used strain no. 2 in our collection as an indicator of bacteriocin activity because it is very sensitive to all of the bacteriocins utilised in typing. For use in bacteriocin titration, the indicator strain is grown in cooked meat medium for 18 h after which it is subcultured by adding 1 ml to 10 ml of freshly boiled and cooled Brain Heart Infusion Broth. After 3 h of growth at 37°C, this culture is diluted 1:100 in Brain Heart Infusion Broth and swabbed on to the surface of a blood agar plate (Difco Brain Heart Infusion Base plus 10% human blood). The bacteriocin to be titrated is serially diluted in broth and either loopfuls or 10 μl volumes of these dilutions are spotted on to the seeded plate. After 18 h of anaerobic incubation at 37°C the reciprocal of the highest dilution showing any inhibition of bacterial growth is expressed as the bacteriocin titre. An end-point dilution of 1/64 would be expressed as 64 units of activity.

VII. TECHNIQUE OF BACTERIOCIN TYPING

The methodology involves testing the susceptibility of unknown strains to ten bacteriocins of *C. perfringens*. The bacteriocin-producing strains and their origin are recorded elsewhere (Mahony, 1974). Before typing, the activity of these bacteriocins is adjusted to 320 units wherever possible. Although most titres obtainable by ammonium sulphate precipitation exceed 320 units, some strains consistently produce lower titres. Such bacteriocins are used at the highest concentration possible. Blood agar plates are swabbed with the test strains of *C. perfringens* grown as described for the indicator strain. Drops (10 μl) of bacteriocin are then systematically placed on these plates with a Pipetman P20 pipette (Mandel Scientific Co.). Multiple inoculating devices delivering an adequate volume might alternatively be used. After the inocula have dried, the plates are incubated anaerobically for 18 h and examined for zones of growth inhibition.

VIII. INTERPRETATION OF BACTERIOCIN TYPING

A. Reading of plates

The patterns of growth inhibition produced by the bacteriocins are recorded. Very weak zones of growth inhibition are recorded as negative; however, the majority of reactions are unequivocal. Figure 1 demonstrates two different typing patterns.

B. Patterns of susceptibility

In testing 274 strains of *C. perfringens* from many sources (laboratory strains, faecal isolates, food poisoning strains and soil strains), typable strains yielded 49 different typing patterns (Mahony, 1974). Since that time, 16 new typing patterns have been observed and the complete list appears in Table II.

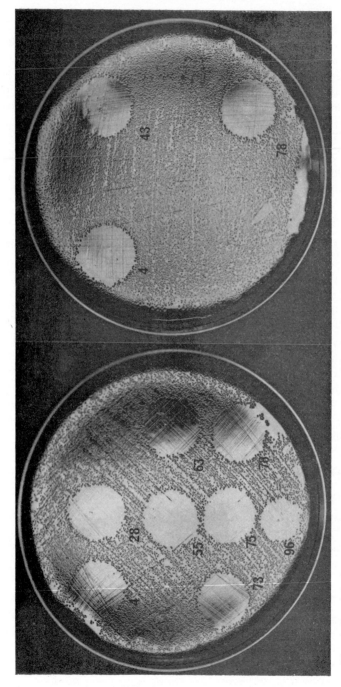

Fig. 1. Bacteriocin typing patterns produced by ten bacteriocins placed on blood agar plates seeded with *C. perfringens*: typing pattern C3 (left) and D1 (right).

The strains in the Group A typing pattern are susceptible to bacteriocins 4, 28, 43, 48, 55 and 63 with variation in response to bacteriocins 75, 78 and 96. The other groups are largely defined by sensitivity to bacteriocins 4, 28 and 43. The largest percentage of strains falls into Group C (52·9%) with types C3, 7 and 8 representing the major subgroups (Mahony, 1974). More recently, in a survey of human faecal specimens (Mahony and Swantee, 1978), bacteriocin type C3 was the dominant type of *C. perfringens* found.

It should be noted that the ability to produce bacteriocins or to be susceptible to bacteriocins is not correlated with the classical toxin typing scheme of the organism. Toxin types C and D were inhibited by bacteriocins produced by toxin type A strains and vice versa. This has also been demonstrated by Sasarman and Antohi (1963, 1968, 1970) for *C. perfringens*.

C. Variability

Under standardised conditions, the results of bacteriocin typing are quite reproducible. Cultures should be typed as soon as possible after isolation since repeated subculturing is undesirable for any typing system. We have reported (Mahony, 1974) that while serially subculturing isolated colonies ten times, and typing four colonies from each subculture a change in the typing pattern through the loss of susceptibility to a particular bacteriocin was occasionally noted; however, this loss might be reversed on further subculture. Of four strains of *C. perfringens* examined in this manner, only one became consistently susceptible to a bacteriocin to which it was originally resistant. We have observed differences in typing patterns when the organisms to be typed are plated on commercial blood plates as opposed to those prepared in our laboratories and at present would not recommend a departure from the methods and materials described above.

IX. POTENTIAL USE OF BACTERIOCIN TYPING OF *C. PERFRINGENS*

Bacteriocin typing of *C. perfringens* may determine the source of clinical infections or of food poisoning. In infections with *C. perfringens* the typing patterns of the organisms isolated from the environment, from necrotic tissue and from the patient's faeces or vagina might indicate the source of the infection. In food poisoning outbreaks bacteriocin typing may supplement or replace serological typing of the organisms. We have looked at 12 of the Hobbs agglutinating types with respect to bacteriocin typing and found that these strains fell into nine different bacteriocin typing patterns. This might indicate that the antigenic typing system is more sensitive in terms of differentiating strains.

TABLE II
Bacteriocin typing scheme for *C. perfringens* as of 1978

Type		4	28	43	48	55	63	73	75	78	96
		Sensitivity to bacteriocin from strain									
A	1	+	+	+	+	+	+	+	+	+	+
	2	+	+	+	+	+	+	+	−	+	+
	3	+	+	+	+	+	+	+	−	−	+
	4	+	+	+	+	+	+	+	+	−	+
	5	+	+	+	+	+	+	−	+	−	+
B	1	+	+	+	−	+	+	+	+	+	+
	2	+	+	+	−	+	+	+	+	−	+
	3	+	+	+	−	+	+	+	−	+	+
	4	+	+	+	−	+	+	+	−	−	+
	5	+	+	+	−	+	+	−	+	+	+
	6	+	+	+	−	+	+	−	+	−	+
	7	+	+	+	−	+	+	−	−	+	+
	8	+	+	+	−	+	−	−	−	+	−
C	1	+	+	−	+	+	+	+	+	+	+
	2	+	+	−	+	+	+	+	−	+	+
	3	+	+	−	−	+	+	+	+	+	+
	4	+	+	−	−	+	+	+	+	−	+
	5	+	+	−	−	+	+	+	−	+	+
	6	+	+	−	−	+	+	+	−	−	+
	7	+	+	−	−	+	+	−	+	+	+
	8	+	+	−	−	+	+	−	−	+	+
	9	+	+	−	−	+	+	−	−	−	+
	10	+	+	−	−	+	+	−	+	−	+
	11	+	+	−	−	+	+	−	+	+	−
	12	+	+	−	−	−	−	+	+	+	−
	13	+	+	−	−	+	−	+	+	+	+
D	1	+	−	+	−	−	−	−	−	+	−
	2	+	−	+	−	−	−	+	−	−	−
	3	+	−	−	−	−	−	−	−	+	−
	4	+	−	−	−	−	−	−	−	−	−
	5	+	−	−	−	−	−	−	+	+	−
	6	+	−	−	−	−	−	+	−	+	−
	7	+	−	−	−	−	−	+	−	−	−
	8	+	−	−	+	−	−	+	−	−	−
	9	+	−	−	−	+	+	+	+	+	+
	10	+	−	−	−	−	−	+	+	+	−
	11	+	−	+	+	+	−	+	+	+	−
	12	+	−	−	−	−	+	+	+	+	+
	13	+	−	−	−	−	−	+	+	+	+
	14	+	−	−	−	+	−	−	+	+	−
	15	+	−	+	−	−	−	+	+	+	−
	16	+	−	−	−	+	+	−	+	+	+

TABLE II (*continued*)

Type		Sensitivity to bacteriocin from strain								
	4	28	43	48	55	63	73	75	78	96
E 1	−	+	+	+	+	+	+	−	−	+
2	−	+	+	+	+	+	+	−	+	+
3	−	+	+	−	+	+	+	−	−	+
4	−	+	+	−	+	+	+	−	+	+
5	−	+	+	−	+	+	+	+	−	+
6	−	+	+	−	+	+	−	+	+	+
7	−	+	+	−	+	+	−	−	−	+
8	−	+	+	−	−	−	−	+	+	−
9	−	+	−	−	+	+	−	+	+	+
10	−	+	−	−	+	+	+	−	−	+
11	−	+	−	−	+	+	−	−	−	+
12	−	+	−	−	+	+	−	−	+	+
13	−	+	−	−	−	+	−	+	+	−
14	−	+	−	−	+	+	+	+	+	+
15	−	+	−	+	+	+	−	+	+	−
16	−	+	+	+	+	+	+	+	+	+
17	−	+	+	−	+	+	−	+	−	+
F 1	−	−	+	−	−	−	+	−	−	−
2	−	−	+	−	−	−	+	+	+	−
G 1	−	−	−	+	−	−	−	−	−	−
2	−	−	−	−	−	+	−	−	−	−
3	−	−	−	−	−	−	−	+	+	−
4	−	−	−	−	+	+	+	+	+	−

+, Susceptible to bacteriocin; −, resistant to bacteriocin.

X. OTHER EPIDEMIOLOGICAL METHODS OF TYPING CLOSTRIDIA: SEROLOGY, TOXINS

Serological typing has been applied to some species of *Clostridium* with varying degrees of success. Henderson (1940) pointed to the wide diversity of somatic antigens of the classical type A strains of *C. perfringens*. Hobbs (1953, 1965) has divided a number of food poisoning strains of *C. perfringens* into 13 serologically distinct types. Specific antisera were obtained after injecting formalin-killed strains. Slide agglutination was performed by mixing antisera and growth from blood agar. In England 70% or more strains are serotypable. The food poisoning strains of *C. perfringens* reported by Hobbs in England had heat-resistant spores (100°C for 1 h or more) and were poor producers of alpha and theta toxins.

Food-poisoning strains isolated in the United States do not conform to this description. Hall *et al.* (1963) studied 83 strains of *C. perfringens* using

Hobbs' 13 antisera. Of 30 strains obtained from food poisoning cases in England, Europe and Asia, 20 agglutinated with the antisera. These strains resembled those described by Hobbs. None of 25 American strains of *C. perfringens* un-associated with food poisoning agglutinated with Hobbs' antisera and none of these strains produced heat resistant spores. Of 28 American strains associated with food poisoning, only 12 were agglutinated with Hobbs' sera, many were without heat-resistant spores, but all were haemolytic. Subsequently, Hobbs *et al.* (1973) also reported heat-sensitive strains implicated in food poisoning and two sets of antisera —Hobbs type 1–24 for heat-resisting strains and types i–xviii for heat-sensitive strains have been developed.

Hughes *et al.* (1976) have revised the latter typing scheme and have assigned arabic numerals to Hobbs' Roman numeral scheme in addition to employing a total of 57 different antisera for typing *C. perfringens*. Sixty-five per cent of isolates from 153 food-poisoning outbreaks were typable as well as 59% of the isolates from 32 cases of gas gangrene and other clinical infections. In 55% of the food-poisoning outbreaks, the causative serotypes were established. From the same laboratory, Stringer *et al.* (1976) have proposed an international serotyping system for *C. perfringens* type A which would utilise specific antisera from England, Japan and the United States.

Dowell and Hawkins (1974) described the slide agglutination test used at the Centre for Disease Control, Atlanta, Georgia, U.S.A. Essentially a strain of *C. perfringens* to be typed is grown in a fluid "antigen" medium for 18 h which is then centrifuged at $12,350 \times g$ for 10 min. The sediment is resuspended in 0·4% formolised 0·85% saline to prepare a turbid suspension for slide agglutination. Drops of *C. perfringens* antisera (Hobbs types 1–13) are placed on marked segments of a glass slide and a drop of antigen is added to each segment and mixed with the antisera by tilting the slide. After 30 s, agglutination intensity is recorded ranging from $1+$ to $4+$ reactions. The Anaerobic Bacteriology Laboratory, Centre for Disease Control, Atlanta, has 74 specific *C. perfringens* antisera in addition to the 13 Hobbs' antisera which are used in analysing strains of *C. perfringens* obtained in outbreaks of food poisoning.

Typing of *C. perfringens* by the fluorescent antibody (FA) technique should be discussed. Klotz (1965) prepared antisera to 56 different strains of *C. perfringens* including 13 Hobb's strains. The antisera were labelled with fluorescein isothiocyanate (FITC) and used to type 79 strains, 34 of which gave strong reactions. Twenty-one strains agglutinated with the Hobbs' antisera; 18 were untypable and 27 gave equivocal results. In spite of the inability to type 57% of the strains, the FA method closely matched the slide agglutination test and was a more sensitive indicator of antigenic similarities. The FA technique has some advantages in that crude material can be stained and an appropriate estimation of cell numbers may be obtained. Also, formolised material can be kept for at least two months without impairing results. It would seem that the antigenic structure of *C. perfringens* resides in the polysaccharide structures of the capsules (Cherniak and Henderson, 1972; Paine and Cherniak, 1975).

Immunofluorescence has been applied to a variety of clostridia (Batty and Walker, 1965) and provides a rapid means of identification; however, typing the species into strains has not been developed. The serology of other species of *Clostridium* has been insufficiently developed to permit more than crude typing of organisms. Such serology is useful in identification of these species, but, as an epidemiological tool, has little value. Much of this information is provided by Mandia (1955) where serotypes of proteolytic clostridia are considered.

Similarly, toxin typing of clostridia may be of importance in species identification but is less useful for strain differentiation. Although five toxin types of *C. perfringens* exist (types A–E), the vast majority of human strains are type A; therefore, some other means must be available for further differentiation of strains for epidemiological purposes. The final identification of *C. botulinum* requires determination of specific toxin type. Although this is of considerable importance, clinically speaking, it tends to obscure the fact that the biological or physical properties of these organisms bear little relationship to their toxin production as has been discussed under the bacteriophages of *C. botulinum*. Typing methods for this species have a wider base than toxigenicity.

XI. BACTERIOPHAGE AND BACTERIOCIN TYPING OF *CLOSTRIDIUM*
SUMMARY AND OVERVIEW

To date no bacteriophage typing scheme exists for any species of *Clostridium*. With respect to *C. perfringens*, it has been our experience that the host range of phages examined is too narrow to make the development of a typing system likely. This view is also shared by Dr S. Hirano (Kagoshima University, Japan) and Dr M. Sebald (Institute Pasteur, Paris, France) (pers. comms). None the less, one must not overlook the promising presentation of Imbert (1968) with respect to typing this species. It is Dr Sebald's opinion that phage typing of *C. histolyticum* might be feasible.

Many bacteriocins of varying properties have been described for *C. perfringens* and the first paper on this subject by Sasarman and Antohi (1963) described 12 bacteriocin types amongst the 81% typable strains tested. The work described by Mahony (1974) and Mahony and Swantee

(1978) and in this Chapter further demonstrates the typability of *C. perfringens* using bacteriocins. One is cautioned, however, that this methodology has not been widely tested and that the concentration of bacteriocin and the number of indicator cells used are critical aspects of this test. The technique is simple and, if proved to be completely reliable and of suitable specificity through extensive trials, would make typing of virtually all strains of *C. perfringens* a relatively easy task. This high degree of typability is not yet possible with serological tests involving a large battery of antisera, although the proposed international typing system would seem to show considerable promise (Stringer *et al.*, 1976).

With respect to *C. botulinum*, phage typing has not been performed, but phage host ranges seem to be restricted to non-proteolytic types B, E and F and non-toxigenic type E (Sugiyama and King, 1972). Bacteriocins seem to have a variety of host ranges: those active only on non-proteolytic strains (Anastasio *et al.*, 1971; Lau, 1974); those active only on proteolytic strains (Ueda and Takagi, 1972); and those which do not seem to be restricted to one class or the other (Beerens and Tahon, 1967). A potential typing system resides in the work of Beerens and Tahon (1967).

Not enough is known about phages of *C. novyi* or *C. tetani* to predict whether phage typing of these organisms is feasible. The rare occurrence of many anaerobic infections or intoxications may discourage development of bacteriocin or bacteriophage typing methodology and this may well explain why the frequently encountered *C. perfringens* has received most attention from an epidemiological viewpoint.

ACKNOWLEDGEMENTS

The author wishes to thank Dr S. Hirano, Dr A. Sasarman and Dr M. Sebald for correspondence, literature and expression of opinions with regard to bacteriocins and bacteriophages of clostridia and Dr R. Lewis for his contribution to the literature search of bacteriocins as presented in his doctoral thesis "Characterisation of a Bacteriocin of *Clostridium perfringens*", Dalhousie University, 1972.

REFERENCES

Ackerman, H. W. (1974). *Path. Biol.* **22**, 909–917.
Anastasio, K. L., Soucheck, J. A. and Sugiyama, H. (1971). *J. Bact.* **107**, 143–149.
Batty, I. and Walker, P. D. (1965). *J. appl. Bact.* **28**, 112–118.
Beerens, H. and Tahon, M. M. (1967). *In* "Botulism 1966" (Eds M. Ingram and T. A. Roberts), pp. 424–428. Chapman and Hill, London.
Betz, J. V. (1968) *Virology* **36**, 9–19.
Betz, J. V. and Anderson, K. E. (1964). *J. Bact.* **87**, 408–415.
Bradley, D. E. and Hoeniger, J. F. M. (1971). *Can. J. Microbiol.* **17**, 397–402.
Bychkov, K. Ya. (1964). *J. Microbiol. Epidem. Immunobiol.* **41**, 39–41.
Cherniak, R. and Henderson, B. G. (1972). *Infect. Immun.* **6**, 32–37.

Clarke, D. J. and Morris, J. G. (1975). *Biochem. Soc. Trans.* **3**, 389–391.
Clarke, D. J. and Morris, J. G. (1976). *J. gen. Microbiol.* **95**, 67–77.
Clarke, D. J., Robson, R. M. and Morris, J. G. (1975). *Antimicrob. Ag. Chemother.* **7**, 256–264.
Cowles, P. B. (1934). *J. Bact.* **27**, 163–164.
Dolman, C. E. and Chang, E. (1972). *Can. J. Microbiol.* **18**, 67–76.
Dowell, V. R. and Hawkins, T. M. (1974). "Laboratory Methods in Anaerobic Bacteriology", CDC laboratory manual, U.S. Dept. of Health, Education and Welfare, Publication No. (CDC), 74–8272.
Elford, W. J., Guélin, A. M., Hotchin, J. E. and Challice, C. E. (1953). *Ann. Inst. Pasteur* **84**, 319–327.
Eklund, M. W. and Poysky, F. T. (1974). *Appl. Microbiol.* **27**, 251–258.
Eklund, M. W., Poysky, F. T. and Boatman, E. S. (1969). *J. Viriol.* **3**, 270–274.
Eklund, M. W., Poysky, F. T., Reed, S. M. and Smith, C. A. (1971). *Science, N.Y.* **172**, 480–482.
Eklund, M. W. Poysky, F. T. and Reed, S. M. (1972). *Nature New Biol.* **235**, 16–18.
Eklund, M. W., Poysky, F. T., Meyers, J. A. and Pelroy, G. A. (1974). *Science, N.Y.* **186**, 456–458.
Eklund, M. W., Poysky, F. T., Peterson, M. E. and Meyers, J. A. (1976). *Infect. Immun.* **14**, 793–803.
Ellison, J. S. and Kautter, J. A. (1970). *J. Bact.* **104**, 19–26.
Ellison, J. S., Mattern, C. F. T. and Daniel, W. A. (1971). *J. Bact.* **108**, 526–534.
Frenkel, H. M. (1940). *Microbiol. Zh. (Kiev)* **7**, 181–187.
Gáspár, G. (1960). *Acta microbiol. Hung.* **7**, 269–276.
Gáspár, G. and Tolnai, G. (1959). *Acta microbiol. Hung.* **6**, 275–281.
Grant, R. B. and Riemann, H. P. (1976). *Can. J. Microbiol.* **22**, 603–610.
Guélin, A. (1949). *Ann. Inst. Pasteur* **77**, 40–46.
Guélin, A. (1950a). *Ann. Inst. Pasteur* **78**, 392–401.
Guélin, A. (1950b). *Ann. Inst. Pasteur* **79**, 447–453.
Guélin, A. (1953). *Ann. Inst. Pasteur* **84**, 562–575.
Guélin, A. and Kréguer, A. (1950). *Ann. Inst. Pasteur* **78**, 532–537.
Guélin, A., Beerens, H. and Petitprez, A. (1966). *Ann. Inst. Pasteur* **111**, 141–148.
Hall, H. E., Angelotti, R., Lewis, K. H. and Foter, M. J. (1963). *J. Bact.* **85**, 1094–1103.
Hariharan, H. and Mitchell, W. R. (1976). *Appl. environ. Microbiol.* **32**, 145–158.
Henderson, D. W. (1940). *J. Hygiene* **40**, 501–512.
Hirano, S. and Imamura, T. (1972a). *Acta Med. Univ. Kagoshima* **14**, 91–101.
Hirano, S. and Imamura, T. (1972b). *Acta Med. Univ. Kagoshima* **14**, 103–110.
Hirano, S. and Imamura, T. (1972c). *Acta Med. Univ. Kagoshima* **14**, 115–118.
Hirano, S. and Yonekura, Y. (1967). *Acta Med. Univ. Kagoshima* **9**, 41–56.
Hirano, S., Imamura, T. and Nakagono, K. (1972). *Acta Med. Univ. Kagoshima* **14**, 111–114.
Hobbs, B. C. (1965). *J. appl. Bacteriol.* **28**, 74–82.
Hobbs, B. C., Smith, M. E., Oakley, C. L., Warrack, G. H. and Cruickshank, J. C. (1953). *J. Hygiene* **51**, 75–101.
Hobbs, B. C., Clifford, W., Ghosh, A. C., Gilbert, R. J., Kendall, M., Roberts, D. and Wieneke, A. A. (1973). *In* "Sampling—Microbiological Monitoring of Environments" (Eds R. G. Board and D. W. Lovelock), pp. 233–252. Academic Press, London.

Hoeniger, J. F. M. and Bradley, D. E. (1971). *Can. J. Microbiol.* **17**, 1567–1572.

Holdeman, L. V. and Moore, W. E. C. (1972). "Anaerobe Laboratory Manual". The Virginia Polytechnic Institute and State University Anaerobe Laboratory, Blacksburg, Virginia.

Hongo, M., Murata, A. and Ogata, S. (1969). *Agr. biol. Chem.* **33**, 337–342.

Hughes, J. A., Turnbull, P. C. B. and Stringer, M. F. (1976). *J. med. Microbiol.* **9**, 475–485.

Imamura, T. (1966a). *Med. J. Kagoshima Univ.* **18**, 157–167.

Imamura, T. (1966b). *Med. J. Kagoshima Univ.* **18**, 168–176.

Imamura, T. and Nakama, R. (1974). *Acta Med. Univ. Kagoshima* **16**, 69–74.

Imbert, S. (1968). *C. r. hebd. Séanc. Acad. Sci. Paris* **267**, 1238–1240.

Inoue, K. and Iida, H. (1968). *J. Virol.* **2**, 537–540.

Inoue, K. and Iida, H. (1970). *Jap. J. Microbiol.* **14**, 87–89.

Inoue, K. and Iida, H. (1971). *Jap. J. med. Sci. Biol.* **24**, 53–56.

Ionesco, H. and Bouanchaud, D. H. (1973). *C. r. hebd. Séanc. Acad. Sci. Paris* **276**, 2855–2857.

Ionesco, H. and Wolff, A. (1975). *C. r. hebd. Séanc. Acad. Sci. Paris* **281**, 2033–2036.

Ionesco, H., Wolff, A. and Sebald, M. (1974). *Ann. Microbiol. Inst. Pasteur* **125B**, 335–346.

Ionesco, H., Bieth, G., Dauguet, C. and Bouanchaud, D. (1976). *Ann. Microbiol. Inst. Pasteur* **127B**, 283–294.

Kautter, D. A., Harmon, S. M., Lynt, R. K. and Lilly, T. (1966). *Appl. Microbiol.* **14**, 616–622.

Kiritani, K., Mitsui, N., Nakamura, S. and Nishida, S. (1973). *Jap. J. Microbiol.* **17**, 361–372.

Klotz, A. W. (1965). *Publ. Hlth Rep.* **80**, 305–311.

Kréguer, A., Guélin, A. and LeBris, J. (1947). *Ann. Inst. Pasteur* **73**, 1038–1039.

Lau, A. H. S., Hawirko, R. Z. and Chow, C. T. (1974). *Can. J. Microbiol.* **20**, 385–390.

Lee, W. H. and Riemann, H. (1970a). *J. gen. Microbiol.* **60**, 117–123.

Lee, W. H. and Riemann, H. (1970b). *J. gen. Microbiol.* **64**, 85–90.

MacPherson, J. N. and Gillies, R. R. (1969). *J. med. Microbiol.* **2**, 161–164.

Mahony, D. E. (1973). *Can. J. Microbiol.* **19**, 735–739.

Mahony, D. E. (1974). *Appl. Microbiol.* **28**, 172–176.

Mahony, D. E. (1977). *Antimicrobiol. Ag. Chemother.* **11**, 1067–1068.

Mahony, D. E. and Butler, M. E. (1971). *Can. J. Microbiol.* **17**, 1–6.

Mahony, D. E. and Easterbrook, K. B. (1970). *Can. J. Microbiol.* **16**, 983–988.

Mahony, D. E. and Kalz, G. G. (1968). *Can. J. Microbiol.* **14**, 1085–1093.

Mahony, D. E. and Swantee, C. A. (1978). *J. clin. Microbiol.* **7**, 307–309.

Mahony, D. E., Butler, M. E. and Lewis, R. G. (1971). *Can. J. Microbiol.* **17**, 1435–1442.

Mandia, J. W. (1955). *J. infect. Dis.* **97**, 66–72.

Mitsui, N., Kiritani, K. and Nishida, S. (1973). *Jap. J. Microbiol.* **17**, 353–360.

Oguma, K. (1976). *J. gen. Microbiol.* **92**, 67–75.

Oguma, K., Iida, H. and Inoue, K. (1973). *Jap. J. Microbiol.* **17**, 425–426.

Oguma, K., Iida, H. and Shiozaki, M. (1976). *Infect. Immun.* **14**, 597–602.

Oguma, K., Iida, H. and Shiozaki, M. (1977). *Jap. J. Microbiol.* **30**, 40–43.

Ohtomo, J., Imamura, T. and Hirano, S. (1966). *Acta Med. Univ. Kagoshima* **8**, 129–136.

Paine, C. M. and Cherniak, R. (1975). *Can. J. Microbiol.* **21**, 181–185.

Paquette, G. and Fredette, V. (1977). *Rev. Can. Biol.* **36**, 205–215.

Popovitch, M. and Sebald, M. (1967). *Bull. Off. Inst. Epiz.* **67**, 1183–1194.

Prescott, L. M. and Altenbern, R. A. (1967a). *J. Bact.* **93**, 1220–1226.

Prescott, L. M. and Altenbern, R. A. (1967b). *J. Virol.* **1**, 1085–1086.

Roseman, D. and Richardson, R. L. (1969). *J. Virol.* **3**, 350.

Sames, R. W. (1956). "Studies on the Bacteriophages of *Clostridium perfringens*". Ph.D. Thesis, Indiana University, Bloomington, Indiana, U.S.A.

Sames, R. W. and McClung, L. S. (1953). *Bacteriol. Proc.* **G52**, 40.

Sasarman, A. and Antohi, M. (1963). *Arch. Roum. Path. exp. Microbiol.* **22**, 377–381.

Sasarman, A. and Antohi, M. (1968). *In* "The Anaerobic Bacteria. Proceedings of an International Workshop held October 16–20, 1967" (Ed. V. Fredette), pp. 125–131. Institute of Microbiology and Hygiene of Montreal University.

Sasarman, A. and Antohi, M. (1970). *In* "Culture Collections of Microorganisms" (Eds H. Iizuka and T. Hasegawa), pp. 541–546. University Park Press, Baltimore.

Sasarman, A. and Antohi, M. (1971). *Rev. Can. Biol.* **30**, 183–189.

Schallehn, G. (1975). *Zbl. Bakt. Abt. Orig. I* **233**, 542–552.

Schallehn, G. and Krämer, J. (1976). *Can. J. Microbiol.* **22**, 435–437.

Schallehn, G. and Lenz, W. (1975). *Zbl. Bakt. Abt. Orig. I* **232**, 100–104.

Sebald, M. and Ionesco, H. (1974). *C. r. hebd. Séanc. Acad. Sci. Paris* **279**, 1503–1506.

Sebald, M. and Popovitch, M. (1967). *Ann. Inst. Pasteur* **113**, 781–789.

Shannon, R. (1957). *J. med. Lab. Techn.* **14**, 199–214.

Smith, H. W. (1959). *J. gen. Microbiol.* **30**, 201–221.

Smith, L. Ds. (1975). "The Pathogenic Anaerobic Bacteria". Charles C. Thomas, Springfield, Illinois, U.S.A.

Smith, L. Ds. and Holdeman, L. V. (1968). "The Pathogenic Anaerobic Bacteria" (Ed. A. Balows). Charles C. Thomas, Springfield, Illinois, U.S.A.

Solomon, H. M., Lynt, R. K., Kautter, D. A. and Lilly, T. (1971). *Appl. Microbiol.* **21**, 295–299.

Spencer, M. C. (1953). "Gas Gangrene and Gas Gangrene Organisms 1940–1952. An Annotated Bibliography of the Russian Literature, 1940–1952, and the Non-Russian Literature for 1952". Armed Forces Medical Library Reference Division, Washington, D.C.

Stewart, A. W. and Johnson, M. G. (1977). *J. gen. Microbiol.* **103**, 45–50.

Stringer, M. F., Turnbull, P. C. B., Hughes, J. A. and Hobbs, B. C. (1976). *Dev. Biol. Stand.* **32**, 85–89.

Sugiyama, H. and King, G. J. (1972). *J. gen. Microbiol.* **70**, 517–525.

Takumi, K. and Kawata, T. (1976). *Jap. J. Microbiol.* **20**, 287–292.

Taylor, D. E. and Guha, A. (1974). *Virology* **59**, 190–200.

Taylor, D. E. and Guha, A. (1975). *J. Virol.* **16**, 107–115.

Tsyp, V. N., Zirbiladze, N. Ia. and Dzhikiia, A. T. (1944). *Tr. Leningrad. inst. vakts. syvor.* **1**, 29–30. (Abstracts in Spencer, 1953, Title 40.)

Tubylewicz, H. (1966a). *Bull. Acad. Pol. Sci. Ser. Sci. Biol.* **14**, 31–36.

Tubylewicz, H. (1966b). *Bull. Acad. Pol. Sci. Ser. Sci. Biol.* **14**, 467–473.

Tubylewicz, H. (1968). *Bull. Acad. Pol. Sci. Ser. Sci. Biol.* **16**, 279–284.

Tubylewicz, H. (1970). *Bull. Acad. Pol. Sci. Ser. Sci. Biol.* **18**, 253–256.

Uchiyama, K. (1966a). *Med. J. Kagoshima Univ.* **18**, 131–144.

Uchiyama, K. (1966b). *Med. J. Kagoshima Univ.* **18**, 145–156.

Ueda, M. and Takagi, A. (1972). *Jap. J. Bacteriol.* **27**, 315. Cited by Mitsui, N.,
    Kiritani, K. and Nishida, S. (1973). *Jap. J. Microbiol.* **17**, 353–360.
Vieu, J. F., Guélin, A. and Dauguet, C. (1965). *Ann. Inst. Pasteur* **109**, 157–160.
Vinet, G., Berthiaume, L. and Fredette, V. (1968). *Rev. Can. Biol.* **27**, 73–74.
Willis, A. T. (1969). "Clostridia of Wound Infection". Butterworth, London.
Wolff, A. and Ionesco, H. (1975). *Ann. Microbiol. Inst. Pasteur* **126B**, 343–356.
Yonekura, Y., Sakamoto, N., Hirakawa, K. and Hirano, S. (1972). *Acta Med.
    Univ. Kagoshima* **14**, 77–83.

CHAPTER II

Serotyping of *Listeria monocytogenes* and Related Species

H. P. R. SEELIGER and K. HÖHNE

University of Würzburg, Federal Republic of Germany

I. MORPHOLOGY AND STRUCTURE

Listeria monocytogenes is a short, Gram-positive, non-acid fast rod without spores. Young cultures consist predominantly of coccoid organisms measuring 0·5 by 1–2 μm. Their ends often appear slightly pointed. The arrangement of cells in smears is quite uncharacteristic. Besides single organisms and pairs, one finds short chains of 3–8 cells and diploforms which are sometimes V- or Y-shaped. Occasionally slender, slightly curved rods, 2–5 μm in length are seen. Club shaped organisms or pleomorphic involution forms have not been observed.

In cultures incubated for 3–6 h at 37°C, the bacillary forms prevail. Thereafter most organisms become coccoid. In older cultures, i.e. after 3–5 days' incubation, filamentous structures often appear, mainly in rough strains. They measure from 6 to 20 μm in length, filaments up to 275 μm have occasionally been observed. By use of immunocytological methods and by immune electron microscopy, Smith and Metzger (1962) demonstrated a capsule-like structure on *L. monocytogenes* surrounding the bacterial cell continuously in a regular size and shape. By histochemical stain techniques this structure was found to consist of mucopolysaccharide material. These "capsules" were demonstrated after cultivating organisms on media enriched with serum and glucose (Smith and Metzger, 1962). The existence of true capsules could, however, not be demonstrated by Seeliger and Bockemühl (1968).

All strains of *L. monocytogenes* hitherto isolated from natural sources have been motile (Seeliger, 1961) in contrast to all species of *Erysipelothrix* and most corynebacteria. Leifson and Palen (1955) stated after investigating 81 stock strains of *L. monocytogenes* that the flagellar arrangement is always peritrichous with at most 4 flagella.

II. CHEMICAL COMPOSITION

A protein substance on the surface of *L. monocytogenes* has been demonstrated (Osebold *et al.*, 1965) by trypsinisation which increased the serologic sensitivity and explained the inagglutinability of some somatic antigens. The material released showed an ultraviolet absorption peak at 200 nm and gave a spot on paper chromatograms compatible with polypeptide.

The cell walls of *Listeria* contain about 20% hexose (glucose and galactose), 5% hexosamine and 5% protein with alanine, glutamic acid, diaminopimelic acid, aspartic acid as well as leucine and are easily broken up by mechanical disintegration. Shaking the cells with glass beads longer than 4 h extensively damages the cell walls. In comparison with other Gram-positive organisms listeriae are also more sensitive to trypsin and nucleases. Roots (1958) obtained homogeneous cell wall preparations after an initial extraction followed by 4 h agitation with glass beads and subsequent enzymatic treatment with trypsin, ribonuclease and deoxyribonuclease. It was concluded that the cell walls of *Listeria* contain great quantities of material specific for the enzymes employed.

All *Listeria* strains freshly isolated from pathological material are capable of lysing red blood cells from most mammalian animals (cf. Seeliger, 1961a; Gray and Killinger, 1966). Certain strains show a pronounced *β*-haemolysis on sheep blood agar plates thereby producing wide halos. The haemolysing activity is most regularly demonstrated using horse blood. Lysis of the red blood cells is effected by a filtrable haemolysin which is also produced in blood-free substrates and which acts on erythrocytes after passage through bacteria-proof filters.

A monocytosis-producing agent (MPA) of *L. monocytogenes* located in the cell wall (Leeler and Gray, 1960) was extracted by Stanley (1949) from dried and ground organisms with organic solvents. Amounts of extractable lipids are highest after extraction with petrol-ether, but chloroform or ether extracts are biologically more active. The isolated lipid MPA is serologically inactive and of low tissue toxicity. A toxic component has been isolated from whole cells of *L. monocytogenes* by aqueous ether extraction and ethyl alcohol precipitation (Srivastava and Siddique, 1974). This component was pyrogenic to rabbits, produced edema and erythema in rabbit skin, and was lethal to chicken embryos. It was also immunogenic to mice against a large homologous challenge dose.

III. ISOLATION PROCEDURES; CULTURAL AND BIOCHEMICAL IDENTIFICATION

A. Isolation from Infected Material

When the organisms are numerous and not mixed with other bacteria, cultivation on bacteriological media does not present any major difficulties. *Listeria* are not fastidious and grow well on glucose-infusion agar, with or without blood or serum added. They also multiply in infusion broth and grow profusely in brain heart-, placenta- or liver broth. Tryptose is preferable to most brands of peptone, and a 2% tryptose agar is an excellent medium for cultivation, propagation and preservation of *L. monocytogenes*.

The various techniques which have proved successful for isolation generally stress the need for homogenisation, when the organisms are to be grown from tissues. The bacteria are often incarcerated in the focal lesions and frequently found intracellularly. The most effective method consists of macerating suspected tissue in a mortar or Waring Blendor together with a few millilitres of sterile distilled water or nutrient broth. Saline should be avoided since it is said to harm the bacterium, especially if the cells are low in number. A portion of the suspension is plated on sheep or horse blood agar, tryptose agar and Trypaflavine Nalidixic Acid (TNA) Agar (Bockemühl *et al.*, 1971). The remainder is stored at 4°C. Body fluids, swabs etc. are plated, and a portion is stored at 4°C. The plates are incubated at 37°C for 18 to 24 h, that of TNA Agar for 48 h, and examined with a scanning microscope, or a hand lens with the plate resting on a laboratory tripod (Bearns and Girard, 1959) and with obliquely transmitted illumination as described by Henry (1933). When viewed in this manner the colonies on tryptose agar are distinctively blue-green and so characteristic that with little practice they can be identified quickly even in highly contaminated cultures (Gray *et al.*, 1948; Gray, 1957; Hartwigk, 1958a) and distinguished from a number of common pathogens (Hartwigk, 1958a). *Listeria* colonies on TNA Agar are of a light green to blue-green colour. The medium has to be clear and must be poured in a thin layer (about 10 to 13 ml per 9 cm plate) in order to allow optimal evaluation (Bockemühl, 1971; Höhne, 1972).

If the initial culture fails to reveal *L. monocytogenes*, the tissue suspension, fluids, swabs etc. which are kept in the refrigerator at 3 to 4°C should be replated after 6 weeks and even after 3 months storage. Usually, a 6 weeks period of refrigeration is sufficient for the growth to appear. Kampelmacher and van Noorle Jansen (1961) found it necessary to refrigerate calf brain material as long as 6 months before the bacterium could be detected.

The mechanism of the enhancing effect at 4°C is not fully understood. When it was first described by Gray *et al.* in 1948, it was suggested that it might involve an inhibitory factor in bovine brain. Attempts to demonstrate such a factor in brains from several different species have been unsuccessful (Murakimi and Kato, 1957). It has been shown repeatedly that delayed growth of listeriae from infected tissue is not limited to bovine brain, but that this is similar with all animal and human tissue and body fluids. The growth at low temperatures may be connected with the psychrophylic nature of the organisms and with slow liberation from its intracellular position. Enhancement by cold storage has been used successfully to isolate the bacterium from such widely diversified sources as pneumonic lungs of infants (Menčikova, 1956), vaginal swabs (Gray, 1960a), silage extracts (Gray, 1960b) and environmental specimens (Weiss and Seeliger, 1975; Höhne *et al.*, 1975).

Unless the organisms grow in pure culture, transfer suspected colonies to fresh blood agar or tryptose agar plates for purification. The characteristic colonial forms are usually apparent after 24–48 h incubation at 37°C and can be distinguished by the naked eye. Then smears are made and Gram stained. If the morphology is characteristic the demonstration of motility is of utmost importance.

For this purpose, a hanging drop preparation is made from a young broth culture incubated at room temperature. In addition semisolid motility agar containing 0·2–0·4% agar is stabbed (about 1 cm). At 20–37°C listeriae inoculated into the agar column swarm through the "motility medium" and produce cloudiness. This can easily be observed by the naked eye. About 0·5 cm below the surface of the agar a layer of increased growth, like an umbrella, is found. In this zone of reduced oxygen tension *Listeria* shows a better development than under aerobic or strictly anaerobic conditions.

The formula of the preferred TNA medium is given below:

Preparation of Trypaflavine-Nalidixic-Acid-Agar:

Tryptose Agar (Difco)	41·05
Nalidixic acid	0·04 g
Distilled water	1000·0 ml

After sterilisation (15 min of 120°C) the medium is cooled to 70°C and the acridines*, dissolved in distilled water, are added to the required concentration (25μg/ml). The medium is mixed thoroughly and poured into Petri dishes. The medium has to be clear and poured in a thin layer (about 10 to 13 ml per plate) in order to allow optimal evaluation (Bockemühl *et al.*, 1971).

B. Biochemical Properties

The demonstration of β-haemolysis is of utmost importance since all virulent *L. monocytogenes* isolates show this property.

The presumptive diagnosis must be confirmed by biochemical and serological studies (see Tables I and II).

Properties of *Listeria monocytogenes*

During the past few years, non-haemolytic *Listeria* isolates have been grown with increasing frequency from faecal and all sorts of environmental

TABLE I
Important properties of *Listeria*

β-haemolysis	+
catalase	+
motility	+
urea	−
nitrate reduction	−
Voges-Proskauer reaction	+
salicin	+
trehalose	+
glucose	+
	acid (without gas)
lactose	$\varnothing$
mannitol	$\varnothing$
aesculin	+

+ = positive reaction
− = negative reaction
$\varnothing$ = variable or slow reaction

*Acriflavin neutral NOKA 565, CASSELLA Farbwerke, Frankfurt (Main) FRG or Proflavinhemisulfat NOKR 565, CASSELLA Farbwerke, Frankfurt (Main) FRG.

TABLE II

Acid production from carbohydrates by species of genus *Listeria*†

	L. monocytogenes	*L. denitrificans*‡	*Murraya grayi*	*M. murrayi*
L–arabinose	−	+	−	−
D–galactose	d	+	+	(+)
glycogen	−	+	−	−
lactose	(d)	+	+	+
mannitol	−	−	+	+
melezitose	d	−	−	−
melibiose	−	(+)	−	−
rhamnose	d	−	−	d
sucrose	(d)	+	−	−
xylose	−	+	−	−

Key: + = acid produced 24–48 h (90% or more strains)
− = no acid produced 21 days (90% or more strains)
d = some strains positive, some negative
(+) = acid produced slowly (3–7 days)
(d) = some strains produce acid slowly (3–7 days), other strains negative

† All species produce acid but no gas in 24–48 h from amygdalin, aesculin, cellobiose, dextrin, fructose, glucose, maltose, mannose, salicin, starch and trehalose.
No acid in 21 days from adonitol, dulcitol, erythritol, inositol, inulin or raffinose.
‡ This species is very likely not a member of the genus *Listeria*.

specimens collected in nature (Weiss and Seeliger, 1975; Seeliger, 1975; Seeliger, 1976). These isolates usually are devoid of any animal pathogenicity and belong to a group of serotypes listed in Table VII.

They are, in the writer's opinion, not true *L. monocytogenes* and may represent a different, closely related species.

In case of doubt, the lack of haemolysing properties should always be demonstrated by the CAMP test as described by Brzin and Seeliger (1974), if horse blood is not available (see below).

As shown in Fig. 1 *L. monocytogenes*—classical serotypes show a very distinct β-haemolysis, which differs from the much stronger haemolytic action of serotype 5 of Ivanov. The serotypes 4f and 4g as well as related varieties completely lack haemolysis and belong to a new spacies, *Listeria innocua* (Seeliger and Schoofs, 1977).

C. The CAMP phenomenon in *Listeria*

The CAMP phenomenon is a simple and accurate method of identifying *Streptococcus agalactiae*. Since the reliability in this aspect is high (only about 2–4% of *Strept. agalactiae* are CAMP negative (Munch-Petersen, 1947)) it was adopted as a routine test as early as 1947 in the New York State Mastitis Control Programme. The explanation for the phenomenon still awaits clarification. The nature of the so-called haemolysis accentuating factor produced by CAMP positive streptococci is as yet unknown. Christie and co-workers stated that the agent is an extracellular filtrable product, quite stable at high temperatures.

Method

A culture of *Staphylococcus aureus* capable of producing a large zone of weak β haemolysis is streaked across the centre of a blood agar medium. A colony of the *Listeria* under examination is streaked perpendicular to and within 3 mm of the line of staphylococcus inoculation. The culture is incubated at 37°C overnight and then checked to see if the haemolysis is accentuated (cf. Brzin and Seeliger, 1974).

Since it is suspected that in some samples of blood (sheep- or calf-blood) staphylococcal haemolysin inhibiting substances are present the erythrocytes are first washed.

As positive control a typical *Strept. agalactiae* strain and a strain of *L. monocytogenes* serotype 1/2a or 4b are included.

A positive CAMP reaction is indicated by a semicircular or arrowhead-shaped zone of complete lysis of the erythrocytes around the staphylococcal streak (Fig. 1). The outer line of the accentuated haemolysis is completely straight (Fraser, 1962).

Performed in the same way with a perpendicular streak, the CAMP test

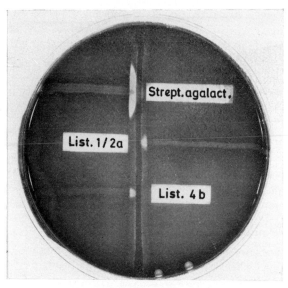

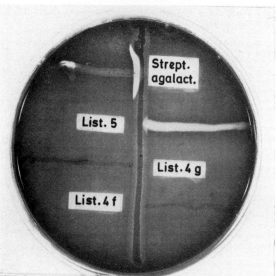

Fig. 1: CAMP test with *Listeria* strains and related serotypes. The classical serotypes of *Listeria monocytogenes* show a very distinct β-haemolysis, which differs from the much stronger haemolytic action of serotype 5. The serotypes 4f, 4g and 6 (*Listeria innocua*) lack haemolysis completely.

with *Listeria* differs from the typical appearance with group B streptococci in that the outside line of the accentuated lysis does not run straight,
i.e. it does not run parallel to the streak of *Staphylococcus*, but has an
arrowhead shaped appearance. This means that it does not follow the
straight line of β haemolysis of the staphylococcal streak but is curved
rather toward the streak itself. The strains formerly regarded as not
haemolysing, uniformly failed to show even a trace of haemolytic activity
in the CAMP test (Brzin and Seeliger, 1974).

IV. ANTIGENIC STRUCTURE AND SEROLOGICAL IDENTIFICATION

A. Historical development

Seastone (1935) was the first to perform serological studies with different
strains of *L. monocytogenes*. In agglutination and absorption experiments
he found a close relationship between cultures from human and animal
sources, with the exception of the original type strain of Murray *et al.*
(1926) which reacted quite differently. Similar results were obtained by
Carey (1936). Webb and Barber (1937) noted a great similarity in the
serological behaviour of *Listeria* strains tested by them. Schultz, Terry,
Brice and Gebhardt (1938) as well as Julianelle and Pons (1939) first drew
the conclusion that *Listeria* should be divided into two distinct serological
groups, which seemed to be connected with the origin of the strain. In
consequence a "rodent group" and a "ruminant group" were distinguished.
Listeria strains isolated from human specimens were found to belong
regularly to one or other of these two serogroups. These studies were
amplified by the fundamental work of Paterson using 54 *Listeria* strains
(Paterson, 1939). By serological analysis of somatic and flagellar antigens
and by use of the modern methods as employed by White and Kauffmann
on the *Salmonella* group, Paterson succeeded in recognising serologically
different O and H antigen patterns in *L. monocytogenes*.

Thus the species *L. monocytogenes* could be subdivided into four serotypes.
These were in no way connected to any particular host species. According
to Paterson "there does not appear any relationship between the bacteriological type and the zoological host species nor do the types appear to be
associated with a particular geographical origin. It will be noted that
members of each type have been isolated from man and that poultry
strains fall into types 1 and 4". Three of the four different varieties could be
distinguished on the basis of their O antigens and the fourth (serotype 2)
was recognised by its different H antigen. After investigating serologically
several thousand *Listeria* strains from a wide variety of sources during the
past 25 years, Paterson's original antigenic scheme of the species

L. monocytogenes is still valid although it had to be extended after the demonstration of further antigenic factors.

Paterson's antigenic scheme was confirmed repeatedly, i.e. by Robbins and Griffin (1944, 1945a, 1945b), who studied the effect of disinfectants on the flagellar antigens with particular reference to the production of antibodies against the various components of the antigenic mosaic. The results are summarised in the following: among the O antigens the factors 1, 2, 4 and 5 are thermostable. Their antigenic power is not influenced by high concentrations of alcohol. After treatment with heat or alcohol many, but not all, strains lose the ability to stimulate production of agglutinin against O-factor 3. The agglutinability, however, is not influenced. Robbins and Griffin ascribe the variability of factor 3 to its particular chemical structure as well as to the arrangement and distribution of factor 3 molecules within the bacterial cell. Possibly factor 3 represents a surface antigen analogous to L antigens of Enterobacteria. It may be recalled that serotypes 1/2a, 4a and 4b are usually found in pathological material from diseased animals and man, while the other serotypes were rarely recognised and mostly originated from faecal specimens of *Listeria* carriers. In consequence, the serological identification of serotypes other than 1/2a (or b) and 4b is not necessary for routine purposes. The same may hold true for serotypes based on the presence or absence of H-factor C which is difficult to demonstrate. So far, serotypes of *L. monocytogenes* have never been found to be correlated to any particular clinical syndrome nor to any particular host, although their distributions, particularly of serotypes 1/2a and 4b show remarkable geographical differences (*vide infra*).

B. Methods for preparation of *Listeria* antigens O-antigens

O-antigens are prepared from smooth cultures incubated on 1% glucose tryptose agar for 24 h at 37°C. The harvested growth is suspended in physiological phosphate-buffered saline (pH 7·2) and boiled in the steamer in order to destroy the H-antigens. Then phenol is added to a final concentration of 0·5%. In order to eliminate the tendency to spontaneous agglutination, all O suspensions are treated by ultrasonic oscillation for 5–10 min. Frequency and duration of ultrasonic treatment depend on the density of suspension and the degree of spontaneous clumping. At 1mHz and 32–38 W 5–10 min treatment yields the desired results.

H-Antigen

H-antigens are prepared by adding equal amounts of 0·6% formolised saline to highly motile cultures in phosphate-buffered 1% glucose tryptose broth after 24-48 hr incubation at 22–24°C. The presence of very active motility should be checked by hanging drop preparations. Cultures showing filaments indicating the presence of rough forms, should not be used for the production of H-antigens. For unknown reasons H-antigens from broth cultures are not always easily agglutinable by homologous immune sera. Thus it may be advisable to use H-antigens grown on the surface of semi-solid agar medium, for example Gard agar supplemented by 1% glucose and phosphate buffer. Its surface must be kept moist. On such media H-antigens of *Listeria* develop well and motility is pronounced.

The growth is harvested in 0·3% formolised saline and is adjusted to pH 7·2 after killing the organisms by allowing the suspensions to stand for 3–4 days at room temperature.

The test for sterility is carried out as follows: 1 ml of suspension is plated on sheep-blood-agar and incubated for 48 h.

C. Production of antisera

Before immunisation is started the animal ought to be tested for the possible presence of *Listeria* agglutinins, which are frequent, but usually found in low titres only. Whether they are *Listeria*-specific is not known. Rabbits with serum titres against *Listeria* of 1 : 80 and above should not be used for the production of *Listeria* antisera.

For the production of antisera rabbits are immunised intravenously with increasing quantities of antigen at 3–4 day intervals. The initial immunising dose consists of 0·5 ml of an antigen suspension corresponding to a density of McFarland 3 ($0·5 \times 10^9$ organisms/ml). This dose is increased by 0·5 ml every time until 2·5 ml are reached. The injections are usually tolerated well. After five injections the animals are allowed to rest for a week. If sufficiently high titres (about 1 : 1600 for O-antisera and 1 : 6400 for H-antisera) are found the animals are exsanguinated.

Sera are divided into small aliquots and are preserved in the deep-freeze without preservative. For use the sera may also be kept in the refrigerator at $+4°C$. Then it is advisable to preserve O-sera by addition of 0·5% phenol and H-sera by 10% glycerine or merthiolate (final concentration 1 : 10,000).

OH-antisera may easily be produced by injection of formalised broth cultures or agar suspensions as described. For routine work OH-antisera may be used in the study of O-antigens if the flagellar antigens of the whole cells to be tested are destroyed by previous boiling. Injections of living organisms do not seem to offer any advantage.

Pure O-antisera are obtained by injection of suspensions boiled for at least 1 h.

Pure H-antisera are produced by absorption of OH-antisera with homologous O-antigen.

Kunkel and Naumann reported in 1974 on the preparation of a pure H-antigen by adsorbing *Listeria* flagella to latex particles. Their efforts did not result in an agglutination between the flagella-latex-mixture and the H-antisera.

D. Production of factor sera

Factor sera may be produced with strains selected and recommended by Seeliger and Donker-Voet (see Table III) which are available from the National Collection of Type Cultures London, The ATCC, Rockville, Md., USA, and the Listeria Culture Collection of the Institute for Hygiene and Microbiology, Würzburg, FRG. The procedure for obtaining factor sera by CASTELLANI's absorption test is outlined on Table V.

TABLE III

List of strains used for serological characterisation of *Listeria mono-cytogenes* and related species

Serotype	Collection No Würzburg SLCC	Also listed
1/2a	2371	NCTC 7973/ATCC 19111
1/2b	2755	DONKER-VOET 1684
1/2c	2372	NCTC 5348/ATCC 19112
3a	2373	NCTC 5105/ATCC 19113
3b	2540	DONKER-VOET L471
3c	2479	DONKER-VOET 1916
4a	2374	NCTC 5214/ATCC 19114
4b	2375	ATCC 19115
4c	2376	ATCC 19116
4d	2377	ATCC 19117
4e	2378	ATCC 19118
4f	3379	
4g	3423	
5	2379	ATCC 19119
6 (sub judice)	2480	DONKER-VOET 1383
7	2482	DONKER-VOET 1627
M. grayi	2861	ATCC 19120
Listeria denitrificans	2836	Collection Institut Pasteur Paris

SLCC: Special Listeria Culture Collection, Würzburg, F.R.G.

NCTC: National Collection of Type Cultures, London, G.B.

DONKER-VOET: C/O Listeria Collection, Rijks Instituut voor de Volksgezondheid, Biltoven, Netherlands.

TABLE IV

Scheme for preparation of O-factor sera for *Listeria monocytogenes*

O-factor	*Listeria monocytogenes* rabbit O-antiserum	Absorbed with O-antigens from serotype
I	1	3
I/II	1 part 1/2 a mixed with 1 part 1/2 b	4
IV	3	1
V/VI	4b	1
VI	4b	4c
VII	4c	4b
VIII	4d	4b
IX	4a	4c
X	5	4d
XI	6	4c + 4d
XII/XIII	7	
XIII	7	*M. grayi*
XIV	*M. grayi*	7
XV	4f	4b

TABLE V

Absorption procedure for *Listeria* H-factor sera

H-factors	*Listeria monocytogenes* rabbit OH-antiserum from serotype	Absorbed with O-antigens from serotype	Absorbed with OH-antigens from serotype
A	1/2a	1/2a	1/2c
AB	1/2a	1/2a	—
C	4b	4b	1/2a
D	1/2c	1/2c	1/2a

For the preparation of OH-factor sera, two controls are necessary.
1. Control with O-antigen used for absorption: no agglutination allowed.
2. Control with OH-antigen of serotype 1/2a, 1/2c and 1/2b, agglutination as follows:

H-factors	OH-antigen from serotype 1/2a	OH-antigen from serotype 1/2c	OH-antigen from serotype 1/2b
A	+	∅	+
AB	+	+	+
C	∅	∅	+
D	∅	+	∅

If the reaction is significantly positive, serum has to be diluted with two parts of 0·85% saline containing 0·5% phenol.

Preparation of O-factor sera

1. Immunise rabbits with boiled antigen and test for presence of desired O- or H-agglutinins
2. Absorb sera, containing the desired factors:
   Mix one part of undiluted serum with two parts of centrifuged antigen sediment.
   Mix thoroughly
3. Incubate for 24 h at 4°C
4. Separate serum from antigen by high speed centrifugation (30 min at about 10,000 rev/min)
5. Save supernatant
6. Centrifuge again
7. Check by slide agglutination with absorbing antigen suspension whether complete absorption is achieved (no agglutination is allowed!)
8. If not, a second absorption will be necessary
9. Check specificity of factor sera by slide agglutination.

For this purpose dense antigen suspensions must be used. After treatment the O-antigen is centrifuged and the supernatant is decanted but for a little residuum which is used for resuspending the sediment. A drop of the concentrated antigen suspension is mixed with a drop of O-factor serum on a glass slide. The slide is tilted between the fingers for 1–2 min and the occurrence of agglutination is watched against a dark background with a good source of light. Two or three factor sera may be examined at the same time on one slide and compared with a saline control.

E. Conservation of factor sera

The factor sera preserved with phenol (0·5% final concentration) or any other suitable preservative, may be kept for years. To 0·9 ml factor serum 0·1 ml of a 5% phenol-saline solution is added. Factor sera in vials may be kept in a refrigerator at 4° to 6°C. Shaking the contents of the vials should be avoided. When withdrawing sera, care should be taken not to tip up the vial abruptly in order to avoid transferring the fine deposit which tends to form after prolonged storage. The reading of the reaction is not affected by an occasional flake.

The antigenic formulae as ascertained by Paterson have been fully confirmed by Seeliger (1975). They proved useful for sero analysis of listeriae and could be simplified for routine work (Table VI).

F. Results of serological analysis

Serological differentiation of *L. monocytogenes* as a routine procedure was introduced by Seeliger (1954, 1958, 1961). Results of almost 3500 typings

TABLE VI

Serotypes of *Listeria monocytogenes*, *Muraya grayi* and related species

Paterson	Seeliger-Donker-Voet	I	II	III	IV	V	VI	VII	VIII	IX	X	XI	XII	XIII	XIV	XV	H-Antigens
1	1/2a	I	II	(III)													A B
	1/2b	I	II	(III)													A B C
2	1/2c	I	II	(III)													B D
3	3a		II	(III)	IV												A B
	3b		II	(III)	IV												A B C
	3c		II	(III)	IV												B D
4	4a			(III)		V		VII		IX							A B C
	4ab			(III)		V	VI	VII		IX							A B C
	4b			(III)		V	VI										A B C
	4c			(III)		V		VII									A B C
	4d			(III)		(V)	VI		"VIII"								A B C
	4e			(III)		V	VI		"(VIII)"	(IX)							A B C
Listeria sp.†	4f ⎫ Complex 4g ⎭			(III) (III)		V V	VI	VII			X	XI					XV	A B C A B C
	5			(III)		(V)	VI	VII	"VIII"		X						A B C
Listeria sp.†	6 (Sub judice)			(III)				VII	"VIII"			XI					A B C
	7			(III)									XII	XIII			A B C
M. grayi (ssp. *grayi*)				(III)									XII		XIV		E
(ssp. *murrayi*)				(III)									XII		XIV		E

"VIII" still under investigation.

† Non-haemolytic strains, additional antigen combinations are known but not listed, probably representing a species or subspecies different from *L. monocytogenes* for which the name *Lysteria innocua* has been proposed.

so far have shown that only a few cultures are unidentifiable, mainly due to roughness. Seeliger (1958, 1961) divided Paterson's O group 3 into 4a and 4b, on the basis of differences in the somatic antigens. A further subdivision on the basis of different somatic antigens by Donker-Voet (1954, 1957, 1959) and Seeliger (1972) resulted in the distinction of further antigen combinations (see Table VI). Serotype 5 was described by Ivanov (1962) who studied listeriosis in sheep. This serotype 5 (Cooper *et al.*, 1973) possesses an O-factor which was not present in serotypes 1 to 4a through 4e. It differs also from other listeriae by its very pronounced haemolytic action. Additional serotypes 3c, 6, 7 and *Murraya murrayi* were described by Donker-Voet (1966) and Stuart and Welshimer (1974).

The present knowledge on the antigenic structure of *L. monocytogenes* is summarised in Table VI.

The situation became complicated when by the detection of non-haemolysing, avirulent strains with the biochemical make-up of *Listeria* and showing a close serological relationship, serotypes 4f, 4g and 6 were recognised jointly by Seeliger and Donker-Voet.

For many reasons not to be discussed here it appears that these serotypes which so far have never been isolated from pathological material of human or animal origin, but frequently from faecal material and environmental sources, may not belong to *L. monocytogenes* (Table VII), but rather to a new species *Listeria innocua*.

TABLE VII

Undesignated serotypes of non-haemolysing biotypes of *Listeria*

O-antigens								
V	VI	VII						
V	VI	VII	VIII					
V	VI			IX				XV
V	VI			IX	X	XI		
V	(VI)	VII				XI	XII	
V	VI	VII		IX				XV
V		(VII)		IX				XV
V		VII		IX	X			XV
V	VI		VIII		X	XI		
V	VI				X	XI		
V		VII						XV
V			VIII					
V			VIII		X	XI		
V					X	XI		
	VI	VII			X			
		VII			X	XI		

This also seems true for many other isolates from similar sources which are also non-haemolysing (as shown by the CAMP-test as the most sensitive method) and not virulent and which show a similar antigenic make-up. Such isolates have not been designated any further, since they are obviously closely related to serotypes 4f and 4g. A survey of antigen combinations so far determined in the authors' laboratory is given in Table VII.

These strains are *Listeria incertae sedis* (*L. innocua*) and should not be confused with classical *L. monocytogenes*.

L. monocytogenes isolates from pathological sources almost invariably belong to serotypes 1/2a, 1/2b and 1/2c, 3a, 3b, and 3c, 4b and 5 (*L. monocytogenes* sub judice!). They represent over 98% of several thousand cultures studied by the senior author since 1951.

The distribution of serotypes throughout the world is not uniform. In the U.S.A. and Canada serotype 4b prevails at a proportion of 65% to 80% of all strains (King and Seeliger, 1959). In the Eastern European countries, in West Africa, in Central Germany, Finland and Sweden serotype 1/2a is most frequently found, while in Western Europe, particularly in France (Seeliger and Linzenmeier, 1955; Lucas and Seeliger 1957; Höhne *et al.*, 1975) and in the Netherlands (Kampelmacher *et al.*, 1966) serotypes 1/2a and 4b are being isolated in about the same proportion. In Western Germany serotype 1/2a prevailed until 1958. Since then a definite shift toward serotype 4b has been noted which in some outbreaks prevailed.

Serological differentiation has been of great help in the identification of suspected cultures, but no progress in the elucidation of the epidemiology and epizootiology of listeriosis has resulted from this work. As a matter of fact serotypes 1/2a and 4b have been isolated even from the same animal, or the mother has excreted a serotype different from that of her listeric child. These observations are indicative of mixed infections rather than of serological instability of the organisms.

G. Cross-reactions

Overlapping serological reactions between the *Listeria* serotypes and other species of bacteria have repeatedly been observed. Comparative studies have constantly demonstrated a serological dissimilarity between *L. monocytogenes* and strains of the *Erysipelothrix* group (Barber, 1939; Julianelle and Pons, 1939). Seeliger found some enterococcus strains which were strongly agglutinated by *Listeria* serotypes 4a and 4b antiserum, but not by sera of the other serotypes. Some serological cross-reactions were also reported between *Listeria* and certain *E. coli* strains (Jaeger and Myers, 1954). The same applies to *L. monocytogenes* and *Staphylococcus aureus*

(Seeliger 1961). There also seem to be overlapping antigens connecting *Listeria* with motile corynebacteria. Haemosensitising substances of *Listeria* and β-haemolytic streptococci may also be interrelated and are considered to be an important source of cross-reactions (Seeliger, 1961). The serological cross-reactions between *L. monocytogenes* and a great number of Gram-positive bacterial species may be explained by the presence of a non-species specific antigen which is common to many Gram-positive organisms (cit. Seeliger, 1961).

These findings call for attention and caution. Serological tests alone are not a sufficient criterion for the diagnosis of *Listeria*, but a reliable help only in combination with cultural and biochemical reactions.

REFERENCES

Barber, M. (1939). *J. Path. Bact.*, **48**, 11–23.

Bearns, R. E. and Girard, K. R. (1959). *Am. J. Med. Technol.*, **25**, 120–126.

Bockemühl, J., Seeliger, H. P. R., and Kathke, R. (1971). *Med. Microbiol. Immunol.*, **157**, 84–95.

Brzin, G., and Seeliger, H. P. R. (1974). "Proc. 6th Int. Symp. on Listeriosis", Nottingham.

Carey, B. W. (1936). *J. Pediat.*, **8**, 626–629.

Christie, R. *et al.* (1944). *Austr. exp. Biol. and med. Sci.*, **22**, 97.

Cooper, R. F., Dennis, S. M., and McMahon, K. J. (1973). *Am. J. Vet. Res.*, **34**, 975–978.

Donker-Voet, J. (1954). *Tschr. Diergeneesk.*, **79**, 743–765.

Donker-Voet, J. (1957). *Tschr. Diergeneesk.*, **82**, 341–350.

Donker-Voet, J. (1959). *Am. J. Vet. Res.*, **20**, 176–179.

Donker-Voet, J. (1966). "Proc. 3rd Int. Symp. on Listeriosis", 13–16 July, pp. 133–137.

Fraser, G. (1962). *Vet. Rec.*, **74**, 50–51.

Gray, M. L., Stafseth, H. J., Thorp, Jr. F., Sholl, L. B., and Riley, Jr., W. F. (1942). *J. Bacteriol.*, **55**, 471–476.

Gray, M. L. (1957). *Zentr. Bakteriol. Parasitenk. Abt. I. Orig.*, **169**, 373–377.

Gray, M. L. (1960). *Lancet*, **2**, 315–317.

Gray, M. L. (1960b). *Science*, **132**, 1767–1768.

Gray, M. L., and Killinger, A. H. (1966). *Bacteriol. Rev.*, **30**, 309–382.

Hartwigk, H. (1958). *Berlin, Munch. Tierärztl. Wschr.*, **71**, 82–85.

Hartwigk, H. (1958a). *Zentr. Bakteriol. Parasitenk. Abt. I. Orig.*, **173**, 568–580.

Henry, B. S. (1933). *J. Infect. Diseases*, **52**, 374–402.

Höhne, K. (1972). Inaug. Diss. Giessen, F.R.G.

Höhne, K., Loose, B., and Seeliger, H. P. R. (1975). *Ann. Microbiol. (Inst. Pasteur)*, **126A**, 501–507.

Ivanov, J. (1962). *Monatsh. Vet. Med.*, **17**, 729–736.

Jaeger, R. F., and Myers, D. M. (1954). *Canad. J. Microbiol.*, **1**, 12–21.

Julianelle, L. A., and Pons, C. A. (1939). *Proc. Soc. Exp. Biol. Med. (N.Y.)*, **40**, 362–363.

Kampelmacher, E. H., and Van Noorle Jansen, K. M. (1961). *Tierärztl. Monatsschr.*, **48**, 442–448.

Kampelmacher, E. H., Donker-Voet, J., and Van Noorle Jansen, L. M. (1966). "Proc. 3rd Int. Symp. on Listeriosis", 13–16 July, pp. 335–340.

King, E. O., and Seeliger, H. P. R. (1959). *J. Bacteriol.*, **77**, 122–123.

Kunkel, M., and Naumann, G. (1974). *Dtsch. Ges. wesen*, **29**, 1611–1613.

Leifson, E., and Palen, M. I. (1955). *J. Bacteriol.*, **70**, 233–240.

Lucas, A., and Seeliger, H. P. R. (1957). *Recueil Méd. Vét.*, **133**, 373–378.

Mencikova, E. (1956). Abstr. 10th Meeting Czech. Soc. Microbiol.

Much-Petersen, E. (1947). *J. Path. Bact.*, **L IX**, 367.

Murakimi, T., and Kato, H. (1957). *J. Fac. Agr. Iwate Univ.*, **3**, 262–267.

Murray, E. G. D., Webb, R. A., and Swann, M. B. R. (1926). *J. Pathol. Bacteriol.*, **29**, 407–439.

Ortel, S. (1972). *Acta microbiol. Acad. Sci. hung.*, **19**, 363–365.

Osebold, J. W., Aalund, O., and Crisp, C. E. (1965). *J. Bacteriol.*, **89**, 84–88.

Paterson, J. St. (1939). *J. Pathol. Bacteriol.*, **48**, 25.

Robbins, M. L., and Griffin, A. M. (1944). *J. Immunol.*, **48**, 63–68.

Robbins, M. L., and Griffin, A. M. (1945a). *J. Immunol.*, **50**, 237–245.

Robbins, M. L., and Griffin, A. M. (1945b). *J. Immunol.*, **50**, 247–254.

Roots, E. (1958b). Beiheft I, Zbl. Vet. Med. Paul Parey, Berlin und Hamburg, pp. 45–48.

Schultz, E. W., Terry, M. C., Brice, A. T., and Gebhardt, L. P. (1934). *Proc. Soc. Exp. Biol. Med.*, **31**, 1021–1023.

Seeliger, H. P. R. (1954). *Z. Hyg.*, **139**, 389–392.

Seeliger, H. P. R., and Linzenmeier, G. (1955). *Ann. Inst. Past.*, **88**, 127–128.

Seeliger, H. P. R. (1958). Johann Ambrosius Barth Verlag, Leipzig.

Seeliger, H. P. R. (1961). S. Karger Verlag, Basel–New York.

Seeliger, H. P. R., and Bockemühl, J. (1968). *Zbl. Bakteriol. Abt. I. Orig.*, **206**, 216–227.

Seeliger, H. P. R. (1975). *Acta microbiol. Acad. Sci. hung.*, **22**, 179–181.

Seeliger, H. P. R. (1976). Colloque de la societé française de pathologie infectieuse. *Med. Mal. infect.*, **9**, 6–14.

Seeliger H. P. R. and Schoofs, M. (1977). "Seventh International Symposium on the Problems of Listeriosis", Varna, Bulgaria, 23–27 September 1977, pp. 3–4.

Smith, C. W., and Metzger, J. F. (1962). *Path. Microbiol.*, **25**, 499–506.

Srivastava, K. K., and Siddique, H. I. (1974). *Am. J. Vet. Res.*, **35**, 561–565.

Stanley, N. F. (1949). *Australian J. Exp. Biol.*, **27**, 123–131.

Stuart, Sarah E., and Welshimer, J. J. (1974). *Int. J. Syst. Bacteriol.*, **24**, 177–185.

Webb, R. A., and Barber, M. (1937). *J. Pathol. Bacteriol.*, **45**, 523–539.

CHAPTER III

Methods of Typing *Erysipelothrix insidiosa*

Iv. Stoev

Institute Against Swine Diseases, Vratza, Bulgaria

I. INTRODUCTION

A. Antigen structure

Studies of the antigens of *Erysipelothrix insidiosa* date from an early stage in the development of microbiology. Watts (1940) was able to differentiate two groups of *Erysipelothrix* by using agglutinogens. Some strains gave rise to antisera which showed cross agglutination with heterologous organisms. This could be avoided by boiling the antigen used for injections and Watts concluded that the organisms showed thermostable group specific antigens and some thermolabile antigens of lower specificity.

Atkinson (1941) also established two major groups on a basis of antigenic structure together with a third group of strains showing a variable pattern of cross reaction. Gledhill (1945) recognised four groups and Dedié (1949, 1950), using the Lancefield method of precipitin detection, established the presence of two acid-soluble type specific antigens, A and B. Strains devoid of these acid-soluble antigens he placed in a further group, N.

Roots and Venske (1952a, b) used a rapid method for the determination of agglutination to demonstrate a further non-specific antigen and Ewald (1955), using agglutination, established that in every group A strain there was a B antigen present and vice versa. Precipitation using acid extracts detected only type specific acid-soluble antigens. Further antigens have subsequently been demonstrated in *Erysipelothrix*. Heuner (1958) detected an antigen, C, in strains which were non-pathogenic for pigs and a further antigen, D, from various animals. Further antigens have been demonstrated; antigen E by Kucsera (1963, 1964), F by Murase *et al.* (1959) and H by Ewald (1967).

The A and B group strains are pathogenic for pigs while the rest of the groups are non-pathogenic, some of them being isolated from the tonsils of pigs, cows, fish, etc.

Serological grouping of *E. insidiosa* was first carried out by agglutination and tube precipitin reactions. Agglutination is complicated because of spontaneous agglutination and non-specific cross reactions. Cross reactions also occur on occasions with the precipitin method, which, in addition, can be difficult to read. Ouchterlony (1953) introduced the agar–gel diffusion precipitin reaction which opened up new possibilities for the typing of *E. insidiosa*. Truszczynski (1961) and Ewald (1962) used this method for the typing of *Erysipelothrix* and demonstrated that the method was reliable, providing results in 18 h whilst precipitin lines were still visible after several days. Mansi (1958) modified the Ouchterlony method to produce a micro-test using layers of agar poured on glass plates and Nikolov (1964) used this method to type *Erysipelothrix* using microscope object glasses.

B. Bacteriophage

Brill and Politynska (1961) proposed the use of specific *Erysipelothrix* phages for typing purposes. They isolated phages specific for serogroup A and described their morphology. They considered that the phages could be used for group-specific diagnosis and similar results were reported by Hammer (1951). Revenko (1968) isolated group N specific phages and used them to type *Erysipelothrix* strains.

C. Haemagglutination

Dinter (1949) showed that some strains of *E. insidiosa* had the capacity to agglutinate hen erythrocytes. The more highly immunogenic strains of serogroup B showed haemagglutinating ability while group A strains did not do so. Strauch and Nitzschke (1958) claimed that haemagglutination activity of strains was dependent upon the composition of their growth media. It was Dinter's intention to use haemagglutination as an indicator of highly immunogenic strains suitable for the production of inactivated vaccines.

D. Haemadsorption

Stoev (1969) established that *Erysipelothrix* has the ability to adsorb erythrocytes and that the reaction can be suppressed by specific immune antisera thus extending the technique of haemadsorption, already well known in the virological field, to the study of these micro-organisms.

E. Distribution of *Erysipelothrix* serogroups

Hubrig (1962) showed that acute *Erysipelothrix* infections are caused mainly by strains belonging to the A serogroup. Serogroups B and N are detected in less acute infections. *Erysipelothrix insidiosa* isolated from the tonsils of healthy pigs in the German Democratic Republic belong to serogroup N.

Murase *et al.* (1959) found that 22 *Erysipelothrix* strains isolated from diseased pigs showing a septic form of infection belonged to serogroup A. Nineteen strains isolated from pigs showing urticarial forms of infection were distributed as follows; 16 serogroup B, 3 serogroup A. Fifty strains of *E. insidiosa* isolated from the tonsils of pigs belonged to different types; 3 to serogroup A, 22 to serogroup B and the rest to serogroups C, D and F. The group antigens are polysaccharide in nature (Gralheer, 1951; Brill *et al.*, 1959; Erler, 1968).

In view of this association of specific serogroups with different forms of the disease and with non-specific carriers, the development of rapid and simple methods of typing *E. insidiosa* has assumed some importance in the study of the epidemiology of this organism. The rest of this Chapter will be concerned with a comparative study of the methods available and the results of their application.

II. METHODS FOR THE TYPING OF *E. INSIDIOSA*

A. Bacterial strains

Eighty-five strains of *E. insidiosa* were used to examine the effectiveness of different typing methods and the relationships between them. A group of control organisms consisting of previously typed strains were obtained from the Dessau Institute in the German Democratic Republic and from other organisations.

B. Serotyping

The method of Atkinson as modified by Nikolov (1964) was used to produce an acetic acid extract from cells. The method involves the use of dilute acetic acid held in a thermostat during extraction. Group specific A and B antisera were produced by immunisation of rabbits.

The microprecipitation method was carried out by Nikolov's modification of Mansi's method (Nikolov, 1964). Both the agar base and the specific antisera were stored at 4–6°C after the addition of merthiolate (1 in 10 000).

C. Bacteriophage

One to two drops of serogroup A specific *Erysipelothrix* bacteriophage (10^7 plaque-forming units per ml) were added to young (1 to 2 h incubation) broth cultures of the strains to be studied. Lytic action of the phage was determined after 18–24 h cultivation at 37°C.

D. Haemadsorption

The ability of strains to adsorb erythrocytes was determined by the examination of colonies grown on the surface of agar. Broth cultures were inoculated on to serum agar (pH 7·8) in Petri dishes. After cultivation for 24, 48, 72 and 96 h at 37°C the agar surface was flooded with a 2% chicken erythrocyte suspension using a dropper and applying the reagent with great care at the edge of the dish. The flooded plate was allowed to stand for 15–20 min at room temperature after which the erythrocytes were gently decanted and the agar suface carefully washed with physiological saline, again using a dropper. Washing was carried out by slow shaking and rotary movement of the dishes. Sharp or careless movements can lead to the separation of *Erysipelothrix* colonies from the surface of the agar. The agar was then dried in an inverted position for 5–10 min and the colonies observed under a microscope at a magnification of 80 diameters.

Erythrocytes for the test were prepared from poultry, rabbit, sheep and guinea pigs and they were carefully washed several times with physiological saline.

In order to check the specificity of haemadsorption, a section of the agar surface carrying *Erysipelothrix* colonies was overlayered with specific anti-*Erysipelothrix* serum derived from pigs and destined for therapeutic use. Colonies were exposed to the serum for 60 min at room temperature, after which time it was carefully poured off and the whole surface of the agar covered with erythrocyte suspension and the haemadsorption characteristics of both the treated and untreated colonies studied by the methods described above.

E. Haemagglutination

Haemagglutination activity of the strains studied was determined by the standard methods.

III. RESULTS OF THE VARIOUS TYPING PROCEDURES

A. Typing by microprecipitation

Using acetic acid extracts, the characteristic precipitation lines appeared within 2 h and were fully developed after 6–8 h. At 24 h they were still characteristic but became more diffuse and more difficult to interpret later.

Of 80 strains of *E. insidiosa* studied, 30 reacted positively with serogroup A specific antiserum, 42 with serogroup B specific antiserum and eight failed to react. The latter have been designated the N group.

B. Typing by bacteriophage

Only two strains of the serogroup B type and none of serogroup N were lysed by the serogroup A *Erysipelothrix* phage. All strains of serogroup A were lysed.

C. Typing by haemadsorption and haemagglutination

The colonies of some strains of *E. insidiosa* adsorb poultry erythrocytes intensely on to their surface. The erythrocytes form a compact layer which cannot easily be disintegrated by washing. The layer of erythrocytes ends sharply at the edge of the colony and does not spread on to the agar surface. It is more compact at the edges and sometimes on the top of the colonies while on the sloping surfaces the layer is more or less broken. This type of adsorption is designated as $+ + + +$ or $+ + +$. The colonies of other strains adsorb only solitary erythrocytes most often on their periphery. Such adsorption is indicated by $+$ or $\pm$. Some strains show no ability to adsorb and this is recorded as a negative $(-)$ reaction. Colonies produced after 48, 72 or 96 h incubation were suitable for demonstrating haemadsorption with hen erythrocytes but 24 h colonies were too young, washing readily from the surface of the agar. Erythrocytes of sheep, guinea pig and rabbit were not adsorbed and were unsuitable for tests with *E. insidiosa*. Adsorption persisted at 4–6°C for at least 96 h.

The pretreatment of *E. insidiosa* colonies with specific anti-*Erysipelothrix* serum completely blocked haemadsorption. Normal pig serum had no such inhibitory action.

Strains which were typed by microprecipitation as serogroup A showed

TABLE I

Results of typing *Erysipelothrix insidiosa* strains by microprecipitation, phage typing, haemadsorption and haemagglutination

Strain	Serogroup	Specific serum		Normal rabbit serum	A-specific anti-*Erysipelothrix* bacteriophage†	Haem-adsorption	Haemagglu-tination
		A	B				
42	A	++++	−	−	Lyses	+	1:8
57	A	+++	−	−	Lyses	+	1:8
40	A	+++	−	−	Lyses	+	1:16
48	A	++++	−	−	Lyses	±	1:4
203	A	+++	−	−	Lyses	±	1:8
34	A	+++	−	−	Lyses	+	1:8
70	A	+++	−	−	Lyses	±	1:8
532	A	+++	−	−	Lyses	+	1:16
16–89	A	+++	−	−	Lyses	+	1:8
14–65	A	+++	−	−	Lyses	+	1:16
B–5	A	+++	−	−	Lyses	+	1:16
OK	A	+++	−	−	Lyses	+	1:16
PI	A	+++	−	−	Lyses	−	1:8
Polianovo	A	+++	−	−	Lyses	−	1:8
43	A	+++	−	−	Lyses	+	1:8
Dren	A	+++	−	−	Lyses	−	1:4
120	A	+++	−	−	Lyses	+	1:4
Ostrov	A	+++	−	−	Lyses	−	1:4
Dimovo	A	++++	−	−	Lyses	−	1:4
Monastiriste	A	−	−	−	Lyses	−	1:4
Mitchurin	A	+++	−	−	Lyses	+	1:8
428	A	+++	−	−	Lyses	+	1:8
Belene	A	+++	−	−	Lyses	−	1:8
Zidarovo	A	+++	−	−	Lyses	+	1:4

Strain	Type					Lysis			Titer
Boichinovtsi	A	+ + + +			−	Lyses		+	1:8
A	A	+ + + +			−	Lyses		±	1:8
773	A	+ + + +			−	Lyses		±	1:4
201	A	+ + + +			−	Lyses		−	1:4
KI	A	+ + + +			−	Lyses		+	1:8
P-10	A	+ + + +			−	Lyses		+	1:8
62	B	−	+ + + +		−	Does not lyse	+ + +	+ + +	1:8
0-1	B	−	+ + + +		−	Does not lyse	+ + +	+ + +	1:8
29	B	−	+ + + +		−	Does not lyse	+ + +	+ + +	1:8
30	B	−	+ + + +		−	Lyses	+ + +	+ + +	1:32
60	B	−	+ + + +		−	Does not lyse	+ + +	+ + +	1:8
72	B	−	+ + + +		−	Does not lyse	+ + +	+ + +	1:32
69	B	−	+ + + +		−	Does not lyse	+ + +	+ + +	1,8
67	B	−	+ + + +		−	Does not lyse	+ + +	+ + +	1:8
9	B	−	+ + + +		−	Does not lyse	+ + +	+ + +	1:16
25	B	−	+ + + +		−	Does not lyse	+ + +	+ + +	1:16
B-8	B	−	+ + + +		−	Does not lyse	+ + +	+ + +	1:16
Gaber	B	−	+ + + +		−	Lyses	+ + +	+ +	1:16
D-64	B	−	+ + + +		−	Does not lyse	+ + +	+ +	1:32
118	B	−	+ + + +		−	Does not lyse	+ + +		1:64
48-Z	B	−	+ + + +		−	Does not lyse	+ + +	+ + +	1:32
28-Z	B	−	+ + + +		−	Does not lyze	+ + +	+ +	1:16
St-55	B	−	+ + + +		−	Does not lyse	+ + +	+ + +	1:8
10-T	B	−	+ + + +		−	Does not lyse	+ + +	+ + +	1:32
B-22	B	−	+ + + +		−	Does not lyse	+ + +	+ + +	1:16
M-330	B	−	+ + + +		−	Does not lyse	+ + +	+ + +	1:16
31	B	−	+ + + +		−	Doqs not lyse	+ + +	+ + +	1:16
2	B	−	+ + + +		−	Does not lyse	+ + +	+ + +	1:8
3	B	−	+ + + +		−	Does not lyse	+ + +	+ + +	1:8
119	B	−	+ + + +		−	Does not lyse	+ + +	+ + +	1:16
Vr-23	B	−	+ + + +		−	Does not lyse	+ + +	+ + +	1:16
Dulevo	B	−	+ + + +		−	Does not lyse	+ + +	+ +	1:16

TABLE I (continued)

Strain	Serogroup	Specific serum		Normal rabbit serum	A-specific anti-Erysipelothrix bacteriophage†	Haem-adsorption	Haemagglutination
		A	B				
G. Damianovo	B	−	++++	−	Does not lyse	++++	1:32
191	B	−	++++	−	Does not lyse	+++++	1:16
Kneza	B	−	++++	−	Does not lyse	+++++	1:8
Levski	B	−	++++	−	Does not lyse	+++++	1:16
117	B	−	++++	−	Does not lyse	+++++	1:16
121	B	−	++++	−	Does not lyse	++++	1:32
24	B	−	++++	−	Does not lyse	+++++	1:32
Vratza	B	−	++++	−	Does not lyse	+++++	1:8
Novoseltzi	B	−	++++	−	Does not lyse	+++++	1:16
D. Mitropolia	B	−	++++	−	Does not lyse	+++++	1:16
Gigen	B	−	++++	−	Does not lyse	+++++	1:16
Brest	B	−	++++	−	Does not lyse	+++++	1:16
Malchika	B	−	++++	−	Does not lyse	+++++	1:8
774	B	−	++++	−	Does not lyse	+++++	1:16
23	B	−	++++	−	Does not lyse	+++++	1:8
194	B	−	+++	−	Does not lyse	+++++	1:16
16	N	−	−	−	Does not lyse	+++++	1:16
1	N	−	++++	−	Does not lyse	++++	1:32
Vr–22	N	−	−	−	Does not lyse	++++	1:16
193	N	−	−	−	Does not lyes	+++++	1:16
4	N	−	−	−	Does nto lyse	++++	1:8
116	N	−	−	−	Does not lyse	++++	1:16
VR$_2$	N	−	−	−	Does not lyse	+++++	1:8
196	N	−	−	−	Does not lyse	+++++	1:16

+++/++++, positive; +, ±, weakly positive; −, negative.
† The phage used was F$_1$ obtained from the Veterinary Institute of Warsaw, Poland.

only a weakly positive or a negative haemagglutination reaction. Strains belonging to serogroup B or N showed strong haemadsorption abilities.

Using the haemagglutination reaction, serogroup B and N strains gave high haemagglutination titres in comparison with strains of serogroup A but there was no sharp distinction between the three groups.

The results of typing by the various methods are summarised in Table I.

IV. DISCUSSION

Of the 85 strains of *E. insidiosa* studied 30 reacted with A group antiserum, 47 with B group serum and eight did not react with either. Naturally the serological approach is considered the most reliable and these results form the basis of allocation of strains to groups.

The results with bacteriophage show a close correlation with serological grouping. Brill and Politynska (1961) explain the occurrence of exceptional lysis seen with old laboratory cultures of serogroup B strains in terms of a change of cell specific antigen. These authors propose that it is necessary only to use the A specific bacteriophage for typing *E. insidiosa*. Revenko (1968) has also reported the existence of B and N group phages. These remain to be studied for their application to typing. Representatives of all three groups are found to have pathological and immunological significance.

The haemadsorption reaction proves to be particularly significant for the typing of *E. insidiosa*. All the strains serologically typed as group A failed to adsorb poultry erythrocytes to a significant extent while those of groups B and N show a markedly positive reaction.

Bacteriophage and haemadsorption tests provide useful auxiliary typing methods. They are simple to perform and can be carried out in most laboratories. Undoubtedly, however, the most useful method is serological typing based on microprecipitation in agar. This method is accurate and differentiates clearly between the three serogroups. The extraction method is simple and the specific antisera are highly potent, such that 1 ml of serum will enable hundreds of strains to be typed. Results can conveniently be read in one working day and the interpretation of the precipitin lines is normally unequivocal.

REFERENCES

Atkinson, N. (1941). *Aust. J. exp. Biol. med. Sci.* **19**, 45–50.
Brill, J. and Politynska, E. (1961). *Zbl. Bakt. I Orig.* **181**, 473–477.
Brill, J., Mikulaszek, E. and Truszezynski, M. (1959). *Zbl. Bakt. I Orig.* **176**, 468–475.
Bukrinskaia, A. G. (1960). "Woprosi virusologii" N2.

Dedié, K. (1949). *Mh. Vet. Med.* **4**, 7–10.
Dedié, K. (1950). *Arch. exp. Vet. Med.* **2**, 56–66.
Dinter, Z. (1949). *Zbl. Bakt. I Orig.* **153**, 281–284.
Erler, W. (1968). *Arch. Exp. Veterinaermed.* **22**, 1139–1145.
Ewald, F. W. (1955). *Mh. Tierheilk.* **7**, 109–119.
Ewald, F. W. (1962). *Berl. Münch. Tierärztl. Wschr.* **75**, 71–73.
Ewald, F. W. (1967). *Berl. Münch. Tierärztl. Wschr.* **80**, 335–339.
Gledhill, A. W. (1945). *J. Path. Bat.* **57**, 179–189.
Gralheer, H. (1951). *Exp. Vet. Med.* **3**, 65–68.
Hammer, D. (1951). Dissertation, München.
Heuner, F. (1958). *Arch. exp. Vet. Med.* **12**, 40–61.
Hubrig, Th. (1962). *Ztbl. Bact. I Orig.* **186**, 344–354.
Kucsera, Gy. (1963). *Acta Vet. Acad. Sci. Hung.* **13**, 61–63.
Kucsera, Gy. (1964). *Acta Vet. Acad. Sci. Hung.* **14**, 293–298.
Mansi, W. (1958). *Nature, Lond.* **181**, 1289–1290.
Murase, N., Suzuki, K., Nakahara, T., Araumi, W. and Hashimoto, K. (1959). *Jap. J. Vet. Sci.* **21**, 113–120.
Nikolov, P. (1964). *Izv. Mikrobiol. Inst., Sofia.* **XVI**, 75–84.
Ouchterlony, Ö. (1953). *Acta path. Microbiol. Scand.* **32**, 231–240.
Revenko, I. P. (1968). *Veterinaria* **7**, 25–27.
Roots, E. and Venske, W. (1952a). *Berl. Münch. tierärztl. Wschr.* **65**, 184–187.
Roots, E. and Venske, W. (1952b). *Berl. Münch. tierärztl. Wschr.* **65**, 223–225.
Sato, S., Matsui, K. and Yoschida, Y. (1965). *Nat. Inst. Anim. Hlth Qt., Tokyo* **5**, 45–46.
Stoev, I. (1969). *Res. Prod. Inst. Swine Dis. Vratza* **II**, 131–135.
Strauch, D. and Nitzschke, E. (1958). *Zbl. Vet. Med.* **5**, B, 968–976.
Truszczynski, M. (1961). *Am. J. vet. Res.*, **22**, 836–838.
Vogel, J. and Shelokov, A. (1957). *Science, N.Y.* **126**, 358–359.
Watts, P. S. (1940). *J. Path. Bact.* **50**, 355–369.

CHAPTER IV

Corynebacterium diphtheriae:
Microbiological Methods Used in Clinical
and Epidemiological Investigations

ALICE SARAGEA, PAULA MAXIMESCU AND EUGENIA MEITERT

The Cantacuzino Institute, Bucharest, Romania

I. GENERAL CONSIDERATIONS

The object of this Chapter is to discuss the background of the practical methods applied by the bacteriologist in diagnosing diphtheria in a suspected case and/or in helping the epidemiologist to uncover epidemic relationships in diphtheria outbreaks or in endemic conditions, rather than to deal with the theoretical descriptions of taxa. It will be based on a complex methodology applied by the authors in controlling diphtheria in Romania in the course of 20 years. The same system has been successfully applied by colleagues, who asked the authors for scientific and technical assistance, in different reference centres of the world.

It seems that few other diseases of man have been as successfully studied as diphtheria (Davis *et al.*, 1973) and marked progress is continuously recorded in different theoretical and practical problems concerning this infection.

A dramatic decline of diphtheria morbidity and mortality has been recorded during the decades since the introduction of modern prophylactic and therapeutic agents. There are, however, many convincing arguments which support the investigation and control of diphtheria as a mass infection.

The high levels of specific immunity acquired through the vast immunisation programmes have on the one hand deeply modified the biological properties of the causative agent, and on the other have brought about many atypical forms of diphtheria; the clinical and bacteriological diagnosis thus present marked difficulties for the young practitioner who has not himself had the opportunity to see classical diphtheria.

Due to the fact that the number of isolated organisms in a well-immunised community is decreasing, the waning interest in this disease usually contributes to an increasing number of non-immunised or partially immunised individuals in the community.

Under these circumstances a recurrence of sporadic cases of the disease or even long-lasting outbreaks occurring in different parts of the world, is important. In the U.S.A., for instance, before the introduction of mass immunisation programmes, more than 10 000 cases, were reported annually. Since 1952 a marked drop in the morbidity of diphtheria has been observed.

However, 50 years after the introduction of diphtheria toxoid, substantial segments of the United States population are still not immunised against diphtheria, and large, sporadic diphtheria outbreaks still occur (Marcuse *et al.*, 1973; Zalma *et al.*, 1970; McCloskey *et al.*, 1972).

General experience also shows that occasional outbreaks may still occur everywhere, suggesting that the organism exists in endemically diverse forms, and when conditions are appropriate, these may initiate either typical or atypical infections of the upper respiratory tract or cutaneous infections (Wilson and Toshach, 1957; Belsey *et al.*, 1969).

Cutaneous diphtheria and skin carriage frequent in native populations of tropical countries represent a possible connection between epidemics. All these facts, the intense tourist traffic throughout the world, as well as the pathogenicity of diphtherial toxaemia which calls for a rapid clinical and microbiological diagnosis, explain the present interest for practical and standardised methods in controlling diphtheria.

II. MICROBIOLOGICAL CHARACTERS OF *CORYNEBACTERIUM DIPHTHERIAE*

A. Definition of the genus *Corynebacterium* and of the type species *Corynebactherium diphtheriae*

"Bergey's Manual of Determinative Bacteriology" (Bergey, 1974) defined this group of micro-organisms.

Straight to slightly curved rods with irregularly stained segments, and sometimes granules. Frequently show club-shaped swellings. Snapping division produces angular and palisade (picket fence) arrangement of cells. Generally, non-motile. Gram-positive, although some species (e.g. *C. diphtheriae*) lose the stain easily, especially in old cultures, while others are more tenacious; granules, however, are strongly Gram-positive. Not acid-fast.

Cell wall composition is characterised by the presence of meso-diaminopimelic acid as the diamino-acid of the peptidoglycan, and by a polysaccharide containing arabinose, galactose and often mannose.

Chemo-organotrophs: carbohydrate metabolism mixed fermentative and respiratory.

Do not produce soluble hemolysins but on solid media containing blood lysis may occur if the red cells are in contact with the colonies. Some pathogenic species produce exotoxins.

Aerobic and facultatively anaerobic; grow best aerobically often with a surface pellicle.

Catalase positive.

Type species: *Corynebacterium diphtheriae* (Klebs, 1883; Lehmann and Neumann, 1896).

Vernacular name: Klebs–Loeffler bacillus.

Corynebacterium diphtheriae: straight or slightly curved rods, frequently swollen at one or both ends, $0 \cdot 3$–$0 \cdot 8$ by $1 \cdot 0$–$8 \cdot 0$ μm; usually stain unevenly and often contain metachromatic granules (polymetaphosphate) which stain bluish purple with Methylene blue. Gram-positive, but rather easily decolourised, especially in old cultures.

Three main distinct cultural types recognised (McLeod, 1943) called *gravis, intermedius* and *mitis*, in accordance with the clinical severity of the cases are most frequently isolated.

Typical strains produce acid from glucose and maltose, but not from sucrose; however, a few strains produce acid from sucrose.

Aerobic and facultative mixed fermentative and respiratory metabolism, does not produce indole, urease or gelatinase, optimum temperature 37°C, range 15–40°C.

Characteristic cell wall sugars are arabinose, galactose and mannose; diamino-acid of peptidoglycan is meso-DAP (Cummins and Harris, 1956).

A toxic glycolipid is present which is a 6-6′-diester of trehalose, containing corynemycolic acid and corynemycolenic acid in equimolar proportions (Kato, 1970).

Pimelic acid, nicotinic acid and beta-alanine required by almost all strains.

The guanidine + cytosine content of the DNA varies from $51 \cdot 8$ to 60%.

Most strains of distinct cultural types (*gravis, intermedius, mitis*) produce an identical exotoxin, highly lethal; however non-toxigenic strains occur which are typical in all other respects. Ability to produce toxin is determined by the presence of prophage (Freeman, 1951) carrying specific determinant called *tox+*, and toxigenicity is induced in non-toxigenic strains by making them lysogenic (Holmes and Barksdale, 1969).

Within the genus *Corynebacterium* we shall deal especially with those members of the group, which are common in the rhinopharynx and hence present problems in differential microbiological diagnosis from the diphtheria bacillus, i.e. (1) *C. diphtheriae* as type species, the causative agent of diphtheria, parasite of man; (2) *C. ulcerans*—the second pathogenic species with human and/or animal habitat; (3) non-pathogenic saprophytic corynebacteria.

Of the saprophytic corynebacteria, from the upper respiratory tract, the following species will be considered: *C. hofmannii*, *C. xerosis*, atypical corynebacteria and some variants with "dwarf" colonies.

B. Characteristics of *C. diphtheriae*

1. *Habitat*

C. diphtheriae, the causative agent of diphtheria, is usually found in the upper respiratory tract, on the skin, conjunctiva and vagina. Its capacity to produce toxin and the immunity state of the individual decide whether diphtheria will evolve. At the site of entry it produces a greyish false membrane—constituted of a fibrinoleukocytic exudate. From the false membrane, where diphtheria bacilli are located, they produce an exotoxin, which invades the tissues by the blood stream.

In individuals with a protective level of 0.03 AU/ml (antitoxin units) the location of *C. diphtheriae* on the mucosa is uneventful or results only in mucosal erythema. The individual may become a symptomless carrier.

C. diphtheriae spreads only among human beings; there is no intermediate host. Cats, dogs and other domestic animals have rarely been incriminated. From sores on the teats of cows, virulent *C. diphtheriae* have been isolated (Henry, 1920; Scott, 1934), but as a rule they have come from sores on the hands of milkers who were carriers of the organism.

2. *Morphology and structure*

Non-motile, non-capsulated, non-flagellate club-shaped bacillus (Fig. 1). Gram-positive. A strongly characteristic feature is the readiness with which it decolourises. Morphologically, characteristic cell forms are obtained on smears prepared from an overnight Loeffler slant culture. With Loeffler's alkaline Methylene blue, Toluidine blue, or double staining (for instance by Neisser or Del Vecchio's method) characteristic granules (Babes, 1886; Ernst, 1888) may be differentially stained.

By double staining, the cells have blue–black granules on a brown

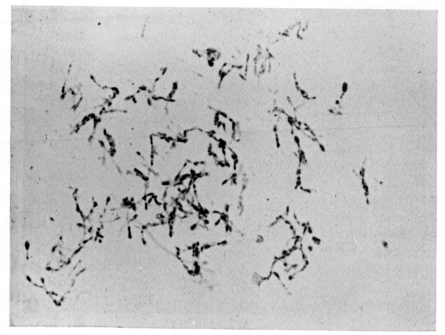

Fig. 1. *C. diphtheriae*

background. The granules are composed of long chain inorganic poly-phosphates (Van Wazer, 1958) and are found most abundantly in cells with retarded growth. They are thought to be storage material for the actively growing cells.

Another characteristic feature of corynebacteria is the arrangement of the cells in the smear. This is often compared to Chinese letters or matches which have been randomly distributed (Fig. 1).

3. *Metabolism and growth requirements*

(a) *Animal protein requirements.* The growth of all members of this group and especially of *C. diphtheriae* is improved by the presence of animal protein, e.g. bovine or horse serum, sheep blood, rabbit blood. Accordingly, all media used for *C. diphtheriae* must be supplemented by animal protein.

For several decades Loeffler's inspissated serum slants represented the medium of choice for *C. diphtheriae*.

However, the difficulties in isolating and recognising any differential colonial morphology on this white medium and its lack of selectivity makes it mainly of value for secondary cultivation of selected colonies.

The exclusive use of Loeffler's medium for primary cultivation of *C. diphtheriae* in many laboratories probably explains the delay and difficulties encountered in the isolation and identification of *C. diphtheriae* by many earlier workers.

(b) *Oxygen requirements* of different species within the genus vary widely. *C. diphtheriae* develops freely and better under aerobic conditions. Diphtherial toxin is also better produced aerobically. Strains which have to be primarily isolated under anaerobic conditions are now referred to the genus *Propionibacterium* (e.g. *C. acnes*).

(c) *Growth factors*. Most of the strains lack certain synthetic capacities in connection with the formation of nicotinic and pantothenic acids. Thus, nicotinic acid becomes an essential component for a suitable culture medium as does pantothenic acid or in some cases beta-alanine from which certain strains form their own pantothenate molecule.

Similarly, certain cultures require biotin, others pimelic acid, whereas others are independent of these substances.

Growth and toxin production are favoured by the addition of amino acids to a synthetic medium and of those aspartic acid and tryptophan are apparently indispensable. Drew and Miller (1951) worked out a completely defined medium for the growth and toxin production of strain PW$_8$, the highly toxigenic strain used throughout the world for commercial toxin production. The medium contains beta-alanine, nicotinic acid, pimelic acid, cystine, glycine, valine, leucine, methionine, biotin, glutamic acid, tryptophan, ammonium ions and some salts such as magnesium, copper, zinc, manganese and iron salts. For everyday use in research work the casein hydrolysate medium of Mueller and Miller (1941) is more practical. For research purposes Barksdale (1970) recommends this medium supplemented with pantothenate, glutamate and tryptophan and therefore calls it PGT medium. He also draws attention to the fact that certain strains do not grow unless thiamine is also present.

Generally, for primary isolation, enrichment, selection of the organism, as well as for routine diagnostic procedures and commercial manufacture of diphtherial toxin, complex media are used, supplied with animal protein and L-cystine. All the other necessary amino-acids are contained in the complex media. The formulae of all media used in the routine standardised diagnostic scheme will be given in Section XIII.

(d) *The optimum growth temperature* is 37°C. For testing phage–host bacteria relationships as well as for toxin production 35°C is optimal.

(e) *The reduction of tellurium salts* is a characteristic feature of *C. diphtheriae* and to a lesser extent of other members of this group. The observations of Klett (1900) concerning the capacity of some bacteria to reduce tellurium salts was used by Conradi and Troch (1912) to develop for *C. diphtheriae* a relatively selective medium containing potassium tellurite.

Morton and Anderson (1971) found needle-like crystals in cells of *C. diphtheriae* and *C. xerosis* grown on plates of tellurite agar. They noticed that most of the crystals were contained wholly within cells, and suggested that the tellurite ion is probably able to diffuse through the cell wall, where it is reduced to tellurium metal, precipitated within the cell affording the black colour to the colonies.

Tucker *et al.* (1962) by applying X-ray diffraction, supplied the first indication that tellurite crystals accumulate in *C. diphtheriae*. The black hue which corynebacteria develop on tellurite blood agar or on Tinsdale agar facilitates the selection of corynebacteria among a mixed pharyngo-nasal flora.

4. *Cultivation and cultural characteristics*

(a) *Types of* C. diphtheriae. A previous section mentioned that potassium tellurite has little effect on *C. diphtheriae* in contrast to most other bacteria. By means of such media, Anderson *et al.* (1931) differentiated between two colony types of *C. diphtheriae*. Initially, they had indications that colony type was correlated to the severity of disease. One type, that occurred dominantly in severe cases, was called the *gravis* type and another, which came from milder cases, the *mitis* type. Anderson *et al.* (1933) differentiated a third type—now generally known as type *intermedius*. The three colony types, with their associated biochemical characters and their different phage types (Saragea and Maximescu, 1964, 1969), are now generally recognised as very stable, and hence still used in current practice. Today, however, there is by no means general agreement concerning correlation of the type with the clinical severity of the case. All three types may be either toxigenic or non-toxigenic. As regards their epidemic behaviour, present evidence indicates that diphtheria in its epidemic form is frequently due to the *gravis* type, *intermedius* coming next.

In Romania, for instance, during the last 20 years *gravis* was the only type isolated in epidemics or later in sporadic cases. The same situation occurred in the main European countries. However, in the U.S.A. in some recently investigated outbreaks (McCloskey *et al.*, 1972) the toxigenic *intermedius* type prevailed.

Mitis is more frequent in endemic areas, and fatality with *mitis* strains is believed to be due more often to tracheal or laryngeal obstruction than to toxic injury. For a thorough discussion of the matter, McLeod's work (1943) may be consulted.

Some strains display various characters which make it difficult to place them in any one type; so-called *minimus* strains (Frobisher *et al.*, 1945) appear to be identical with *intermedius* (Johnstone and McLeod, 1949; Freeman and Minzel, 1950).

Other atypical strains were cited by different authors, among which the most important seem to be the atypical *gravis* corresponding to type IV Wright and Christison (McLeod, 1943; Saragea *et al.*, 1962) and variants with "dwarf" colonies described by Maximescu (pers. comm.).

(b) *Growth characteristics: media used.* The following media are used for routine microbiological diagnosis:

A liquid enrichment medium ECST (egg yolk, cystine, serum, tellurite) (Calalb *et al.*, 1961).

Tinsdale as a selective solid medium (Meitert and Saragea, 1967).

Cystine–tellurite–blood agar (CTBA), a medium used for differentiating the three biotypes: *gravis, mitis, intermedius.*

Blood agar (BA).

Loeffler slants for stock culture.

Plain or glucose broth.

(c) *Cultural aspects.* In the enrichment medium a positive or suspect swab develops black points attached to the cotton of the swab, within 10–14 h. However, this cannot account for more than a suspicion of positivity.

On Tinsdale plates (Tinsdale, 1947; Meitert and Saragea, 1967), *C. diphtheriae* develops in most cases after 24 h incubation, small circular greyish black colonies, surrounded by a specific well-delineated brown halo, representing the most characteristic feature on which the high specificity of this medium is based (Fig. 2).

In a few cases, colonies develop only after 48 h incubation period during which the brown halo becomes intensely brown. The occurrence of brown haloed colonies is specific only for *C. diphtheriae* and *C. ulcerans*, both high cystinase producers. Even one colony on the plate can be picked out. None of the diphtheroids and/or of the variants of this group presents this feature. The brown halo results from the reaction between potassium tellurite and hydrogen sulphide (H_2S) produced by cystinase, owing to the L-cystine contained in the medium. This reaction is represented by a local tellurite sulphur precipitate. The sodium thiosulphate provides a good reducing agent. Welsch and Thibaut (1948), analysing the phenomenon of the specific halo described by Tinsdale (1947), demonstrated that it represents one of the two main biochemical properties of *C. diphtheriae*: cystinase and the capacity to acidify the medium. The association of these two

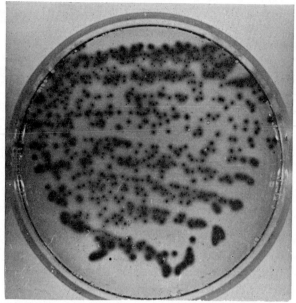

FIG. 2. Growth characters of *C. diphtheriae* colonies on Tinsdale plates. Brown haloes around greyish black colonies.

properties is not to be found in corynebacteria other than *C. diphtheriae* and *C. ulcerans*, which explains the specificity and practical interest of this halo.

On cystine–tellurite–blood agar (CTBA) plates. The three types *gravis*, *mitis* and *intermedius* present very characteristic colonial features.

Gravis strains—large plate, daisy-headed colonies, grey to black, with a rough, matt surface, displaying radial striations and crenated edges (Fig. 3). When transilluminated in a dissection microscope (stereoscope) they show greyish transparent margins and a dense black opaque centre. Consistency: friable colonies, tendency to crumble when touched by the loop.

Mitis colonies are middle sized or large, convex with regular margins and a glossy smooth surface (Fig. 4); they appear black and opaque when examined in transmitted light; consistency: butyrous.

Intermedius type—small or middle sized colonies with a flat granular surface—or sometimes with a raised centre, and regular edges (Fig. 5). When examined in transmitted light they show a translucent margin. Thus, by its colonial appearance, the *intermedius* type occupies an intermediary position between the rough granular *gravis* and the smooth *mitis*.

On blood agar (BA) plates, *C. diphtheriae* colonies are pearl greyish in

FIG. 3. *C. diphtheriae* type *gravis* colonies on CTBA plates.

FIG. 4. *C. diphtheriae* type *mitis* colonies on CTBA plates.

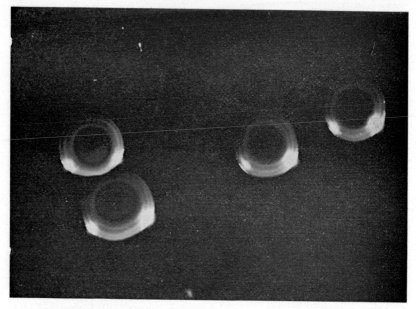

FIG. 5. *C. diphtheriae* type *intermedius* colonies on CTBA plates.

colour, with granular or shiny surface, with crenated or regular edges; (Fig. 6) they are friable or butyrous.

On Loeffler slants, the granular culture is white, pale or deeper cream in colour, butyrous in consistency and easily emulsifiable in saline; it sometimes shows spontaneous agglutination.

On nutrient broth growth readily occurs. Type *gravis*: veil on surface and a small deposit at the bottom, the remaining broth clear. Type *mitis*: homogeneous, slightly turbid. Type *intermedius*: granular deposit on the walls and at the bottom.

5. *Enzymes*

(a) *Cystinase*. Hydrogen sulphide production from cystine is one of the most characteristic enzymatic features of *C. diphtheriae* (Pisu and Guarnacci, 1971). Cystinase is produced also by *C. ulcerans* and *C. ovis*. This enzyme is not to be found in any other of the diphtheroids.

(b) *Glycoside hydrolases*. A general characteristic of almost all members of the group is production of acid from carbohydrates without gas. *C. hofmannii* is the only member of the group which does not split any carbohydrate. Glycoside hydrolases are very important "markers", useful for the diagnosis of the type species of *C. diphtheriae*, as well as its biotypes,

FIG. 6. *C. diphtheriae* type *gravis* colonies, on BA plates.

gravis, intermedius, and *mitis*. *C. diphtheriae* produces acid from glucose, maltose and levulose but not from sucrose, apart from the rare exceptions described by Frobisher *et al.* (1945) and Mauss and Keown (1946). Sucrose fermentation has to be examined with great care, as it is used for eliminating other corynebacteria as candidates for the diagnosis of *C. diphtheriae*, although sucrose positive *C. diphtheriae* do occur. Among polysaccharides *C. diphtheriae* type *gravis* splits dextrose, starch, and glycogen. Type *mitis* does not form acid from polysaccharides, and *intermedius* splits only dextrose. *C. ulcerans* may ferment like *gravis* or *intermedius* and also possesses trehalose-1-glucohydrolase.

(c) *Urease*. Püschel (1936) underlined the ability of most diphtheroid bacilli to hydrolyse urea, with production of ammonia, in contrast to *C. diphtheriae*, which is uniformly negative. The great majority of *C. hofmannii*, i.e. 90%, and about 60–70% of *C. xerosis* strains possess this enzyme.

All *C. ulcerans* produce urease. Evidence of this enzyme represents a valuable marker.

(d) *Phosphatase*. First described by Bray (1944) is present in some diphtheroids (*C. xerosis* and some *C. hofmannii*), atypical corynebacteria, and always absent in *C. diphtheriae*.

(e) *Nitrate-reductase* is present in all corynebacteria except a variant of the *mitis* type, *C. belfanti*, isolated from cases of ozena (Bezjack, 1958; Oehring, 1963a, b), *C. ulcerans* and *C. ovis*.

(f) *Haemolysin*. On sheep blood agar and still better on rabbit blood agar, the *mitis* type is haemolytic, *intermedius* non-haemolytic, and *gravis* variable in this respect.

(g) *Neuraminidase* (N-acetylneuraminate glycohydrolase) has been found in some *C. diphtheriae* strains, both toxigenic and non-toxigenic (Morijama and Barksdale, 1967). This enzyme has been shown to be intracellular, possibly associated with the cytoplasmic membrane. Diphtherial toxin and diphtherial neuraminidase have the same molecular weights, despite the fact that neuraminidase can be produced by the PW_8 toxin-producing cells grown in conditions unsuitable for toxin production (Barksdale, 1970). According to the above authors, neuraminidase appears to be distinct immunochemically and electrophoretically (Blumberg and Warren, 1961; Heide and Haupt, 1962).

(h) *Catalase* is a property of all corynebacteria.

(i) *Gelatin* is liquefied only by *C. ulcerans*.

6. *Resistance and sensitivity*

Diphtheria bacilli may live for as long as 5 weeks in dust.

They are readily killed by heat (10 min at 58°C) and by the common antiseptics. *Corynebacterium diphtheriae* has been isolated from dust around beds in diphtheria wards (Christie, 1969). Even normal saline will kill diphtheria bacilli within a short time, a point of importance in laboratory practice. The organism is fully sensitive towards most antibiotics. It is moderately sensitive to sulphonamides *in vitro*.

7. *Pathogenicity*

(a) *Pathogenicity*. Diphtheria is a typical *tox* infection, often fatal, and its lesions are mainly produced by the action of a powerful exotoxin diffusing rapidly throughout the body from the growing bacilli at the site of infection.

The problem of pathogenesis is mainly that of understanding the origin, nature and mode of action of the toxic protein, elaborated by diphtheria bacilli.

(b) *Experimental models*. Experimental infections have been demonstrated in guinea pigs and rabbits which are most susceptible; dogs, cats, and pigeons are moderately susceptible, and rats and mice very resistant.

In this respect the interested research worker is referred to Loeffler's (1884) early review concerning the pathogenicity in animals.

In guinea pigs, the following clinical symptoms may occur: the animal appears to be suffering and sits crouched in its cage; a soft swelling occurs at the site of inoculation, within 24 h, and death usually occurs within 48–92 h. The time to death may be shorter with larger doses of toxin.

At necropsy of a guinea pig that died within 4 days, the following three typical signs were noted: (i) gelatinous haemorrhagic oedema at the site of inoculation (subsequently developing to necrosis); (ii) swelling and congestion of the adrenal glands; (iii) serous or haemorrhagic exudates in the pleural and abdominal cavities with intensive congestion of the abdominal viscera. Degenerative changes can always be demonstrated in the heart, liver and kidneys. Sublethal doses cause late paralysis similar to that observed in man.

All toxic effects described, as well as death, can be specifically neutralised and prevented by a previously administered specific diphtheria antitoxin.

By intradermal inoculations on the shaved skin of a rabbit, diphtherial toxin or toxigenic strains promote a typical erythematous reaction which may also be neutralised by a previously administered antitoxin.

An experimental model was achieved (Calalb *et al.*, 1963) by conjunctival infection of the guinea pig with *C. diphtheriae* strains. Bonciu *et al.* (1965, 1966), and Meitert *et al.* (1972a, b) established clear-cut differences in the pathogenic effects of the three biotypes: *gravis, intermedius, mitis*, tox^+ and tox^-.

(c) *Clinical aspects.* In man the first obvious sign of infection is the occurrence of a greyish, extensive, adherent membrane, in most cases on the fauces. It may also be localised on the nasal mucosae, where there is a thick, sero-sanguinous rhinorrhoea, sometimes on skin wounds, leg ulcers, on the fingers, ears, eyes, vagina.

Diphtheria bacilli multiply at the site of infection and the toxin produced is carried by the blood stream to the rest of the body. It produces lesions mainly in the heart and nervous system, possibly leading to death. Paralysis of the palate muscles and of other muscles are common complications. How toxin reaches the nervous system is as yet uncertain.

When a general specific level of immunity has been acquired in large sections of the population, circulating toxinogenic *C. diphtheriae* may give rise to atypical clinical forms of diphtheria, characterised by pharyngeal redness and slight fever, situations that have to be differentiated from common colds.

8. *Virulence and toxinogeny*

With Freeman's discovery (1951) that pairs of toxigenic (tox^+) and non-toxigenic (*tox⁻*) strains of *C. diphtheriae* could be obtained, differing from one another by only one prophage and by their toxigenic capacity, it became obvious that "invasiveness" (virulence) and toxigenicity were separable properties (Barksdale *et al.*, 1960; Van Heyningen and Arseculeratne, 1964; Murata *et al.*, 1959).

For example, rabbits infected with the original strain (*tox⁻*), developed pseudomembranous lesions, but later recovered, whereas rabbits infected with the toxigenic pair strain (*tox⁺*) developed necrotic lesions and died. This separation of invasiveness from toxigenicity was consistent with the reported cases of mild diphtheritic infections in man caused by non-toxigenic diphtheria bacilli, as well as with those records of atypical clinical cases of diphtheria occurring in individuals who possess circulating antitoxin (Belsey *et al.*, 1969; Barksdale, 1970; Brooks *et al.*, 1974).

Easily distinguishable properties, "invasiveness" and "toxigencity", are also evident in the wide spread (in Romania) during the present post-epidemic period, in very well immunised children, of a large number of non-toxigenic strains in healthy carriers. Sometimes cause, e.g. sore throat, chills or erythematous angina from which patients usually recover without any treatment.

Virulence (invasiveness) is obviously under the genetic control of the bacterium, while toxigeny is under that of the temperate corynebacteriophage. Thus, "invasiveness" is expressed by the growth of bacilli at the site of infection (pseudomembrane) and is related to certain protein and lipid components associated with their surfaces (Huang, 1942a, b; Lautrop, 1950, 1955; Kato, 1970) while toxinogeny is related to and dependent on the presence of prophage in the bacterial genome.

III. DIPHTHERIAL TOXIN

A. Phage–bacteria relationships in *C. diphtheriae*

1. *Lysogeny and toxin production*

(a) *General considerations.* Lysogeny is the perpetuation of temperate bacteriophages, prophages, as a part of the bacterial replicating system (Barksdale and Arden, 1974).

When a non-lysogenic bacterium like *C. diphtheriae* incorporates the genome of a temperate bacteriophage into its chromosome it becomes lysogenic—it has been lysogenised; it carries a phage genome as a prophage.

In the change from the non-lysogenic to the lysogenic state, bacteria acquire the capacity to release phage either after ultraviolet light induction, by chemical induction or by other physical conditions as, for instance, shaking cultures, thus acquiring a state of immunity against the liberated phage.

The acquisition of these two properties is obligatory for all lysogenic conversions. In some cases bacteria acquire new properties through lysogenisation. One of the most striking effects of genotypic modifications brought about in the cell by lysogenisation is the toxigenic conversion in *C. diphtheriae*. By exposing a phage-sensitive, non-toxigenic (*tox⁻*) strain to a filtered lysate of phage B (previously isolated, described and referred to by Toshach (1950) as phage C/13) Freeman (1951) made the remarkable discovery that the *tox⁻* strain exposed to phage resulted in the outgrowth of *C. diphtheriae* toxigenic (*tox⁺*) strain. The same change was induced in countless experiments by various investigators and by ourselves using the temperate mutant β isolated by Barksdale from phage B in different *C. diphtheriae* strains. Similar results were also obtained by using other temperate bacteriophages. The non-toxigenic strains were always demonstrated as not producing toxin prior to exposure to phage. Passage of the phage-modified strains through antiphage serum in no way altered the capacity to produce toxin (Freeman and Morse, 1952). This signified that the lysogeny and toxigeny were both stable properties. Groman (1955) succeeded in curing the C_4 (β) *tox⁺* strain of its prophage and toxigeny by superinfection with a virulent mutant of phage β, thus demonstrating that elimination of prophage from lysogenic toxigenic strains resulted in loss of toxigenicity.

Later, in analytical experiments, it became obvious that converting genes are expressed during vegetative replication of phage. Matsuda and Barksdale (1967) using a virulent mutant of β were able to obtain yields of toxin in one cycle of virus growth equal to the best yield of the PW_8 strain. Corynebacteriophage β-*tox⁺* is a DNA virus. Groman (1955) found a temperate phage γ-*tox⁻* which was related to β-*tox⁺* and recombinants of the two were obtained. Holmes and Barksdale (1969, 1970) elaborated a mating system and achieved the mapping of the *tox* gene. They showed that gene *tox⁺* is distributed among a considerable variety of corynebacteriophages, differing both serologically and genetically. Thus *tox⁺* occurs even in phages not closely related to β.

2. The gene tox of corynebacteriophages

(a) *Stability of integration of* tox *prophages.* There are *tox⁺* and *tox⁻* bacteriophages.

Even when the *tox⁺* gene is present in a bacteriophage it cannot always be expressed. Experiments of skilled workers (Barksdale, 1970) as well as our own data, obtained during the surveillance of the pathogenic agent, showed that, irrespective of the method of storing of the organism, prophages and toxigeny are stably integrated and that only in pseudolysogenic carrier cultures is the loss of toxigeny observed. In this connection the existence of the PW$_8$ highly toxigenic strain should be emphasised. This strain, isolated from a case in 1896 by Park and Williams, is used as a high toxin producer throughout the world and all its variants maintained at different reference centres of the world are still toxigenic and lysogenic (Maximescu, 1968).

(b) *Expression of gene* tox. *Role of iron in toxin production.* The tox^+ gene is essential for toxin synthesis, but the expression of gene *tox* is possible only in certain *C. diphtheriae* strains. Even with the PW$_8$ strain, a very high toxin producer, important variations with regard to toxin were recorded in different laboratories. Although long recognised, the phenomenon did not receive a satisfactory explanation until Pappenheimer and Johnson (1936) showed that inorganic iron concentration in the medium played an important and unexpected role in connection with toxin formation. At a concentration of about 100 μg of iron per litre of medium, the peak of toxin production was reached and further additions, although improving growth, resulted in a rapid fall of toxin, up to 500 μg of Fe per litre, when toxin could no longer be demonstrated.

Barksdale (1955) and Barksdale *et al.* (1961) checked the conditions in which the *tox* character may be expressed, and also found that the physiological state of the host bacteria was important. Toxin is synthesised in high yield only by bacteria with an abnormally low iron content. Thus, the classic Park Williams 8 strain (PW$_8$) possesses the unusual capacity of increasing in mass five- to six-fold after depleting its exogenous supply of iron.

The mechanism by which iron controls toxin production has not yet been elucidated. Recent evidence (Pappenheimer and Gill, 1973) suggests that the repressor of the *tox* gene may be an iron-containing bacterial protein.

In complex media containing chelating agents, such as phosphorus and calcium ions, the effect of iron on toxin production was not dramatic. In Pappenheimer's refined medium far less iron was needed to repress the yields of toxin because it did not contain chelating agents. Since then, only complex media have been used for the commercial production of toxin.

It is thus certain that while the structural information for toxin biosynthesis is now known to be carried by the phage genome (Uchida *et al.*, 1971), its expression is controlled by the nature and physiological state of the bacterial host. In particular, high yields of toxin are only synthesised by lysogenic bacteria with low iron content. Maximescu *et al.* (1968, 1974a, b) showed that gene *tox* can be expressed and may induce toxin also in strains of *C. ulcerans* and *C. ovis*, the latter representing a species which has been associated with infections in sheep (apart from some very rare exceptions in humans). When any one of these corynebacteria were infected with certain *tox*[+]-carrying bacteriophages, they produced a typical diphtherial toxin.

(c) Tox: *structural gene, for toxin biosynthesis.* Genetic crosses between toxigenic and related non-toxigenic diphtherial phages have revealed that the capacity to convert to toxigeny, the *tox* character, maps in a single region of the phage genome. Until 1971 it was not established, however, whether this region contains the structural information for toxin synthesis, or whether *tox* acts indirectly to permit the expression of a host structural gene.

Having this in mind, Holmes and Barksdale (1969) claimed the need for phages which would code for altered toxins. At an earlier date Matsuda and Barksdale (quoted by Barksdale and Arden, 1974) had screened a large number of *tox*⁻ wild diphtheria bacilli, for altered toxin, but found none (unpublished data).

Uchida *et al.* (1971) and Matsuda, *et al.* (1972) succeeded in obtaining nitrosoguanidine mutants of phage β which induced production of altered toxin proteins (cross-reacting materials), thus furnishing the first certain evidence that *tox* is the structural gene for diphtherial toxin. The protein thus obtained, while not toxic, had many of the characteristics of toxin. It was immunologically identical and cross-reacted with pure and specific diphtherial toxin.

As regards the cellular site of toxin production the data of Uchida and Yoneda (1967) suggested that it is the cytoplasmic membrane.

B. Physiology of diphtherial toxin

1. *Molecular structure*

Diphtherial toxin, the product of gene *tox*, is released extracellularly as a single polypeptide chain of molecular weight of 62 000 daltons (Gill and Dinius, 1971; Collier and Kandel, 1971). This is lethal for man, guinea pigs, rabbits and birds in doses of 130 ng per kilogram body weight (Barksdale, 1970).

In rabbits only a few picograms will give a visible skin reaction. Mice and rats are singularly resistant to toxin. The only known biological activities of the intact toxin molecule are its toxicity and its ability to fix to mammalian cells.

Intact, newly synthesised diphtherial toxin is enzymically inactive *in vitro*. When subjected to a brief treatment with trypsin in the presence of a thiol the molecule is "nicked" by hydrolysis of a single peptide bond located in the loop, formed by the N-terminal end of its two disulphide bridges. The resulting enzymically active "nicked" and reduced toxin consists of two fragments—A, amino-terminal and B, carboxyl-terminal of

24 000 and 38 000 daltons respectively, that are held together by weak interactions (Gill and Dinius, 1971; Collier and Kandel, 1971) (Fig. 7).

When the fractions are still connected by disulphide bridges the product is called "nicked toxin". The two fragments can be separated from one another by gel filtration in 6 M urea solution, or 0·1% sodium dodecyl sulphate.

Both fragments play a different but essential role in toxicity. Fragment B is required for fragment A to reach the cytoplasm of susceptible animal cells. Mutations of the phage *tox* gene, affecting enzyme activity (fragment A) or attachment to the cell (fragment B) result in a non-toxic but serologically related protein.

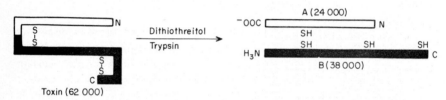

Fig. 7. Molecular structure of intact toxin molecule, before and after reduction with dithiotreitol, followed by mild hydrolysis with trypsin, according to Uchida *et al.* (1971).

2. *Biological activity of diphtherial toxin*

The mode of action of diphtherial toxin at the molecular level has been clarified in the last few years. Strauss and Hendee (1959) showed that low toxin concentrations (10^{-8} M or even less) completely block amino acid uptake by cultured human cells. Subsequently, Kato and Pappenheimer (1960) furnished evidence of the effect of toxin on the synthesis of mammalian cell protein. Collier and Pappenheimer (1964) found that NAD (nicotinamide dinucleotide) was required for the inhibition by toxin of protein synthesis in a cell-free system (from HeLa cells and rabbit reticulocytes). Later Collier (1967) and Goor and Pappenheimer (1967) showed that suitably activated toxin preparations inhibit protein synthesis by inactivating specifically the translocating enzyme of eukaryotes—aminoacyl transferase II. Toxin preparations catalyse reactions in certain susceptible animal cells and in the extracts of eukaryotic cells, containing transferase II (Honjo *et al.*, 1968; Gill *et al.*, 1969; Pappenheimer *et al.*, 1972) according to the following equation:

NAD$^+$ + transferase II (or EF$_2$) → ADPR–transferase II + nicotinamide + H$^+$ where ADPR = adenosine diphosphate ribose.

These observations show that toxin inactivates EF$_2$ (transferase II) by ADP ribosylation in the living animal as well as in cell-free extracts. The intact toxin molecule is enzymically inactive *in vitro*, and the presence of certain degradation products accounts for the enzymic activity of most toxin preparations; activation signifies reduction of one of the disulphide bridges and hydrolysis of a peptide bond. This change unmasks the 24 000 daltons heat stable fragment A, on which the enzymic activity is located. It is suspected (Uchida *et al.*, 1971) that such modifications of the toxin occur at the cell membrane where fragment A is split off and gains access to the cell cytoplasm, inactivating transferase II intracellularly.

The remaining fragment B (38 000 daltons) has no known enzymic activity but seems to be required for attachment of the toxin to the sensitive cell membrane. A single molecule of fragment A is sufficient to convert the entire cell content of EF$_2$ to its inactive derivative within a few hours.

By mutations of the structural gene *tox* it has become possible to prepare diphtherial lysogenic strains producing mutant, non-toxic proteins of various types. Some possess an altered fragment A and thus have become enzymically inactive, others contain a normal fragment A which cannot gain access to the cell cytoplasm because of an altered B fragment. There is a further exciting problem to be underlined in this connection.

It is known that formol toxoid immunises and prevents the lesions produced by diphtherial toxin. It reacts specifically with diphtherial antitoxin to form toxoid–antitoxin precipitates. *In vivo* the antibodies neutralise toxin. Toxoid is incapable of blocking adsorption of toxin to HeLa cells. Until the discovery of the molecular structure of diphtherial toxin it was assumed that formol conversion of toxin to toxoid renders the toxic site of toxin non-toxic, but leaves it antigenic. According to the new data, it seems that the toxoid induces antibodies, which block adsorption of the carboxy-terminal B fragment of the toxin to the animal cell.

3. *General properties of diphtherial toxin*

Studies on diphtherial toxin have been promoted by its availability in large amounts and by the accuracy and sensitivity of the methods for its bioassay, originally worked out by Ehrlich (1897).

Diphtherial toxin is a typical thermolabile exotoxin, readily inactivated by heat (5 min at 70°C) as well as by acid solutions of pH below 6 and by sunlight. The isoelectric point is 4·1. The toxin should therefore be stored in the cold and in the dark, but even so, it slowly and gradually loses its toxicity, without parallel loss of its serological activity. The toxins produced by all naturally occurring strains appear to be immunologically identical.

Immunisation with toxoid, plain or purified and adsorbed, results in protection against all *C. diphtheriae* strains capable of causing toxaemic diphtheria.

Toxin was obtained in a crystalline form by Pope and Stevens in 1958. The minimum lethal dose (MLD), which for the most active purified preparations is less than 40 ng, was originally defined by Ehrlich as that amount of toxin required to kill a 250 g guinea pig on the fourth day, following subcutaneous injection with three characteristic signs: gelatinous and/or haemorrhagic oedema and local swelling at the site of inoculation, adrenal congestion and haemorrhagic or clear exudate in the pleural and peritoneal cavities. A few picograms injected into rabbit skin will produce a visible reaction. As detailed above the toxin is very antigenic and is neutralised by antitoxin.

Treatment with formalin converts toxin into an antigenic but non-toxic product: toxoid (English), anatoxin (French). The specific intoxication produced by the exotoxin *in vivo* can be specifically neutralised by anti-toxin, and this reaction is important both for performing the Schick test in man and as a rapid method of toxin–antitoxin titration in animals.

Large quantities of toxin are used for the preparation of toxoids as prophylactics and as antigens, to obtain antitoxin in horses for passive immunisation and for testing individual susceptibility to diphtheria by means of the Schick test.

During the last decades highly immunogenic prophylactics have been obtained by purification of diphtherial toxoid, and adsorption on aluminium salts (aluminium phosphate or aluminium hydroxide). These prophylactics today represent the principal means of controlling diphtheria. The toxoids are standardised in Lf (Limes flocculans) according to Ramon's (1923, 1928) flocculation method for *in vitro* titration. Lf is defined as that quantity of toxin or toxoid that flocculates most rapidly when mixed with 1 AU (antitoxic unit of the standard serum). It is common practice everywhere to immunise infants with a combined vaccine containing either diphtheria and tetanus toxoids, or diphtheria, tetanus toxoids and *Bordetella pertussis* vaccine.

Routine methods of assay of toxinogeny in *C. diphtheriae* strains will be treated in Section IX.

C. Schick test

1. *General*

Susceptibility to diphtheria may be checked by means of the Schick test (1913), which determines the specific action of diphtheria toxin on the skin. Intradermal injection of a small dose of toxin (1/50 MLD) in 0·2 ml into the right forearm of non-immune individuals results in an erythematous reaction around the site of injection (Table I). If the individual is immune, the antitoxin present in his blood and tissue fluids will neutralise the toxin and block the reaction. A similar amount of boiled toxin ($100\,^{\circ}$C for $\frac{1}{2}$ h) in-oculated in the other forearm should not elicit any reaction unless the individual is sensitive to the foreign proteins in the preparation (pseudo-reactions).

The non-specific skin reactions usually reach a maximum within 48–72 h and then fade, in contrast to the specific positive reaction which becomes more intense after 48 h and persists up to 72–96 h after inoculation in the case of a non-immune individual.

The various types of responses are summarised in Table I.

TABLE I

Various responses to Schick test

Type of reaction	Schick test After 48 h	72 h	Schick control 48 h	72 h	Results, probable content of antitoxin	Interpretation of results
Redness > 10 mm in diameter	+ †	+ +	−	−	Positive < 0·03 AU/ml	Non-immune
Lack of reaction	−	−	−	−	Negative > 0·03 AU/ml	Immune
Partial reaction	+	−	+	−	Pseudo-negative > 0·03 AU/ml	Immune but sensitive to foreign proteins
Partial combined reaction	+	+ +	+	−	Pseudo-positive (combined) reaction < 0·03 AU/ml	Non-immune and sensitive to foreign proteins

+, + +, + + + = different degrees of sensitivity. The reaction is usually erythematous and of different sizes—a function of the degree of lack of immunity. A positive reaction is considered starting from an area of 10 mm diameter (+) up to 20 (+ +) or 30 (+ + +); sometimes the centre of the area is raised, erythematous, and the periphery looking like a cockade.

− (negative reaction) indicating a level of about 0·03 AU/ml or more, in the circulating blood.

2. *Various responses to the Schick test*

The Schick test is a semiquantitative reaction showing toxin activity (Table I). The level of circulating antitoxin which is required to protect against diphtheria in all ordinary circumstances is 0·03 AU/ml serum. More accurate estimations of serum antitoxin levels may be achieved by tissue culture tests or better by intradermal titrations in rabbit, according to Jensen's method (1933).

The Schick test is a reliable susceptibility test useful in the investigation of the immunity status as a result of mass immunisation. It is used to control the antigenic potency of vaccines and of different immunisation schemes.

To control the disease in the community it is considered that 90–95% should be Schick negative.

IV. ANTIGENIC STRUCTURE OF *C. DIPHTHERIAE*

A. General

Increasing importance has been ascribed during the last three decades to the bacterial cell components of *C. diphtheriae.*

Interest has been correlated with the need to develop a method of typing as in other bacterial species, and to explain the following:

(1) The phenomena of virulence and invasiveness.
(2) The occurrence of mild symptoms (atypical clinical form) in immunised carriers (Favorova and Kostjukova, 1966).
(3) The spontaneous disappearance from healthy carriers harbouring atoxigenic bacilli.

In this connection it was important to develop a serological typing method and to clarify whether there is any antibacterial immunity and its role in protection against diphtheria.

B. Serological typing

A series of findings led to the assumption that *C. diphtheriae* contains some cellular factors of virulence, besides toxin (Niggemeyer, 1955; Solovieva and Ananieva, 1945), but further verification is needed. Before Anderson *et al.'s* (1931) discovery concerning the existence of the three biotypes (*gravis, mitis* and *intermedius*) a classification scheme for *C. diphtheriae* was lacking. This led several authors to work on serological classification. Progress in research on the serological relationships within *C. diphtheriae* encountered difficulties when attempts were made to obtain stable suspensions of all *C. diphtheriae* strains. The first who dealt with seroagglutination in *C. diphtheriae* was Nicholas (1896).

Early workers reported that the diphtherial bacillus was an antigenically heterogeneous species, falling into two to eight types (Langer, 1916; Durand, 1918, 1920a, b; Havens, 1920; Smith, 1923; Eagleton and Baxter, 1923). After the discovery of the three biotypes, Ewing (1933), while investigating the relationships of type *gravis* strains, found four groups, A, B, C, D. Robinson and Peenay's (1936) results confirmed their existence, discovered an additional serological type and named them types I–V.

Tarnowski (1942), subsequently found 11 *gravis* types, of which five were identical to those of Robinson and Peenay. Like Murray (1935), he found clear-cut serological differences between the three classical biotypes and was able to classify a further five serological *intermedius* and 13 serological *mitis* types. Hewitt (1947) described 13 *gravis* types, two *intermedius* and 40 different *mitis* types and observed a strict relationship between serology and virulence.

Ferris (1950) by typing 794 strains in Australia was able to classify them in 14 serotypes, namely: four *gravis*, two *intermedius*, and eight *mitis* serotypes. He also made the general remark, subsequently confirmed by many other workers, that serotype II, which has a world-wide distribution, contains *mitis*, as well as *gravis* type strains, and that they can be differentiated by carbohydrate fermentation and haemolysis. Kostjukova *et al.* (1971) reports on the works of Delyiagina, Suslova and Pelevina, who also found that a serotype II was prevalent everywhere in far-away regions of the Soviet Union, as well as among strains from Poland, Hungary, Czechoslovakia, and Far Eastern countries. In Great Britain, type I was the most common, in U.S.A. type V, and in Egypt type IV.

It is generally agreed that *gravis* strains display specific and clear-cut agglutination reactions and the great majority of agglutinable strains has been found among toxigenic strains.

Shortcomings of serological typing in epidemiological investigations are the facts that a single serotype could be dominant in a country or region, that the most non-toxigenic strains were inagglutinable, and cross-reactions in the *mitis* and the *intermedius* groups could occur.

At present, it seems that the only recognised serogroup types are those of Robinson and Peenay, inadequate, however, from a practical epidemiological point of view. Due to the antigenic heterogeneity, many combined serological schemes of classification have been proposed, but none has been adopted as an internationally standardised scheme.

C. Investigations on bacterial antigens

Wong (1938) and Wong and Tung (1938, 1939a, b) demonstrated three kinds of antigens in diphtheria bacilli, i.e. a lipid and a polysaccharide substance, both of which reacted type specifically, and an alkali soluble protein, which had group specificity.

These investigations comprised only a small number of strains and did not attempt to use the three antigens in systematic classification. Lautrop (1950, 1955) investigated the occurrence of heat-labile and heat-stable antigens in the diphtheria bacillus. His experiments showed that the heat-labile antigen caused O-inagglutinability of *C. diphtheriae*. Lautrop named the heat-labile antigens "K" antigens, and stated that they are somatic surface antigens, causing O-inagglutinability of living or formalin-treated cultures. The heat-stable antigens were named "O" somatic antigens. Lautrop (1955) considered the O-antigen as a group antigen, but underlined the fact that the O-antigen as a diphtherial group antigen is of limited usefulness as it is present also in *C. ovis* and *C. hofmannii*. It is evident that ordinary typing sera prepared by immunisation with heat-killed or with formalin-killed organisms contain group reacting O-antibodies, able to interfere with typing by giving rise to cross-reactions.

Most of the *gravis* and *intermedius* strains have a more developed "K" antigen, in contrast with the *mitis* strains, which makes it possible to obtain satisfactory results in ordinary serotyping.

More clear-cut results can be obtained using typing sera, from which antibodies have been removed by absorption with an autoclaved culture (Lautrop, 1955).

There is general agreement that protein "K" antigens are surface antigens and O-antigens are polysaccharides (Wong, 1940; Wong and Tung, 1940; Kröger and Thofern, 1952a, b; Cummins, 1954).

Cummins (1954, 1965) presumes that these O-antigens represent modifications of the arabinogalactan component of the cell walls (Barksdale, 1970). Barber *et al.* (1966a, b), studying polysaccharides from different corynebacteria, found that the structure is similar in most, but not all, so that they cannot be employed as taxonomic criteria for the entire group. They also showed (Barber *et al.*, 1963, 1965) that there are type-specific protein antigens, a polysaccharide serologically specific for *C. diphtheriae*, and nucleoprotein antigens common for all corynebacteria. Kostjukova *et al.* (1970) present the results of immunological studies of the surface fractions of *C. diphtheriae* obtained by salt extractions at various pH values and by chloroform treatment. The alkali soluble fraction seemed to be the most antigenic, containing two type-specific antigens. Type specificity, evidenced by precipitation reactions with the alkali-soluble antigens, did not correspond fully to the type specificity determined by agglutination with live cultures. Some of the heat-labile surface antigens are type specific, others biotype specific and some even strain specific. Their possible association with pathogenicity has not yet been clarified. Branham *et al.* (1959) isolated from two *gravis* strains an immunologically active factor which in their opinion, caused swelling and oedema at the site of infection. However, this has not been verified by others. Antibacterial immunity has been demonstrated by several authors in different experimental models. Some of our latest field experiments (unpublished data) have shown that in closed children's communities there is a high percentage of prolonged *tox⁻* diphtheria bacilli carriers 18–20 months after infection. Due to the antibacterial immunity, a precipitous drop in the carrier rate was to be observed thereafter without any application of vaccines or antibiotics. These facts were also confirmed in experimental models on guinea pigs. Animals immunised with living *tox⁻* bacilli have shown a high percentage of protection against a booster dose with *tox⁺* bacilli (Drăgoi *et al.*, 1979). Hence, the conclusions are that antibacterial immunity seems to act as a natural adjuvant for the antitoxic immunity.

Healthy *tox⁻* bacillus carriers, who have high antitoxin levels, should not be treated by antibiotics.

It becomes evident that more investigations are still necessary to elucidate some theoretical problems concerning the antigenic structure of *C. diphtheriae*.

In order to use the serological character for practical purposes, Calalb *et al.* (1966, 1968) experimented with a polyvalent diphtherial antiserum for slide agglutination. Due to its specificity, they proposed the technique as a rapid means of identification in routine cultures of *C. diphtheriae*, but spontaneous agglutination (15%) as well as the lack of agglutinability (10%) limited the value of the method.

Recently, Fîciu and Maximescu (unpublished) improved the stability and the rate of agglutinability by using as diluent a saline-buffered solution, and by introducing in the polyvalent serum new monovalent sera. Using the slide agglutination in the identification of more than 1500 corynebacteria they obtained high specificity, high rate of agglutinability of gravis *tox⁺* strains (98%) and an improved rate of positive reactions (87·3%). These results which agree with those of Suslova (1964) recommend slide agglutination for routine identification of *C. diphtheriae*.

V. PHAGE TYPING OF *C. DIPHTHERIAE*

A. General

The discovery of bacteriophages by Twort (1915) and d'Hérelle in 1917 and that of lysogeny by Bordet and Ciucă (1921a, b) opened new fields of bacteriological research. Among the first were attempts to utilise phage for bacterial classification.

The well-known classification of *C. diphtheriae* into three biotypes *gravis*, *mitis* and *intermedius* (Anderson *et al.*, 1931) is based on biochemical and cultural differentiation. This classification established by the medical staff of Leeds (Anderson *et al.*, 1931) during the big diphtheria epidemics which occurred in Great Britain in 1931 represented remarkable progress, and has since led to considerable advances in the epidemiological study of diphtheria.

However, the experience of those responsible for the surveillance of the epidemiological process in diphtheria subsequently showed that this first classification was not satisfactory for thorough epidemiological investigations. The persistence over a long period and over large geographical areas, of only one biotype as, for instance, the *gravis* type throughout Europe in the severe diphtheria epidemic during the Second World War (Tasman and Lansberg, 1957), that of *mitis* in the U.S.A. (National Communicable Diseases Center Surveillance Rept., 1968), that of *intermedius* in Madagascar and India (Saragea and Maximescu, 1966) and the wide distribution of *gravis*-type strains for about 15 years in Romania (Saragea, 1965; Saragea *et al.*, 1965, 1967a, b, 1968) supports this statement. The differences in clinical severity in various areas contaminated by the same biotype suggested the existence of some biological differences between the strains. This indicated that further subdivision of the biotypes was needed (McLeod, 1943).

Even one of the authors of this first classification, McLeod (1943), foresaw that: "A more analytical research is necessary in order to evidence a series of complex relations and subdivisions not yet discovered".

B. History of research on phage typing

The first attempts to investigate phage sensitivity in *C. diphtheriae* were made in Australia (Keogh *et al.*, 1938). These authors used two phages and were able to characterise a group of strains belonging to the *gravis* and *intermedius* types. They found a close correlation between phage sensitivity and serological type. Toshach (1950), by isolating and adapting three lysogenic phages (one of which, C/13, was subsequently used by Freeman under the name of B phage) succeeded in typing some *C. diphtheriae* strains from Canada, but no definite "pattern" related to type or source was established. Fahey (1952) proposed and discussed a tentative typing scheme, but did not find any correlation between biotype and phage sensitivity.

Thibaut and Frédéricq (1956a, b) found nine phage types among *gravis* strains. Christensen (1957) was able to differentiate strains from two different epidemics on the basis of their lysogeny; Ortalli *et al.* (1956) obtained a phage sample displaying a wide lytic activity against all three diphtheria biotypes and *C. ulcerans*; Gabrilovitch (1966), using ten phage preparations, was able to class 73% of *gravis* strains into nine phage types and observed a close relationship with the biotype and serological properties. Endemann (1960) with the help of two lysogenic phages, and later Rische and Endemann (1962) with three more temperate phages classified a group of 389 *gravis* strains into four phage types. They found only one *intermedius* and no *mitis* phage types.

Starting from studies on the frequency of naturally occurring lysogeny, among about 10 000 *C. diphtheriae* strains, a remarkable number of lysogenic phage preparations were isolated in Romania. They were subsequently selected and grouped with respect to their lytic activity, against a large number of wild *C. diphtheriae* strains belonging to all 3 biotypes. Seventeen lysogenic (endogenous) phages were selected and grouped together with two reference phages (25 Warshaw and Freeman's B phage) (1956–1959). The set was completed in 1962 by five German phage preparations. A group of 24 lysogenic phage preparations, representing the basic set of typing phages, was thus established and has, to date, been used in the classification of about 35 000 *C. diphtheriae* strains in Romania.

The phage typing scheme established was named at first "Provisional Phage Typing Scheme for *C. diphtheriae*" (Saragea and Maximescu, 1964). The scheme was improved in 1969 (Fig. 8) and called by the authors (Saragea and Maximescu, 1969) the "original" phage typing scheme for *C. diphtheriae*. Two of the 24 phage preparations, i.e. nos. 23 and 24, were discarded, as they were not biotype specific. Two new phage types were identified by the same set. At present, by means of 22 lysogenic phages, the three classical biotypes may be classified into 21 phage types, namely four *mitis*, three *intermedius*, and 14 *gravis* types irrespective of the characters of lytic reactions (Figs 9–11). The "original" phage types are stable and specifically correlated to biotype and toxinogenicity.

This scheme has proved its efficiency during the entire epidemic period of diphtheria in Romania (Saragea *et al.*, 1966a, b, 1967a, b, 1968, 1973; Gabrilovitch, 1966; Hajderi *et al.*, 1967; Ejowa, 1967; Saragea and Maximescu, 1969). It has also been used successfully to analyse different epidemic outbreaks in other countries (Saragea and Maximescu, 1966; Saragea *et al.*, 1966b, 1972a; Gibson *et al.*, 1970; McCloskey *et al.*, 1972; Zamiri *et al.*, 1972). During the period 1956–1969, 99–100% of the *C. diphtheriae* type *gravis* tox^+ strains and 75% of the total number of *C. diphtheriae* strains isolated in Romania proved typable. Everywhere abroad where *C. diphtheriae* type *gravis* tox^+ strains were prevalent and

Indicator strains No.	Name	Origin	Biotype	1 (67, Romania)	2 (8, Freeman)	3 (25, Warshaw)	4 (Pc19, Romania)	5 (W, PWgWeissensee)	6 (Mio, Romania)	7 (29A, Romania)	8 (27c, Romania)	9 (23, Romania)	10 (252, East Germany)	11 (121, East Germany)	12 (340, East Germany)	13 (19, Romania)	14 (11S, Romania)	15 (12S, Romania)	16 (19R, East Germany)	17 (17, Romania)	18 (25S, Romania)	19 (7L, Romania)	20 (8L, Romania)	21 (20M, Romania)	22 (16, Romania)	Phage types
1	25	Warshaw	mitis tox+	●	●																					I
2	192	U.S.A.	mitis tox+	●	●	●							O						O							I "a"
3	1180	Freeman	mitis tox-	●	●	●	●	●	●				O						O							II
4	15872	Romania	mitis tox-							●	●															III
5	3522	Romania	intermedius tox+								●															IV
6	4465	Romania	intermedius tox-							●	●	●														V
7	328+ / 5429	East Germany / Romania	intermedius tox+ / gravis tox+												●											VI
8	5492	Romania	gravis tox+							●	●															VII
9	430+ / 9408	East Germany / Romania	gravis tox+											●	●	O										VIII
10	5714	Romania	gravis tox+											●	●							●	●	●	●	VIII "a"
11	19+ / 8390	East Germany / Romania	gravis tox+											●	●	O	●					●	●	●	●	IX
12	304 / 54	Copenhagen / East Germany	gravis tox- / gravis tox+											●	●	●	●	●	●	●	●					XIV
13	5829	Romania	gravis tox+											●		●	●	●	●	●	●					XIV "a"
14	11824	Romania	gravis tox+													●	●	●	●	●	●					XIV "b"
15	1202	Australia	atypical gravis tox+													●	●	●	●	●	●					XIV "c"
16	3148 / 10R / 9253	Romania / East Germany / Romania	gravis tox+ / gravis tox+ / gravis tox-											●		●	●	●	●	●	●	●	●	●	●	XVI
17	951	Romania	gravis tox+													●	●	●	●	●	●	●	●	●	●	XII
18	340	East Germany	gravis tox+												●	●		●	●	●	●	●	●	●	●	XIII
19	215	U.S.A.	gravis tox+											●		●	●	●	●	●	●	●	●	●	●	XIII "a"
20	252+ / A1377	East Germany / Romania	gravis tox+												●	●	●	●	●	●	●	●	●	●	●	XVII
21	121	East Germany	gravis tox+													●	●	●	●	●	●	●	●	●	●	XIX

FIG. 8. Original phage typing scheme for *C. diphtheriae*. ●, constant lytic reactions characterising the phage types; ○, inconstant lytic reactions (not obligatory for the phage types). *tox+* = toxigenic; *tox-* = non-toxigenic.

played an epidemic role, the original phage typing scheme was likewise satisfactory.

In Romania, as the general level of immunity against *C. diphtheriae* increased during the period between 1958 and 1970, from 60% Schick negative reactors to 96% as a consequence of large-scale diphtheria immunisation programmes, significant changes occurred in the distribution of wild *C. diphtheriae* biotypes. A wide-spread surveillance programme of the causative agent revealed that the *gravis tox+* type, formerly so widely distributed, was gradually replaced by *intermedius tox-* and *gravis tox-*

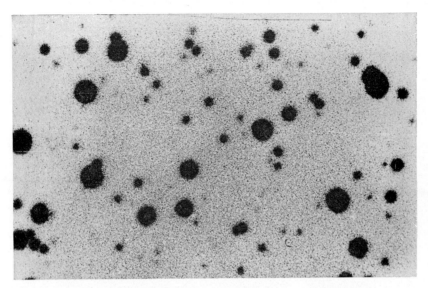

Fig. 9. Large, clear plaques of the original phage No. 2.

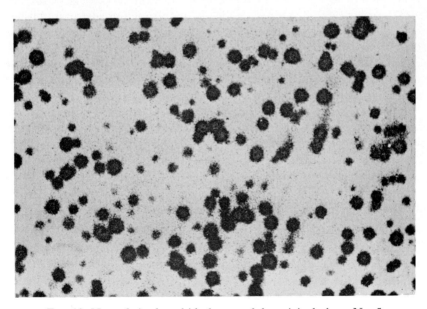

Fig. 10. Normal sized, turbid plaques of the original phage No. 5.

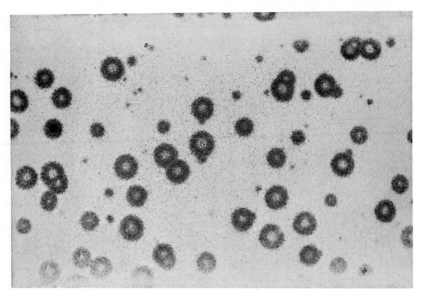

FIG. 11. Normal sized turbid plaques, with an opaque centre, original phage No. 3.

strains. Concomitantly the frequency of phage typable strains decreased from 74·7 to 46·5%. The high incidence of healthy carriers of *tox⁻* *C. diphtheriae* strains in the post-epidemic period in Romania (unpublished data), the sporadic occurrence of asymptomatic forms of sore throat in *C. diphtheriae* carriers, and the steadily diminishing capacity of the original phage-typing scheme to type these strains, determined the authors to undertake further studies in order to develop a second scheme, named for convenience, the "additional" phage typing scheme for *C. diphtheriae* (Maximescu *et al.*, 1972). This new scheme supplemented the original one and is based on different principles. The set of typing phages is represented by two categories of phages, endogenous lysogenic phages and "host restrictions" in different strains of a unique highly virulent phage (951 "L"), isolated in Romania.

C. Original phage typing scheme for *C. diphtheriae*

1. *Principles*

The original scheme (Saragea and Maximescu, 1964, 1969) is based upon the principles of phage typing schemes using lysogenic (endogenous) phages (see Fig. 8).

The phage types are represented either by patterns of lytic reactions (Figs. 12–14) or by isolated reactions. All strains displaying the same lytic

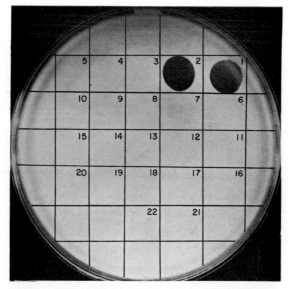

FIG. 12. Lytic pattern of the original phage type I.

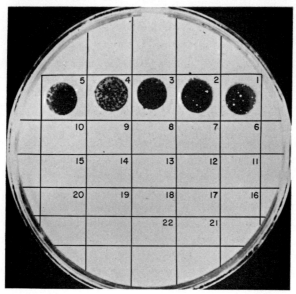

FIG. 13. Lytic pattern of the original phage type II.

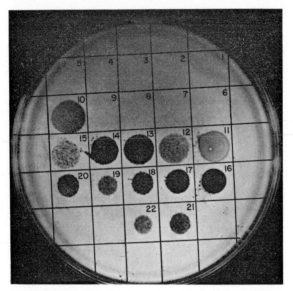

FIG. 14. Lytic pattern of the original phage type XVI.

pattern or single reactions are considered to belong to the same phage type; 22 phage preparations are used in routine typing. Some of the phage preparations, nos. 1, 2; 4, 5; 13–18 and 19, 20, 22, are similar in their lytic activity in each group.

Despite the similarity of their activity, all phages from each group should be retained, because some of the strains isolated during the last few years have partially lost their phage receptors. Thus, by keeping all similar phages the chances of revealing the phage sensitivity of strains and of being able to refer them to known phage patterns is improved.

2. *Basic set of typing phages*

The 22 typing phages (upper horizontal range, Fig. 8), are numerically recorded, their position being indicated by arabic figures. All phage preparations were obtained by induction from their lysogenic hosts, either from overnight shaken cultures or with ultraviolet light. Only one phage (position 4), which is probably also endogenous in origin, was obtained by the authors from a throat gargle in a diphtheria case. Phage no. 5 was isolated by ultraviolet-light induction from the PW_8 strain, variant Weissensee (Berlin).

Six of the 22 phage preparations are reference phages, obtained from abroad, i.e. phages no. 2 (phage B. Freeman); no. 3 (25 Warshaw) and phages no. 10, 11, 12, 16, also of endogenous origin, isolated at the phage typing centre of Wernigerode (Endemann 1960; Rische and Endemann, 1962). The typing phages are arranged in order and grouped according to their lytic activity against the three biotypes, *mitis*, *intermedius* and *gravis*.

The first group of typing phages contains the 1–6 phage preparations which are active against the *mitis* type strains; the second from 7 and 8 are active on *intermedius* and *gravis* tox⁻ and 9 attacks only *intermedius* strains. All phage preparations from position 10–22 possess lytic properties against *gravis* type strains. Only two of the *gravis* phages, i.e. nos 10 and 16, show sporadic and very weak reactions against two of the *mitis* reference strains *C. diphtheriae* 192 and 1180 (phage types I*a* and II).

The number of *gravis* typing phages from the basic set is outstanding (Fig. 8). This is explained by the prevalence and large variety of *gravis* type strains in Romania, during the time this scheme was developed.

3. *Phage sensitive (indicator) strains*

The vertical rows on the left of the scheme (Fig. 8) include the phage-sensitive strains (prototypes of the different phage types) arranged in order and grouped according to the biotype to which they belong. Each phage type is characterised by one, two or more specific lytic reactions, grouped according to a certain "lytic pattern". The second, third and fourth vertical rows record the number, origin and biotype to which the strains belong. Their toxinogenic properties are also shown. For each phage type, one or more sensitive strains of different geographical origin are recorded. Each phage type is designated by a Roman numeral in the right vertical row.

Thus, with the help of the 22 typing phages, biotype *mitis* strains can be divided into four phage types: I, I*a*, II and III; *intermedius* strains into three types: IV, V, VI; and *gravis* strains into 14 types, called for convenience, VII, VIII, VIII "a", IX, XIV, XIV "a", XIV "b", XIV "c", XVI, XII, XIII, XIII "a", XVII and XIX. The strict relationship between phage type and the biotype of each individual strain is remarkable. An exception is that of phage type VI, which, may characterise either *gravis*, or *intermedius* tox⁺ or tox⁻. It is worthy of note that, besides the well-established "lytic patterns" presented by this scheme there are a number of *mitis* and *intermedius C. diphtheriae* strains which are irregularly sensitive against some phages, irrespective of the biotype specificity of the phages.

Such strains were called "allosensitive". They are infrequent.

4. *Properties of the original phage typing scheme*

The principal properties of a good typing system are the following:

The species must be divisible into as many types as possible.

The types and reactions, should be stable.

The technique must be simple, easy to perform and interpret and reproducible.

The method must be standardisable.

The value of the classification must be verified through intensive field applications, i.e. the incidence of types should be interpreted on a sufficiently representative sample of biological material.

Due to its stability, diversity and specificity, the "original" phage-typing scheme meets all of these criteria.

(a) *Stability of the* C. diphtheriae *original phage-typing scheme* has been proven both *in vitro* and *in vivo*. Strains stored *in vitro* (irrespective of storage method, i.e. freeze-dried, or by monthly subcultures) retained their lytic pattern for about 20 years (the period of our observation) and did not change after passage in experimental animals. The phage type stability could also be fully demonstrated in epidemic situations. During an outbreak the same type could be isolated from patients and from their contacts. Bearing this in mind, the whole area of an outbreak could be determined from the spread of the prevalent epidemic phage type and could readily be cleared up, until the last healthy carrier or contact was identified and treated.

In a certain geographical area, the same phage type could persist for a long time if all healthy carriers were not treated.

The same phage type was found over a long period of time in healthy carriers showing pathological modifications of the rhinopharynx.

(b) *The specificity of phage type* is demonstrated by the fact that each biotype is specifically lysed by groups of phages and that other bacterial genera, pseudodiphtheric species or other members of the genus *Corynebacterium* were never lysed by the *C. diphtheriae* phages representing this basic set. Exceptions were *C. ulcerans* and *C. ovis* species which may be lysed by some *C. diphtheriae* phages and even induced to synthesise diphtherial toxin (Maximescu, 1968; Maximescu *et al.*, 1968, 1974a, b). This property was taken as an argument for considering them to be closely related to *C. diphtheriae* (Carne, 1968).

(c) *Diversity of phage types* in most cases is due to the lysogenic state of the receptive strain and expressed in the large number of phage types determined by this scheme capable of subdividing the three classical biotypes.

5. *The genetic basis of the original* C. diphtheriae *phage typing scheme*

The three properties mentioned above which ascertain the value of this scheme have recently been verified (Saragea *et al.*, unpublished data). There were indications that the lytic pattern of the phage types is genetically determined as a result of their lysogenic state. The stability of the lysogenic state determines the phage type stability. The world-wide distribution of *gravis* phage type XVI and the fact that most *gravis* types may be obtained by its lysogenisation with appropriate phages, support the hypothesis (Saragea *et al.*, unpublished).

A "genetic structural formula" was thus established for almost all original phage types. It indicates the stability, specificity and diversity of the original *C. diphtheriae* phage types, and consequently the good reproducibility of the results.

6. *Serological relationship of the typing phages*

During our routine work on *C. diphtheriae* phage typing the differences in lytic activity of various groups of phages were repeatedly correlated with the biotype of the host cell. This raised the question of possible serological differences among typing phages. Saragea and Maximescu (1969) found three principal serological groups among the typing phages closely linked to their host–range activities.

7. *Practical results of the* C. diphtheriae *original phage typing scheme*

The original phage-typing scheme proved its epidemiological efficiency in the surveillance programme of the causative agent in Romania over the whole epidemic period as well as in a number of other countries (Canada, unpublished data; Australia, Gibson *et al.*, 1970; Iran, Zamiri *et al.*, 1972; France, Saragea *et al.*, 1972a; Great Britain, unpublished data; Soviet Union, Krilova *et al.*, 1969; Berezkina and Kravcenko 1970; Albania, Hajderi *et al.*, 1967; U.S.A., McCloskey *et al.*, 1972; Madagascar, Hungary, Jugoslavia, Italy, unpublished data).

By phage typing the origin and geographical spread of epidemics, the role of healthy carriers, the means of transmission and the relationships among different outbreaks may be determined. Yearly geographical maps of the incidence and distribution of phage types in Romania have enabled us to follow the filiation of epidemics, the alternation of phage types, or the penetration of a new type in a region, and to establish possible links with visitors from abroad.

8. *Distribution of the original phage types*

During the first period studied in Romania (1956–1964) the most widespread phage type among *gravis* type strains was phage type XIV which caused moderately severe cases. Subsequently it was followed by XIV "a" phage type, isolated from all outbreaks, occurring in sanatoria where children with poliomyelitis sequelae were concentrated.

Phage type XVI characterised a very large and severe outbreak in 1960, phage type XII two other severe and large outbreaks, and phage type VIII "a", only one outbreak in 1962.

Subsequently phage type IX alone has been prevalent in all epidemics but at present is found in only very rare, sporadic cases (one to two cases yearly) all over the country. Phage types XIII and XVII have not been epidemic in Romania. The latter was identified in 1968 in only two healthy carriers, who had come in contact with tourists from Hungary, where this type was more common. *Mitis* phage types I, Ia, II and III, *intermedius* IV, V, VI and *gravis* non-toxigenic VII were only isolated from healthy carriers and in non-epidemic areas.

In conclusion: by means of this scheme it has been possible to characterise 73% of all *C. diphtheriae* strains isolated in Romania during the epidemic period of 1956–1967, and 62% of strains received from other geographical areas. Among strains isolated from other countries 99% of the *gravis* strains were typable by this scheme.

The prevalence of a certain epidemic phage type could be followed in various parts of the world. For instance phage type XVI has been identified in a collection of strains isolated in Australia as long ago as 1938.

The same phage type was identified during the last decade in Europe (Italy, Jugoslavia, Hungary) and Canada (Vancouver). *Mitis* I "a" has been identified in the U.S.A. and Canada but never in Europe, before it suddenly appeared in Hungary in 1974. Some of the individuals carrying this type had been visiting relatives in the U.S.A. one month before.

9. *Limitations of the method*

Worthy of note is the smaller typing capacity of the biotypes *intermedius tox⁻* and *mitis tox⁻* (39 and 19% respectively). These two biotypes are more frequently isolated from healthy carriers and in non-epidemic areas or periods. The failures of the "original" phage-typing scheme indicate that while complete for *gravis tox⁺* strains the "original" basic set of typing phages has to be supplemented for *mitis*, *intermedius* and *gravis tox⁻* in order to increase the typability among these three biotypes to higher levels.

The limits of this method, which became more evident in the postepidemic period of diphtheria in Romania recently, led to the development of an additional phage-typing scheme (Maximescu *et al.*, 1972).

D. Additional phage typing scheme for *C. diphtheriae*

1. *Principles*

(a) *The additional basic set of typing phages.* This set comprises 33 phages (Fig. 15). It is based upon the principles of both lysogenic and host-restricted preparations of phages. Twenty-one of the 33 typing phages are represented by specific host restrictions of the highly virulent lysogenic phage 951, arising from a *C. diphtheriae gravis tox*[+] strain, recorded as 951 "L". This strain was isolated in 1959 from an outbreak in Romania. At the time of isolation (Ciucă *et al.*, 1960) it consisted of a heterogeneous cell population, some clones being lysogenic and able to release phages active against other clones derived from the same strain. Strain 951 "L" carried a highly virulent phage, endowed with remarkable plasticity (Maximescu and Saragea, 1969), which made it possible to obtain various very specific host restrictions in previously untypable *C. diphtheriae* strains. These phages, obtained by host restriction, differ in their lytic activity both from the original phage and from one another. The other group of 12 phage preparations were obtained, either from *C. diphtheriae* biotype *intermedius tox*[−] strains (five preparations) or from *C. diphtheriae gravis* biotype *tox*[+] strains (seven preparations). Thus, initially, the basic set of 33 phages was used routinely on 3000 *C. diphtheriae* strains which until then had been untypable (Fig. 15). The frequency of typability among these strains increased step by step with the introduction of new phage preparations from 32% in 1967 to 73% in 1971.

Afterwards 19 typing phages were selected according to the frequency of their lytic activity from among the 33 preparations of the initial basic set (Fig. 16). They were designated by the numbers given them in the initial series of the 33 preparations. The following phage preparations are now commonly used in routine work: 1, 2, 6, 7, 8, 9, 10, 11, 13, 15, 17, 19, 20, 21, 23, 25, 26, 29 and 33 (Fig. 16).

If a strain is non-typable by these 19 selected preparations it is subjected to the whole basic set of 33 phages. Figure 16 presents the origin of the 33 phages and the indicator strain for each. Most indicator strains are susceptible to only one typing phage but some strains display sensitivity towards two or more phages. The host restricted phages manifest a strong activity on the propagating indicator host strain, but occasionally show low activities (isolated plaques) against other strains.

(b) *Additional phage types.* After the application of the basic set to some 3000 *C. diphtheriae* strains, 14 main additional phage types were selected. Their phage sensitivity was manifested either by isolated (Fig. 16) lytic reactions, or by "lytic patterns" as in the "original" phage typing scheme. These 14 additional phage types were determined by means of the 19

C. diphtheriae indicator strains; biochemical type, toxinogenesis and origin \ Number and origin of phages				1	2	3	4	5	6	7	8	9	10	11	12	13	14	15	16	17	18	19	20	21	22	23	24	25	26	27	28	29	30	31	32	33
				1023 Copenhague/9284	1023 Copenhague/7111	951 „L"/9571	951 „L"/8853	951 „L"/9717	951 „L"/10789	951 „L"/10823	40 RDA/10269	40 RDA/11557	40 RDA/7748	507 Albania/4465	3148 Romania/6740	951 „L"/524	494 Romania/5492	951 „L"/16262	951 „L"/7762	951 „L"/A 763	951 „L"/554	525 Albania/951	537 Albania/550	792 Ceylon/803	798 Ceylon/810	951 „L"/1334	951 „L"/1336	951 „L"/1359	951 „L"/1381	951 „L"/1410	951 „L"/1416	951 „L"/6104	951 „L"/1563	951 „L"/1581	951 „L"/1681	951 „L"/7508
1	9284	*int. tox*$^-$	Romania	CL	–	–	–	–	–	–	–	–	–	–	–	–	–	–	–	–	–	–	–	CL	–	–	–	–	–	–	–	–	–	–	–	–
2	7111	*int. tox*$^-$	Romania	–	CL	–	–	–	–	–	–	–	–	–	–	–	–	–	–	–	–	–	–	–	–	–	–	–	–	–	–	–	–	–	–	–
3	9571	*int. tox*$^-$	Romania	–	–	CL	–	–	–	–	–	–	–	–	–	–	–	–	–	–	–	–	–	–	–	–	–	–	–	–	–	–	–	–	–	–
4	8853	*int. tox*$^-$	Romania	–	3+n	CL	CL	1+n	–	SCL	–	–	–	–	–	±	–	±	–	±	–	3+n	–	–	–	–	–	–	–	SCL	–	±	±	–	–	–
5	9717	*int. tox*$^+$	Romania	–	3+n	–	–	CL	–	–	–	–	–	–	–	–	3+n	–	–	3+n	–	–	–	–	–	–	–	SCL	±	–	–	–	CL	–	–	–
6	10789	*int. tox*$^-$	Romania	–	–	–	–	–	CL	–	–	–	–	–	–	–	–	–	–	–	–	–	–	–	–	–	–	–	–	–	–	–	–	–	–	–
7	10823	*int. tox*$^-$	Romania	–	–	–	–	–	–	CL	SCL	SCL	±	–	–	–	–	–	–	–	–	±	–	–	–	–	–	–	–	–	–	–	–	–	–	–
8	10269	*gravis tox*$^-$	Romania	–	–	–	–	–	–	–	CL	–	–	–	–	–	–	–	–	–	–	–	–	CL	–	–	–	–	–	–	–	–	–	–	–	–
9	11557	*int. tox*$^-$	Romania	–	–	–	–	–	–	–	–	CL	–	–	–	–	–	–	–	–	–	–	–	–	–	–	–	–	–	–	–	–	–	–	–	–
10	7748	*int. tox*$^-$	Romania	–	–	–	–	–	–	±	SCL	CL	CL	–	–	–	–	–	–	–	–	–	–	–	–	–	–	–	–	–	–	–	–	–	–	–
11	4465	*int. tox*$^-$	Romania	–	–	–	–	–	–	–	–	–	1+n	CL	CL	–	CL	–	–	CL	–	–	–	–	–	–	–	–	SCL	–	SCL	–*)	±	–	–	–
12	6740	*int. tox*$^-$	Romania	–	–	–	–	–	–	–	±	–	3+n	CL	SCL	–	CL	–	–	–	–	–	–	–	–	–	–	–	–	–	CL	–	–	–	–	–
13	524	*int. tox*$^+$	Chesterfield	–	–	–	–	–	–	–	–	–	–	–	–	CL	–	–	–	–	–	–	–	–	–	–	–	–	–	–	–	–	–	–	–	–
14	5492	*gravis tox*$^-$	Romania	–	–	–	–	–	–	–	–	–	OL	–	–	–	CL	–	–	SOL	–	–	–	–	–	–	–	–	–	–	–	–	–	–	–	–
15	16262	*gravis tox*$^-$	Romania	–	–	–	–	–	–	–	–	–	–	–	–	–	–	CL	–	CL	–	–	–	–	–	–	–	–	–	SCL	–	–	–	–	–	–
16	7762	*gravis tox*$^+$	Romania	–	–	–	–	–	–	–	–	–	–	–	–	–	–	–	SCL	–	–	CL	–	–	–	–	–	–	–	–	–	–	–	–	–	–
17	A 763	*int. tox*$^-$	Romania	–	–	–	–	–	–	–	–	–	–	–	–	–	–	–	–	CL	–	–	–	–	–	–	–	–	–	SOL	–	–	–	–	–	–
18	554	*mitis tox*$^+$	Albania	–	–	–	–	–	–	–	–	–	–	–	–	–	–	–	–	–	CL	–	–	–	–	–	–	–	–	–	–	–	–	–	–	–
19	951	*gravis tox*$^+$	Romania	–	–	–	–	±	–	–	SCL	±	±	–	–	±	–	1+n	3+n	–	–	CL	–	–	–	–	–	–	–	–	1+n	1+n	–	–	–	–
20	550	*int. tox*$^-$	Albania	–	–	–	–	–	–	–	–	–	–	–	–	–	–	–	–	–	–	–	CL	–	–	–	–	–	–	–	–	–	–	–	–	–
21	803	*mitis tox*$^-$	Ceylon	–	–	–	–	–	–	–	–	–	–	–	–	–	–	–	–	–	–	–	–	CL	–	–	–	–	–	–	–	–	–	–	–	–
22	810	*mitis tox*$^-$	Ceylon	–	–	–	–	–	–	–	–	–	–	–	–	–	–	–	–	–	–	–	–	–	CL	–	–	–	–	–	–	–	–	–	–	–
23	1334	*int. tox*$^-$	Canada	–	–	–	–	–	–	–	–	–	–	–	–	–	–	–	–	–	–	–	–	–	–	CL	–	–	–	–	–	–	–	–	–	–
24	1336	*int. tox*$^-$	Canada	–	–	–	–	–	–	–	–	–	–	–	–	–	–	–	–	–	–	–	–	–	–	–	CL	–	–	–	–	–	–	–	–	–
25	1359	*gravis tox*$^-$	Moscow	–	–	–	–	–	–	–	–	–	–	–	–	–	–	–	–	–	–	–	–	SCL	–	–	–	CL	–	–	–	–	1+n	–	–	–
26	1381	*mitis tox*$^+$	Moscow	–	–	–	–	–	–	–	–	–	–	–	–	–	±	±	–	CL	–	–	–	–	–	–	–	–	SCL	–	–	–	–	–	–	–
27	4110	*mitis tox*$^-$	France	–	–	–	–	–	–	–	–	–	–	–	–	–	–	–	–	–	–	–	–	–	–	–	–	–	–	CL	–	–	–	–	–	–
28	1416	*mitis tox*$^+$	France	–	–	–	–	–	–	–	–	–	–	–	–	–	–	–	–	–	–	–	–	–	–	–	–	–	–	–	SOL	–	–	–	–	–
29	6104	*mitis tox*$^+$	Romania	–	–	–	–	–	–	–	–	–	–	–	–	–	–	–	–	–	–	–	–	–	–	–	–	–	–	–	–	CL	–	–	–	–
30	1563	*int. tox*$^+$	Iran	–	–	–	–	–	–	–	–	–	–	–	–	–	–	–	–	–	–	–	–	–	–	–	–	–	–	–	–	–	CL	–	–	–
31	1581	*mitis tox*$^-$	Canada	–	±	–	–	–	–	–	–	–	–	–	–	–	–	–	–	–	–	–	–	–	–	–	–	–	–	–	–	–	–	CL	–	–
32	1681	*gravis tox*$^-$	Nigeria	–	–	–	–	–	–	–	–	–	–	–	–	–	–	–	–	–	–	–	–	–	–	–	–	–	–	–	–	–	–	–	CL	–
33	7508	*gravis tox*$^-$	Romania	–	–	–	–	–	–	–	–	–	–	–	–	–	–	–	–	–	–	–	–	–	–	–	–	–	–	–	–	–	–	–	–	CL

FIG. 15. Additional set of typing phages.

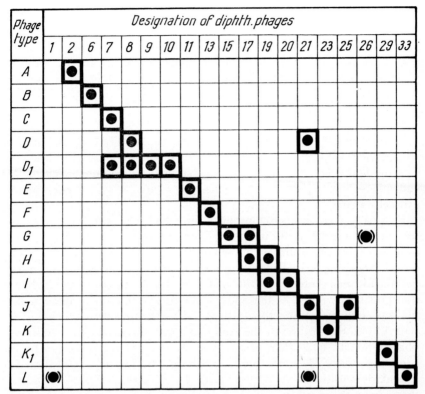

FIG. 16. Additional routine phage typing scheme. ●, lytic usual pattern; (●), inconsistent lytic activity.

selected phage samples mentioned above. Phage types were labelled with capital letters as follows: A, B, C, D, D_1, E, F, G, H, I, J, K, K_1, L. Thereafter a new routine assay with the additional scheme was carried out on about 7000–8000 *C. diphtheriae* strains.

(c) *Characters of the additional phage typing scheme.* The following points have to be stressed:

The additional phage types were stable as proved in repeated tests, irrespective of the means used to store the strains. This was proved during the approximately 3 years' surveillance of closed children's communities, showing high carriage percentages of non-toxigenic diphtheria bacilli. The same phage type was identified throughout the whole period.

While the original phage typing scheme is at present almost complete and further changes may be limited as it has proved its value on about 35 000 *C. diphtheriae* wild strains from Romania and approximately

3000 *C. diphtheriae* strains from abroad, the additional scheme is more likely to undergo further developments. The system on which it is partly based, i.e. the plasticity of the highly virulent 951 "L" phage and the possibility to restrict it in previously phage-resistant bacteria, makes it possible to obtain new specific phages for strains from different areas of the world.

2. *Practical value and efficiency of the additional phage typing scheme*

Both schemes are useful epidemiological tools. The original one for situations of epidemic occurrence of *gravis tox⁺* and/or *intermedius tox⁺*, and of some *intermedius tox⁻ C. diphtheriae* strains, the additional scheme for endemic or post-epidemic situations, when *tox⁻ C. diphtheriae* strains generally prevail.

However, there are circumstances in which the latter phage typing scheme can also be useful in epidemic situations. One particular type, for instance, type "K", displaying sensitivity against host restricted phage no. 23, has characterised a large number of *intermedius tox⁺* strains from an outbreak in San Antonio, Texas, in 1972 (McCloskey *et al.*, 1972). It was thus possible by means of this additional type to differentiate and elucidate the double source of infection of this epidemic (*gravis tox⁺*, original phage type XIV in one section of the population and *intermedius tox⁺* additional phage type "K" in another). Phage type "K" has two receptors, one for the additional phage 23, which produces strong lytic reactions, and the other irregularly sensitive to the original phage 11. This phage type is widespread throughout the south-west U.S.A. and Canada (Toshach *et al.*, 1977). After the outbreak mentioned above, this type was identified by the authors among strains isolated from Indian reservations in New Mexico and Arizona. It seems to be an endemic ancestral type in the U.S.A. rather than a recent import. It was also frequently isolated from cutaneous lesions, leg ulcers, nose and ear suppurations in Canadian eskimoes.

The use of the additional phage typing scheme enabled the authors to follow changes in the distribution of *C. diphtheriae* strains in Romania in the post-epidemic period, to study the distribution of healthy carriers and of atypical clinical forms. Among the additional phage types, the most frequent was *gravis tox⁻* phage type "L".

It is still the prevalent phage type in healthy carriers, and is followed in order of frequency by phage types B, D, J and A.

3. *Subtypes of the original* intermedius *phage type V*

During the development of the additional phage typing scheme it was observed that two *intermedius tox⁻* strains, belonging to original phage type V, commonly isolated from healthy carriers, showed different lytic patterns

Phage subtype	Designation of phages				
	11	12	14	17	26
V a	●	●			
V b	●	●	●		
V c	●	●		●	
V d	●	●	●	●	
V e	●	●			●
V f	●	●	●		●
V g	●	●		●	●
V h	●	●	●	●	●

FIG. 17. Additional subtypes of the original phage type V.

when submitted to the additional basic set of phages. Subsequently eight subtypes in all were recognised and were labelled V "a", V "b", V "c", V "d", V "e", V "f", V "g" and V "h" (Fig. 17).

The strains originating from a different geographical area all belonged to the same subtype.

Thus the additional phage typing scheme in C. diphtheriae, together with the original one, represent useful tools in epidemiological investigation.

Electron micrographs of different C. diphtheriae phages are shown in Figs 18 and 19 (Petrovici et al., 1969).

VI. BACTERIOCINS OF CORYNEBACTERIA

A. General

Thibaut and Frédéricq (1956a, b) were the first to draw attention to bacteriocins in C. diphtheriae type gravis, displaying activity only on homologous strains. This substance was shown not to be sensitive to proteolytic enzymes. Subsequently they were reported by Tashpulatova and Krilova (1967) and referred to, later on, by Krilova (1969) as "cory-

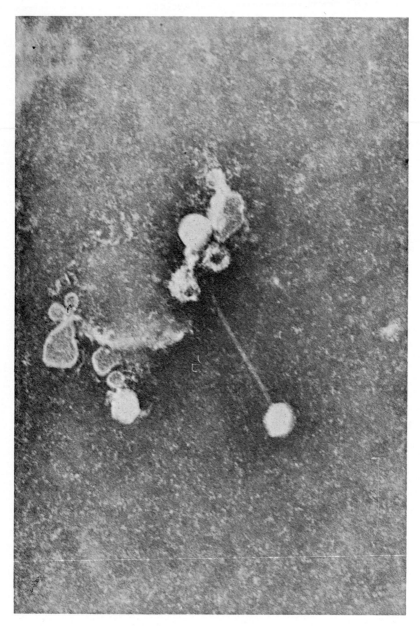

Fig. 18. Electron micrograph of an isolated *C. diphtheriae* phage. Magnification ×86 000.

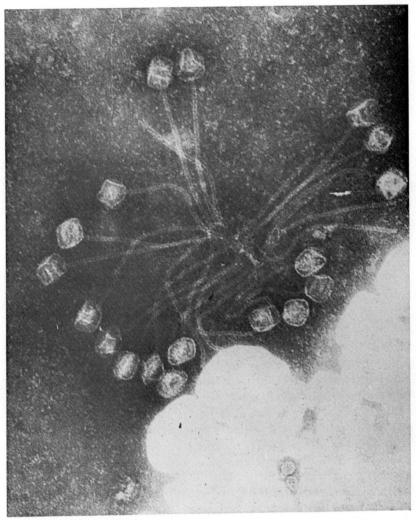

Fig. 19. Electron micrograph of *C. diphtheriae* bacteriophages attached to the surface of a *C. diphtheriae* cell. Magnification × 86 000.

cines" and the producers as "corycinogenic" strains. These authors showed that 65% of the *C. diphtheriae* strains, all *mitis* types, were bacteriocinogenic. Corycins produced by *mitis* are homogeneous, belonging to group A.

In Romania, Meitert (1969a, b) and Meitert and Bica-Popii (1971) examining 592 strains of the genus *Corynebacterium*, found that eight species (i.e. *C. diphtheriae, C. ulcerans, C. hofmannii, C. xerosis, C. murium,*

C. pyogenes, *C. equi*, *C. renale*) and atypical corynebacteria contained bacteriocinogenic strains in different proportions. The frequency of bacteriocinogeny among these strains varied in terms of quantity and sensitivity of the indicator strains. Media composition also influenced the percentage of bacteriocinogenic strains. When indicator strains of the same species were used a lower incidence of bacteriocinogeny was detected in different species, especially in *C. murium*, *C. ulcerans* and *C. equi*, than when heterologous species were also used (*C. diphtheriae*, and *C. renale*).

Similar observations were made by Gibson and Colman (1973), who found that *C. murium* and *C. hofmannii* bacteriocinogenic strains are more active against *C. diphtheriae* strains and bacteriocinogeny of *C. diphtheriae* is more frequently demonstrated against *C. ulcerans* species.

Gibson and Colman (1973), checking bacteriocinogeny in 83 strains, pertaining to five different *Corynebacterium* species, also demonstrated bacteriocinogenic activity in all the species studied by them (i.e. *C. diph-theriae*, *C. belfanti*, *C. xerose*, *C. renale* and *C. bovis*). The bacteriocins obtained from *C. diphtheriae* were referred to by these authors as "diph-thericins" and were demonstrated in 94% of the *C. diphtheriae* studied.

C. diphtheriae type *mitis* was demonstrated in Romania (Meitert, 1969a, b) to show the highest percentage of bacteriocinogeny (42%), followed by *intermedius* (15%) and *gravis* (7%). The same author noted that bacterio-cinogeny is more frequent among *C. diphtheriae* toxigenic strains. Bacterio-cinogeny was constant in only 54% of the bacteriocinogenic strains.

Meitert (1969a, b) and Meitert and Bica-Popii (1971) found that the bacteriocinogenic corynebacteria displayed immunity against their own bacteriocin. They noted that bacteriocins produced by different coryne-bacteria are active against strains of different microbial genera: β-haemo-lytic streptococci, viridans streptococci, enterococci, staphylococci, *Escherichia coli*, *Salmonella*, *Shigella*, *Proteus vulgaris*, *Listeria mono-cytogenes* and *Erysipelothrix insidiosa*. Coexistence of lysogeny and bacterio-cinogeny was observed in 6–48% of strains.

B. *Corynebacterium diphtheriae* bacteriocin typing

1. *History*

The discovery of bacteriocins by Thibaut and Frédéricq (1956a, b) opened prospects for their use in classifying *C. diphtheriae*.

The use of bacteriocins, for epidemiological purposes, was signalled by Emilianov *et al.* (1968). Krîlova (1972) and Krîlova *et al.* (1973) checked bacteriocin typing and phage typing and were able to classify *tox⁻*, *C. diphtheriae mitis* strains.

Meitert (1972, 1975) and Gibson and Colman (1973) used bacteriocin typing as an independent marker to subdivide phage types and serotypes respectively. Strains which were untypable by other methods were also classified.

Zamiri and McEntegart (1973) and Toshach *et al.* (1972) used this method successfully for the classification of untypable *C. diphtheriae* strains.

Meitert (1972, 1975) uses the contact method described by Hamon (1965) and adapted by her to the growth requirements of *C. diphtheriae*. She proposed a bacteriocin typing scheme permitting the classification of different sensitivity patterns of the strains, against a set of 20 bacteriocin producers selected from among different *Corynebacterium* species namely: 14 *C. diphtheriae* strains, one *C. ulcerans*, four *C. hofmannii* and one *C. xerosis* strain.

Rurka (1974) devised a technique whereby producer strains were grown on a quadrant of filter paper placed on the culture medium surface. The bacteriocins would diffuse through this paper which was then removed and the indicator strains applied with a multipoint applicator. This modification allowed all indicator strains to be applied in a single operation.

Following the typing of 588 Canadian strains, and 64 other *C. diphtheriae* strains from different countries (U.S.A., England) the method was found to be reproducible, but still of limited value. Only 24·4% of the strains examined were typable and 86·8% of the typable strains were of the same type.

The bacteriocin typing scheme proposed by Meitert (1975) contains 16 bacteriocin groups A–P, 12 of which are subdivided into types specified by Arabic numerals (Fig. 20a, b). Most of the 1171 *C. diphtheriae* and *C. ulcerans* strains, isolated chiefly from carriers between 1969 and 1972, proved to belong to bacteriocin group A (69·5%). The rest belonged to other bacteriocin groups: E, G, F, D, B, I, C, H, P, O, L, N, J, K and M. According to Meitert, the bacteriocin type is not concordant either with the biotype or species *C. diphtheriae*, for example, the same bacteriocin type may include *gravis*, *mitis* or *intermedius* strains of *C. diphtheriae* and even *C. ulcerans*. However a correlation could be observed by the authors between a certain bacteriocin type and phage type in the case of strains isolated at a given moment and in a certain community.

Gibson and Colman (1973), using Gillies's bacteriocin typing technique (1964), found a correlation between bacteriocin type and phage-, sero-, and biotype.

2. C. diphtheriae *bacteriocin typing scheme*

The bacteriocin typing scheme, established by Meitert (1972, 1973) (Fig. 20) is based upon determination of the sensitivity of strains tested against a set of 20 constant bacteriocin producer strains, labelled 1–20.

Bacterio-cinic group	Bacterio-cine type	Set of bacteriocinogenic strains																			
		1	2	3	4	5	6	7	8	9	10	11	12	13	14	15	16	17	18	19	20
		153	155	198	730	760	828	866	868	871	872	873	876	877	878	882	898	2378	2380	Hoff	XER
A	A1	●	●	●	●					●	●	●						●	●	●	●
	A2	●	●				●	●	●	●	●	●	●	●	●	●	●	●	●	●	●
	A3	●	●	●					●	●	●	●		●			●	●	●		●
	A4	●	●	●							●			●				●		●	●
	A5	●	●		●	●				●							●	●	●	●	●
	A6	●	●		●		●										●	●	●	●	●
	A7	●	●		●			●	●	●	●	●	●				●	●	●	●	●
	A8	●	●		●				●	●	●	●	●	●			●	●	●		●
	A9	●	●		●			●	●		●	●	●				●	●			●
	A10	●	●		●				●	●						●		●	●	●	●
	A11	●	●		●					●			●					●	●	●	●
	A12	●	●		●					●								●	●	●	●
	A13	●	●		●													●	●	●	●
	A14	●	●			●	●		●	●	●	●	●				●	●	●	●	●
	A15	●	●				●	●	●	●	●	●	●	●	●	●	●	●	●	●	●
	A16	●	●				●	●	●	●	●	●	●				●	●	●	●	●
	A17	●	●				●	●	●	●	●	●	●				●	●	●	●	●
	A18	●	●	●			●	●	●	●	●	●		●	●	●	●	●	●	●	●
	A19	●	●				●	●	●	●	●	●		●	●		●	●	●	●	●
	A20	●	●				●	●	●	●	●	●	●	●			●	●	●	●	●
	A21	●	●				●		●	●	●	●	●	●				●	●	●	●
	A22	●	●				●		●	●	●	●	●	●			●			●	●
	A23	●	●				●		●	●	●						●			●	●
	A24	●	●				●		●	●		●			●			●	●		●
	A25	●	●				●				●						●	●	●	●	●
	A26	●	●				●				●						●	●	●	●	●
	A27	●	●				●														
	A28	●	●					●	●	●	●	●	●		●	●	●	●	●	●	●
	A29	●	●					●	●	●	●	●	●		●	●	●	●	●	●	●
	A30	●	●					●	●	●	●	●	●				●	●	●	●	●
	A31	●	●						●	●	●	●	●	●	●	●	●	●	●	●	●
	A32	●	●						●	●	●	●	●	●	●	●		●	●	●	●
	A33	●	●						●	●	●	●	●	●			●	●	●	●	●
	A34	●	●						●	●	●	●	●	●			●	●	●		●
	A35	●	●						●	●	●	●	●				●	●		●	●
	A36	●	●						●	●	●	●	●			●	●	●	●	●	●
	A37	●	●						●	●	●	●	●	●			●	●		●	●
	A38	●	●						●	●	●	●	●	●						●	●
	A39	●	●						●	●	●	●		●			●	●	●	●	●
	A40	●	●						●	●	●	●					●	●	●	●	●
	A41	●	●						●	●	●	●					●	●	●	●	●
	A42	●	●						●	●				●			●	●	●	●	●
	A43	●	●						●	●				●				●	●		●
	A44	●	●						●	●		●						●	●	●	●
	A45	●	●						●		●	●	●				●	●	●	●	●
	A46	●	●						●		●	●	●	●			●			●	●
	A47	●	●						●		●	●	●						●		●
	A48	●	●						●		●	●		●			●				●
	A49	●	●						●		●	●		●			●	●			●
	A50	●	●						●		●			●				●	●	●	●
	A51	●	●						●		●	●	●					●	●	●	●
	A52	●	●						●	●				●			●	●	●	●	●
	A53	●	●						●								●	●	●	●	●
	A54	●	●								●	●		●			●	●	●	●	●
	A55	●	●								●	●						●	●		●
	A56	●	●								●		●				●	●	●	●	●
	A57	●	●								●			●			●	●	●	●	●
	A58	●	●								●						●	●	●	●	●
	A59	●	●								●						●	●	●	●	●
	A60	●	●												●	●	●	●	●	●	●
	A61	●	●													●	●	●	●	●	●
	A62	●	●														●	●	●	●	●
	A63	●	●														●	●	●		●
	A64	●	●														●	●		●	●
	A65	●	●															●	●	●	●
	A66	●	●															●	●		●
	A67	●	●																	●	●
	A68	●	●																		●
	A69	●	●																		●
	A70	●			●		●										●	●	●		●
	A71	●			●					●							●	●	●		●
	A72	●			●																●
	A73	●					●	●	●	●	●	●	●	●			●	●	●	●	●
	A74	●						●	●	●	●	●	●				●	●	●		●
	A75	●															●	●	●		●

FIG. 20. Bacteriocin typing scheme in *C. diphtheriae*. ●, Bacteriocin activity.

FIG. 20 (*continued*)

The sensitive strains could be divided by the set of 20 bacterio-cinogenic strains listed in 16 bacteriocin groups (A to P). The strains belonging to group A are sensitive to bacteriocinogenic strain 1; those contained in group B are resistant to bacteriocins of strain 1 and obligatorily sensitive to the bacteriocin of strain 2 (see Fig. 20). Twelve of the 16 bacteriocin groups were further subdivided into several types. For instance the A_1, A_2, A_3, etc. bacteriocin types all belonged to bacteriocin group A. The 16 bacteriocin groups included 162 types distributed as shown in Table II. Characters of bacteriocin groups and types are presented in Fig. 20.

(a) *The set of bacteriocinogenic strains used for bacteriocin typing of* C. diphtheriae. This is represented to date by 20 different strains of coryne-bacteria (Table III). These were selected from among 396 strains (367 *C. diphtheriae*, 15 *C. hofmannii*, 9 *C. xerosis*, 3 *C. ulcerans* and 2 atypical corynebacteria), which were first checked for bacteriocin production according to Hamon's method (1965) adapted by Meitert (1969a) to the growth requirements of C. *diphtheriae* (Table III).

(b) *Technique for demonstrating bacteriocinogeny*. Drops of overnight liquid

TABLE II

Subdivision of bacteriocin groups into bacteriocin types

Bacteriocin groups	Number of bacteriocin types contained in each group
A	75
B	16
C	6
D	8
E	12
F	10
G	12
H	6
I	7
J	1
K	2
L	2
M	1
N	2
O	1
P	1

cultures (Pope Linggood cystinated broth) of each strain to be tested are spotted by means of a Pasteur pipette on to the surface of brain heart agar, usually eight per plate. After 48 h incubation, at 37°C, growth develops from each drop. These are killed by chloroform vapour for 2 h. Then a drop of a 2½–3 h liquid culture of the sensitive strain is applied to the edges of each patch of growth.

After 18 h incubation at 37°C, the production of bacteriocins is expressed by inhibition of the sensitive test cultures as presented in Fig. 21.

Storing of bacteriocinogenic strains. Bacteriocinogenic strains are cultivated and stored either on Loeffler slants for 2–3 months in the dark at room temperature or in the lyophilised state. These bacteriocins have not shown up in electron microscopy. Bacteriocinogenic strains are available from the Diphtheria Laboratory of the Cantacuzino Institute, Bucharest, Romania. (c) *Technique of bacteriocin typing of* C. diphtheriae. Heavy, overnight Loeffler cultures from each of the bacteriocinogenic strains is plated along a 1 cm wide diametric band across each plate. After 48 h at 37°C, the

TABLE III

Characters of bacteriocinogenic strains used for *C. diphtheriae* bacteriocin typing

Registered number of bacteriocinogenic strains	*Corynebacterium* species	Toxigenic character	Origin
153	*C. hofmannii*	—	Carrier
155	*C. hofmannii*	—	Carrier
198	*C. hofmannii*	—	Carrier
730	*C. diphtheriae intermedius*	*tox⁻*	Carrier
760	*C. ulcerans*	*Di-tox⁻*	Carrier
828	*C. diphtheriae intermedius*	*tox⁻*	Carrier
866	*C. diphtheriae mitis*	*tox⁺*	Case
868	*C. diphtheriae mitis*	*tox⁺*	Case
871	*C. diphtheriae mitis*	*tox⁺*	Case
872	*C. diphtheriae mitis*	*tox⁺*	Case
873	*C. diphtheriae mitis*	*tox⁺*	Case
876	*C. diphtheriae intermedius*	*tox⁺*	Case
877	*C. diphtheriae mitis*	*tox⁺*	Case
878	*C. diphtheriae mitis*	*tox⁺*	Case
882	*C. diphtheriae mitis*	*tox⁺*	Case
898	*C. diphtheriae mitis*	*tox⁺*	Case
2378	*C. diphtheriae intermedius*	*tox⁻*	Carrier
2380	*C. diphtheriae intermedius*	*tox⁻*	Carrier
Hof.	*C. hofmannii*	—	—
Xer.	*C. xerosis*	—	—

growth is treated for 2 h by chloroform vapour. Young liquid cultures (2–3 h incubation at 37° in cystinated Pope Linggood broth) of the strains to be tested are then inoculated with a thin loop, at right angles to the killed bacteriocinogenic growth. The streaks are made on both sides of the diametrical band, from the edges of the bacteriocinogenic culture to the margins of the plate (Fig. 22).

Plates of 12 cm diameter accommodate 16–20 strains to be tested. The results are recorded after overnight incubation at 37°C. The growth of a bacteriocin sensitive strain is inhibited over a smaller or bigger area starting from the central bacteriocinogenic band. The whole range of reactions is recorded. The minimum positive inhibition area is 2 mm, with or without isolated resistant colonies.

Smaller inhibition zones than 2 mm are considered only when the strain would not otherwise be defined.

(d) *Media used.* The medium recommended for bacteriocin typing of *C. diphtheriae* is brain–heart infusion with $1 \cdot 3\%$ agar, $0 \cdot 001\%$ cystine to

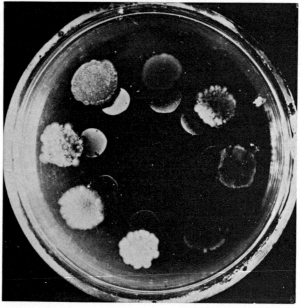

Fig. 21. Assay of bacteriocinogeny, in eight different strains. The inhibition areas, looking like "half-moons" on the bacteriocin sensitive strains, indicate the bacteriocinogenic activity of the strain under test. One of the bacteriocinogenic strains, probably also lysogenic, elaborates its phage on the same sensitive strain; plaques of lysis can be seen on the area.

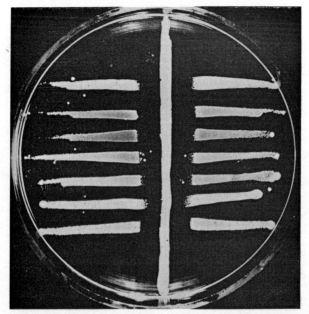

FIG. 22. Bacteriocin typing assay. The diametrical culture is the bacteriocinogenic culture. The cross streaks are the different bacteriocin sensitive strains of *C. diphtheriae*.

which 0.1% $CaCl_2$ ($1\,M$) and 10% bovine serum are added (Meitert, 1969b). This is the same formula as for phage typing (see Section XIII) except that the agar content must not exceed 1.3%.

Reproducible results require that the agar thickness is above 3–4 mm and that the plates should be freshly prepared.

As liquid medium, Pope Linggood broth or any other available commercial tryptic digest broth, supplemented by cystine, may be used.

3. *Distribution of* C. diphtheriae *bacteriocin types*

Bacteriocin typing has been applied to 1150 *C. diphtheriae* strains of different biotypes and phage types as well as to 21 *C. ulcerans* strains. All strains were typable. Type distribution can be seen in Table IV.

Bacteriocin types are not correlated with biotypes or species (*gravis*, *mitis*, and *intermedius* of *C. diphtheriae* or *C. ulcerans*).

4. *Stability of bacteriocin types*

During subsequent transfers on media, bacteriocin types can easily vary

5

TABLE IV

Distribution of *C. diphtheriae* and *C. ulcerans* strains in different bacteriocin groups

Bacteriocin group	Number of strains	Percentage
A	764	65·2
B	27	2·3
C	25	2·1
D	42	3·5
E	78	6·6
F	55	4·7
G	78	6·6
H	21	1·7
I	28	2·3
J	4	0·3
K	4	0·3
L	5	0·4
M	3	0·2
N	5*	0·4
O	13	1·1
P	19	1·6

from one bacteriocin type to another. In these conditions stability is observed only with respect to bacteriocin group.

5. *Relationships of bacteriocin types of phage types*

Phage typing classifies 85·7% of the *C. diphtheriae* strains and 99% of *C. ulcerans* strains. All strains were characterised by bacteriocin typing, although a strict correlation could not be established between phage types and bacteriocin types. A prevalent phage type in a community corresponded always to only one bacteriocin type (Fig. 23). In some communities the prevalent phage type could sometimes be subdivided by bacteriocins (Fig. 24).

6. *Limits and possibilities for further developments in* C. diphtheriae *bacteriocin typing*

The variability of bacteriocin production may be related to differences in growth factors, agar concentration, storage of bacteriocinogenic strains or growth conditions of the strains tested.

Despite the instability of the bacteriocin types the typing method seems to be reliable if the results are interpreted in a local sense and over a relatively short period of time.

Strains	Set of bacteriocinogenic strains (1–20)	Bacteriocin type	Lysotype	Biotype	Toxinogenicity	Date of isolation	Communities
8493	●● ●●●●●●● ●●●●●	A_{30}	V a	Intermed	–		
8494	●● ● ●●●●●●● ●●●●●	A_7	L	Gravis	–		
8495	●● ● ●●●●●●● ●●●●●	A_7	L	Gravis	–	April 1971	Buzău
8496	●● ● ●●●●●●● ●●●●●	A_7	L	Gravis	–		
8497	●● ● ●●●●●●● ●●●●●	A_7	L	Gravis	–		
8498	●● ● ●●●●●●● ●●●●●	A_7	L	Gravis	–		
8499	●● ● ●●●●●●● ●●●●●	A_7	L	Gravis	–		
8500	●● ● ●●●●●●● ●●●●●	A_7	L	Gravis	–	April 1971	Lehliu
8501	●● ● ●●●●●●● ●●●●●	A_7	L	Gravis	–		
8502	●● ●●●●●●●●●●● ●●●●●	A_{15}	V d	Intermed	–		
8503	●● ●●●●●●●●●●● ●●●●●	A_{15}	V d	Intermed	–		
8504	●● ●●●●●●●●●●●● ●●●●●	A_{15}	V d	Intermed	–		
8505	● ●●●●●●●●● ●●●●●	A_{73}	IV	Intermed	+		
8506	● ●●●●●●●●● ●●●●●	A_{73}	IV	Intermed	+		
8507	● ●●●●●●●●● ●●●●●	A_{73}	IV	Intermed	+	April 1971	Brăila
8509	● ●●●●●●●●● ●●●●●	A_{73}	IV	Intermed	+		
8510	● ●●●●●●●●● ●●●●●	A_{73}	IV	Intermed	+		
8511	● ●●●●●●●●● ●●●●●	A_{73}	IV	Intermed	+		
8512	●● ● ●●●●●●● ●●●●	A_{21}	XII	Gravis	+		
8513	●● ● ●●●●●●● ●●●●	A_{21}	XII	Gravis	+		

Fig. 23. Example of the relationships among phage types and bacteriocin types. Bacteriocin typing of 20 *C. diphtheriae* strains, isolated in different communities. ●, Bacteriocin activity (2–15 mm).

Strains	Set of bacteriocinogenic strains (1–20)	Bacteriocin type	Biotype	Toxinogenicity	Date of isolation	Communities
7177	●● ● ●●●●●●● ●●●●●	A_{20}				
7178	●● ● ●●●●●●● ●●●●●	A_{20}				
7193	●● ● ●●●●●●● ●●●●	A_{21}				
7197	●● ● ●●●●●●● ●●●●	A_{21}				
7219	●● ● ●●●●●●● ●●●●	A_{16}	*Gravis*	—	Dec. 1970	Buzău
7220	●● ● ●●●●●●● ●●●●	A_{20}				
7224	●● ● ●●●●●●● ●●●●	A_{20}				
7226	●● ● ●●●●●● ●●●●	A_{20}				
7227	●● ● ●●●●●●●● ●●●●	A_{16}				
7276	●● ● ●●●●●● ●●●●	A_{21}				
7278	●● ● ● ●●●●	A_{27}				
7279	●● ● ●●●●●● ●●●●	A_{21}	*Gravis*	—	Dec. 1970	Constanta
7284	●● ●●● ● ●●●●	A_{54}				
7286	●● ●●● ● ●●●●	A_{54}				
7287	●● ●●● ● ●●●●	A_{54}				
7355	●● ● ●●●●	A_{53}				
7357	●● ● ●●●●	A_{53}	*Gravis*	—	Dec. 1970	Adjud
7358	●● ● ●●●●	A_{53}				
7359	●● ●●●● ● ●●●●●●	A_{36}				

Fig. 24. Example of subdivision of an additional phage type "L" by means of bacteriocin typing. ●, Bacteriocin activity (2–30 mm).

VII. EPIDEMIOLOGY OF DIPHTHERIA

Diphtheria is typically a disease of communities, schools and households with children of susceptible ages (Wilson and Miles, 1946). Closeness and duration of contact play a major part in determining the spread of the disease. Contact during sleeping hours in common dormitories is far more dangerous than casual contact during waking hours (Christie, 1969).

When many clinical cases of diphtheria from an outbreak are admitted to hospital, spread of the disease is chiefly due to carriers and/or, at present, to asymptomatic cases of diphtheria, closely resembling simple cases of tonsillitis (erythema, mild sore throat). According to Wildführ (1949) and to our own experience, diphtheria bacilli are often detectable in the throat during the incubation period and it is likely that early carriers might be more dangerous than convalescent or healthy carriers, owing to the great number of growing bacilli they carry. Observations made by Wright *et al.* (1941) showed that dust is a possible source of infection in hospitals and other institutions. Diphtheria bacilli in dust remain fully virulent for as long as 5 weeks (Christie, 1969). *C. diphtheriae* has no intermediate animal host, although virulent diphtheria bacilli have been isolated from horses and monkeys.

Stănică *et al.* (1968) isolated *C. diphtheriae* in 14·7%, *C. ulcerans* in 7% and other corynebacteria in 35% of horses. The spread of diphtheria by milk is an epidemiological curiosity. It usually spreads from ulcers on the teats of cows with which the milkers (who may be diphtheria bacilli carriers) come in contact. Diphtheria outbreaks are not as a rule explosive in character but develop step by step, by human contact. The intensity of infection is governed by the level of specific immunity of the population and by the success of efforts to control its distribution by the rate spread of the causative agent. One or two decades ago, in the epidemic period, when mass immunisation was not so extensively applied, several outbreaks were recorded in Romania. These normally resulted in 60–100 cases, with approximately 25–30 contacts per case, each carrying the same phage type as the index cases. At that time precise principles of surveillance and control were established and were rapidly applied in order to limit an outbreak as soon as possible (Saragea *et al.*, 1965). Subsequently, with the application of mass immunisation and thorough surveillance of the causative agent in Romania (since 1965–1966), outbreaks have consisted of no more than five to seven cases. During the last 5–6 years, when the general level of immunity is measured at 95–96% Schick negative reactors, only very rare sporadic cases (one to two per year) occur throughout the country and no more than five or six contacts per case are identified.

Phage typing led to marked progress in the follow-up of carriers and/or contacts of an outbreak and their relationship to the cases because only carriers harbouring the same phage type as the case (usually *gravis tox*[+] strains) could be regarded as dangerous. Today, the occurrence during a routine survey of a *C. diphtheriae gravis tox*[+] phage type in a carrier is a curiosity in Romania, representing a rare event, indicating the need for a more analytical epidemiological investigation. The majority of strains isolated at present in communities and/or in an open environment, originate from healthy carriers and are represented by non-toxigenic *C. diphtheriae gravis* or *intermedius* types. No prophylactic or therapeutic measures are necessary for these carriers. In the U.S.A. during recent years, outbreaks accounting for 70–200 cases have been described in Austin, Texas, Phoenix, Arizona, and San Antonio, Texas (Doegge and Walker, 1962; Zalma *et al.*, 1970; Marcuse *et al.*, 1973; Brooks *et al.*, 1974). These outbreaks persisted over periods of months and years, suggesting high levels of endemic carriage and failures in containment in spite of the intensive mass immunisation programmes. This represents a striking example showing that diphtheria cannot be exterminated by immunisation programmes alone. In the best immunised community a lack of immunity is always to be found, stressing the need for thorough and continuous surveillance. This is readily possible by using all modern microbiological typing methods.

The occurrence of cutaneous diphtheria among American Indians in south-west U.S.A. (Belsey *et al.*, 1969) and Canada (Wilson and Toshach, 1957) raises new questions concerning the relationships between these forms, and faucial diphtheria. During the Second World War cutaneous diphtheria appeared in soldiers and large epidemics were reported in troops serving in tropical or subtropical countries. Close personal contact and poor personal hygiene favoured spread of the disease. Lesions occurred most frequently on the lower extremities, scrotum and hands. Few patients died, unless they also suffered from faucial diphtheria, but many (20–25%) experienced neurological complications. These findings underline the need for methods tracing the origin of outbreaks, the paths of spread, connection of cases and determination of the duration of carriage.

VIII. OTHER CORYNEBACTERIA WITH HUMAN HABITAT

A. *Corynebacterium ulcerans*

1. *Definition*

Corynebacterium ulcerans is another pathogenic member of the genus *Corynebacterium*. It produces two different toxins, a species specific toxin present in all strains and a typical diphtherial toxin, immunologically identical with that elicited in *C. diphtheriae*, which, as in *C. diphtheriae*, may or may not be present in different strains.

2. History

First isolated from the nasopharynx of a group of individuals and considered as "diphtheroid" by Gilbert and Stewart (1926) and Wilson and Miles (1946), it was called by the authors by its present name, *C. ulcerans*. Mair in 1928 reported anomalous results in neutralisation experiments with diphtheria antitoxin, in guinea pigs and Barratt (1933) found that it fermented starch and liquefied gelatin. Petrie and McClean (1934) showed that it may produce two kinds of toxin, one immunologically identical to diphtherial toxin and its own toxin related to the toxin of *C. ovis*. Jebb (1948), Saxholm (1951) and Cook and Jebb (1952), compared a series of strains isolated from acute tonsillitis, healthy carriers, or skin lesions. The latter authors suggested some more rapid enzymic reactions (hydrolysis of urea, and nitrate reduction in routine examination) to differentiate them from *C. diphtheriae gravis*.

Maximescu (1967), studying reference and wild *C. ulcerans* strains isolated from healthy human carriers in Romania, developed a complete routine scheme for their bacteriological diagnosis. Maximescu (1968), in an attempt to elucidate the taxonomic position of *C. ulcerans*, showed that it has characters in common with *C. diphtheriae* and *C. ovis*, sharing cystinase, ability to ferment carbohydrates, sensitivity to diphtheria phages and, in spite of major differences in pathogenicity, the capacity to induce diphtherial toxin biosynthesis in all three species by lysogenisation with a diphtherial *tox*⁺ bacteriophage.

The pathogenicity differential test with *C. ovis* in mice was established by the same authors (Maximescu *et al.*, 1968, 1974a, b). *C. ovis* kills mice by any route of inoculation, while *C. ulcerans* strains only cause arthritis.

3. Habitat

Both in human cases of acute tonsillitis or healthy carriers and in animals (horses or cows) producing ulcerative lesions or healthy carriers.

4. Microbiological characters

(a) *Morphology*. Gram-stained smears prepared from overnight cultures on Loeffler slants (Fig. 25) show very characteristic features, i.e. a mixture of pleomorphic club-shaped granular, thin, slightly curved, diphtheria-like bacilli, and pleomorphic, minute cocci, with a great tendency to over-decolourise, as in *C. diphtheriae*. Smears sometimes present only pleo-

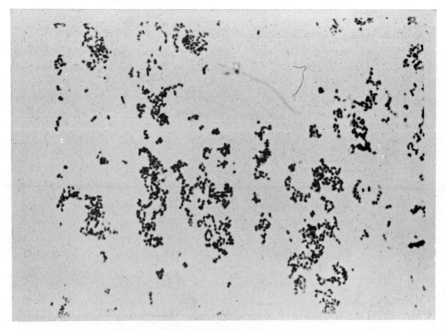

FIG. 25. *C. ulcerans.*

morphic readily decolourised cocci without bacillary forms. Broth cultures display a greater capacity to show bacillary forms.

(b) *Growth*, aerobic, optimum temperature 37°C.

(c) *Cultural characters.* On Loeffler slants—a thick, rich, creamy culture.
   On CTBA large greyish black, dense colonies with regular margins and shiny-smooth surface are produced. On blood agar plates, large, dense, opaque yellowish-white colonies, with a smooth surface develop. The consistency is friable and easily emulsifiable.
   On Tinsdale plates, large black colonies are surrounded by an intense and precocious brown halo which appears in about 16–24 h at 37°C. In broth, slightly turbid cultures, with a veil or ring on the surface are produced. These clarify the next day, forming a powderish deposit at the bottom of the tube.

(d) *Enzymes.* Enzymic activity is very intense and reactions are rapidly produced.
   Carbohydrate fermentations: acid with glucose, maltose, levulose,

dextrin, starch, and glycogen. Some strains resemble the *C. diphtheriae* type *intermedius* (Jebb and Martin, 1965; Maximescu *et al.*, 1974a).

A characteristic feature, to be emphasised, is the fermentation of trehalose, which is split very slowly (up to 16 days). Cystine and urea are rapidly split. The phosphatase reaction is positive within 24 h. Gelatin is liquefied in stab cultures within 10 days at room temperature. Nitrate reductase is not present in *C. ulcerans*.

(e) *Pathogenicity. C. ulcerans* possesses its own toxin, supplemented in some strains by a diphtherial toxin ($Di\text{-}tox^-$).

The whole broth culture of a $Di\text{-}tox^-$ strain inoculated subcutaneously is lethal within a few days for the guinea pig. The post mortem findings are as follows: an intense gelatinous and haemorrhagic oedema at the site of inoculation. Pleural and peritoneal exudates without lesions in the adrenals. All these phenomena are not prevented by diphtheria antitoxin.

By intradermal inoculation in rabbits, severe necrosis and ulcers of the skin occur.

Suppurative lesions are induced in white mice. When endowed with diphtherial toxin, subcutaneously inoculated culture filtrates produce death in about 24–48 h, with lesions characteristic of diphtheria intoxication. These may be prevented by diphtherial antitoxin.

In tissue cultures and Elek–Ouchterlony plates, these strains produce reactions similar to those of tox^+ *C. diphtheriae* strains.

(f) *Bacteriocins* described by Meitert (1969b).

5. *Phage–bacterial host relationships in* C. ulcerans

(a) *General.* Thorough studies on its taxonomic position indicate that *C. ulcerans* is somewhat intermediate between *C. diphtheriae* and *C. ovis* (Barrat, 1933; Jebb, 1948; Rountree and Carne, 1967; Carne, 1968). Further investigations on the phage–bacterial host relationships and their relations with toxinogenicity in these species, performed by Maximescu *et al.* (1968) led to exciting findings on diphtherial toxin production. Repeated investigations showed that each of the two species, *C. ulcerans* and *C. ovis*, when experimentally lysogenised with diphtherial phages, produced besides their species-specific toxin, another one which is biologically and immunologically identical to the *C. diphtheriae* exotoxin.

(b) *Phage typing in* C. ulcerans *and* C. ovis. From the previously described research a question arose: to what extent are *C. ulcerans* and *C. ovis* biologically related to each other and to *C. diphtheriae*? Thus, further studies on the phage–bacterial host relationships in *C. diphtheriae, C.*

ulcerans and *C. ovis* (Maximescu *et al.*, 1974a, b) were carried out with a view to detecting other characters that would contribute to a correct circumscription of each species. The authors tried to find a phage sensitivity "marker" to define the borderline between the species.

During an early period, *C. ulcerans* and *C. ovis* were rarely lysed by *C. diphtheriae* typing phages; *C. ulcerans* could not be classified into known lytic "patterns". In contrast, *C. ulcerans* typing phages were more active. *C. ulcerans* phages are known to be species specific and stable (Henriksen, 1955).

The phages proved useful for lysosensitivity testing since all available *C. ulcerans* and *C. ovis* strains could be typed and classified into well-defined types. As a consequence, six *C. ulcerans* phages and one *C. diphtheriae* phage "restricted" in *C. ovis* were selected for a provisional phage typing scheme by which *C. ulcerans* and *C. ovis* strains have thereafter been routinely tested for phage sensitivity. Maximescu *et al.* (1974a) outlined a provisional scheme by which 13 well-defined phage types could be defined. This scheme yielded satisfactory results in the routine typing of wild and

Host strain	Phage	(a) $15542/234$	(b) $213/40_c$	(c) $16259/40_c$	(d) $210L/16L$	(g) $9304/984$	(h) $984/984$	76 $76c/ovis21$	Phage-type
984	C.ulcerans	●	●	●	●	●	●	●	A_0
9304	,,	●	●	●			●	●	IA
5229	,,	●	●	●		●		●	IA_1
996	C.ovis	●	●		●	●	●	●	IIB_0
990	,,	●	●		●		●	●	IIB_1
16259	C.ulcerans	●	●				●	●	IIB_2
16L	,,	●	●		●	●			$IIIC$
210L	,,	●	●						IVD
318c	,,	(●)	●						IVD_1
213L	,,	●			●	●			VE
15542	,,		●						VIG
4870	,,		●		●				VIG_1
C.0.21	C.ovis		●					●	VIG_2

FIG. 26. Phage typing scheme for *C. ulcerans* and *C. ovis*. ●, confluent lysis; (●), isolated plaques.

TABLE V

Phage-typing of *Corynebacterium ulcerans* and *C. ovis* strains of different origins

Phage-type	Di-tox+	Di-tox−	Ao +	Ao −	IA +	IA −	IA1 +	IA1 −	IIB0 +	IIB0 −	IIB1 +	IIB1 −	IIB2 +	IIB2 −	IIIC +	IIIC −	IVD +	IVD −	IVD1 +	IVD1 −	VE +	VE −	VIG +	VIG −	VIG1 +	VIG1 −	VIG2 +	VIG2 −	Sensitive non-typable	No. of tested strains
C. ulcerans wild type human	44	96	—	—	54	—	4	—	—	—	—	—	1	15	15	—	9	8	22	—	—	5	4	1	—	—	—	—	2	140
C. ulcerans equine	6	29	—	—	9	—	2	—	—	—	—	—	—	—	6	—	16	—	—	—	2	—	—	—	—	—	—	—	—	35
C. ulcerans monkeys	1	32	—	—	—	—	—	—	—	—	—	—	—	—	1	—	—	—	—	—	—	—	—	—	32	—	—	—	—	33
C. ulcerans collection	13	3	4	—	2	—	—	—	—	—	—	—	—	—	1	—	6	—	—	—	1	—	1	—	—	—	—	—	1	16
C. ovis	2	60	—	—	—	—	—	—	26	—	2	—	2	—	—	—	—	—	—	—	—	—	—	—	—	—	—	32	—	62
Collection total	66	220	4	—	65	1	6	—	26	—	2	—	1	25	25	—	21	25	8	24	1	—	5	—	36	1	—	32	3	286

reference strains. Also, taxonomic differentiation between *C. ulcerans* and *C. ovis* was possible.

(c) *Efficiency of phage typing in* C. ulcerans *and* C. ovis. All but 1% of the *C. ulcerans* and *C. ovis* strains could be typed and classified in well-defined "lytic patterns" (i.e. phage types) (Fig. 26). Collection strains of different origins and dates of isolation were typable and their "patterns" were stable (Table V).

The phage types Ao, IA, IIB$_2$, IVD, IVD$_1$, VE, VIG and VIG$_1$ are specific for *C. ulcerans* strains; phage types IIBo, IIB$_1$, VIG were found only within *C. ovis*. Phage type IIIC was found in both *C. ulcerans* and *C. ovis*, only among *tox*+ strains. The most widespread types are IA, IVD and IIIC. In man and horses identical phage types were found. Most of the wild strains isolated in humans occurred in carriers, and only rarely was their incidence associated with sore throat or tonsillitis. Most of the carriers were children from rural areas. Among horses, up to 50% were carriers. A high incidence of *C. ulcerans* strains (43%) was described by Panaitescu *et al.* (1977) among monkeys. The isolated *C. ulcerans* strains belonged to different phage types. Most of the strains were classified in phage type VIG, and were *Di-tox*⁻. These animals were imported from Malaysia. Type IIIC, *Di-tox*+ was isolated from a clinical case of diphtheria in *Maccaca mulata* of Indian origin. The possibility of distinguishing stable types among strains of different origin, isolated at different times, proves the validity of this test in the characterisation of *C. ulcerans* and *C. ovis*. Technical details of the procedures are described in Section XI.

B. *Corynebacterium hofmannii*

1. *Isolated*

Isolated by von Hofmann in 1888.

2. *Habitat*

In the nasopharynx, in the auditive meatus and on the skin of the face of healthy individuals, where it lives as a harmless saprophyte. Because of its frequency it represents the most important diphtheroid with respect to the microbiological diagnosis of diphtheria.

3. *Morphology*

Short rods (1·5–2 μm long) thick, pointed ends, intensely Gram-positive, more tenacious of the Gram stain without granules. There is no tendency to pleomorphism. Bacilli are very regularly arranged in parallel rows (Fig. 27). The trained observer recognises it without difficulty, but

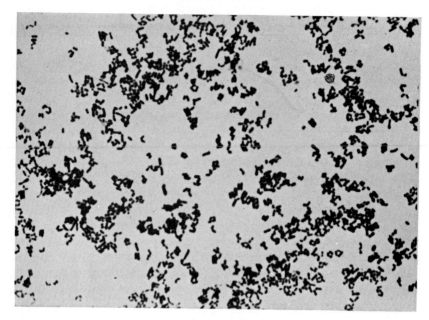

FIG. 27. *C. hofmannii.*

assurance of its diagnosis comes by comparing its morphology with the other characters.

Non-motile, non-sporing, non-capsulated.

4. *Growth*

Aerobic and facultatively anaerobic. Optimal temperature around 37°C.

5. *Cultural characteristics*

Develops readily on conventional laboratory media, including Loeffler medium and cystine–tellurite blood agar. On Loeffler, after 24 h there is a heavy creamy smooth growth, homogeneous in structure and easily emulsifiable. On tellurite media, colonies are smooth with a rather wide greyish opaque or transparent margin, because it reduces tellurite salts very poorly. In broth, growth is uniform, turbid, the culture gradually settling to the bottom as a powdery deposit. No pellicle is formed. On Tinsdale plates greyish black large colonies, without any halo.

6. *Enzymes*

The only diagnostically important enzymes are urease, nitrate reductase, and in some of the strains also phosphatase.

7. Pathogenicity

It is not pathogenic.

8. Antigenic structure

Antigenic structure differs from that of *C. diphtheriae* and *C. xerosis*. It is never agglutinated by antibacterial *C. diphtheriae* sera. Two main precipitating antigens A and B were isolated and characterised (Banach and Hawirko, 1966). Both antigens consist of a major protein with polysaccharide and nucleic acid. Antigen A contains less protein and carbohydrate than antigen B. The principal sugars of antigen A are arabinose and glucose. Both are heat stable, acid resistant, and species specific. Acid extracts of heterologous species fail to cross-react, even with high-titred *C. hofmannii* sera.

9. Lysogenicity and phage sensitivity

Meitert and Bica-Popii (1971) described lysogeny in this species. They isolated four temperate bacteriophages from wild lysogenic strains and studied the morphology and antigenic relationships between them (Meitert and Bica-Popii, 1972). Later, Saragea *et al.* (1975) succeeded in isolating a set of a further 18 preparations and studied their biological

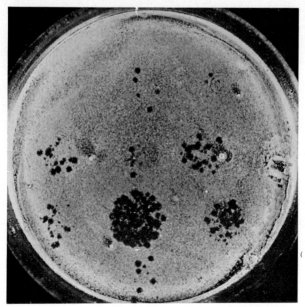

Fig. 28. Phage lysis plaques in *C. hofmannii.*

activity, serological specificity and morphology. They showed that *C. hofmannii* phages do not cross-react with *C. diphtheriae* phages, display large haloed plaques (Fig. 28), a higher multiplicity, and smaller electron microscopic size than *C. diphtheriae* phages, or the phages specific for other corynebacteria. Morphologically (Fig. 29), they resemble only *C. murium* phages (Petrovici *et al.*, 1974). Their strict species specificity (Saragea *et al.*, 1975) provides the possibility of using phage sensitivity of *C. hofmannii* as a taxonomic criterion, or as a valuable marker in genetic studies.

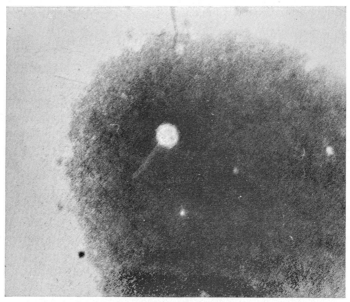

Fig. 29. Electron photograph of *C. hofmannii* phage. Magnification × 125 000.

10. *Bacteriocins*

Meitert (1969a, b) and Meitert and Bica-Popii (1975) while investigating bacteriocinogeny among various members of the *Corynebacterium* genus, found that *C. hofmannii* strains were 57% bacteriocinogenic against other members of the genus as well as against other bacterial genera like haemolytic streptococci, enterococci, and staphylococci.

In order to see whether bacteriocinogenic strains of *C. hofmannii* could make *C. diphtheriae* strains disappear if the two were grown together, Meitert (1973) devised an *in vitro* experimental model and showed that *C. diphtheriae* was almost completely eliminated by *C. hofmannii* after

80 days, in contrast to pure culture controls where each bacterium survived for that period.

Bacteriocinogenic *C. hofmannii* strains are included in the set of bacteriocinogenic strains used in the proposed scheme for bacteriocin typing which was described earlier.

C. Corynebacterium xerosis

Described by Reymond *et al.* in 1881 (cited by Wilson and Miles, 1966).

1. Habitat

C. xerosis is isolated from normal and diseased conjunctiva, ear and occasionally from the throat, although much less frequently than *C. hofmannii*.

2. Morphology

In Gram stained smears prepared from 24 h cultures on Loeffler slants at 37°C, the cell morphology resembles *C. diphtheriae*. However, the cells are arranged in palisades, with a prevalence of barred and segmented forms. Club formation and metachromatic granules are infrequent (Fig. 30).

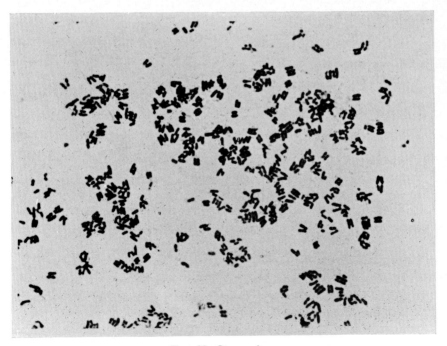

FIG. 30. *C. xerosis*.

Pleomorphism is very rare. Gram stain is more tenaciously retained than by *C. diphtheriae*. It is non-motile, non-spore forming, and non-capsulated.

3. *Growth characteristics*

Aerobic and facultatively anaerobic; optimal temperature, 37°C. Growth is less vigorous than that of the diphtheria bacillus and isolated colonies are smaller on tellurite–blood agar. Their colour is dark black and they have a glossy surface. An inexperienced technician may confuse them with *C. diphtheriae* type *intermedius* colonies. However, differences between the glossy surface of *C. xerosis* and the granular aspect of *C. diphtheriae* type are notable. On Tinsdale plates, it gives dark colonies without a brown halo.

Broth remains clear with a slight granular deposit at the bottom.

4. *Enzymes*

Acid from glucose, sucrose, maltose and fructose. Nitrates are reduced. Urea is hydrolysed by most strains. *C. xerosis* lacks cystinase and is not haemolytic. *C. xerosis* is, in many respects, similar to *C. cutis communis* which however does not ferment maltose.

5. *Antigenic structure*

Not well known. but certainly different from *C. diphtheriae*.

6. *Pathogenicity*

There is no evidence of pathogenicity.

7. *Bacteriophages*

Have not yet been described for *C. xerosis*.

8. *Bacteriocins*

C. xerosis possesses bacteriocins active against other members of the genus *Corynebacterium* and against other genera. A highly bacteriocinogenic *C. xerosis* strain is included in the set of bacteriocinogenic strains used in bacteriocin typing of *C. diphtheriae* (see above).

D. Atypical corynebacteria

Variants of atypical corynebacteria have been reported in Romania by Saragea *et al.* (1962). They were isolated usually towards the end of some of the diphtheria outbreaks during the epidemic period up to 1965. Their occurrence after application of extensive prophylactic procedures and antibiotic treatment of carriers as well as certain similarities with atypical mycobacteria, led the authors to assume that these forms were "variants" representing a degradation of the type species *C. diphtheriae*. These variants are very similar to type VI described by Wright and Christison (1935).

1. *Habitat*

Nasopharynx, skin and conjunctiva in humans.

2. *Morphology*

Variable, either very long or very small, often gigantic, branched rods, club-shaped, containing big metachromatic Gram-positive bodies, which are easily decolourised. Intensely pleomorphic. In smears arranged like Chinese letters (Fig. 31). Slender bacilli are sometimes seen in the same microscopic field.

3. *Growth*

Strictly aerobic. A most characteristic feature is the rough surface of the cultures and the tardy growth on media used for *C. diphtheriae* (2–4 days).

On Loeffler slants, thick, rough, white or intensely yellow culture develops. The growth is not uniformly emulsified. When touched with a loop, the culture crumbles into big, rough, inhomogeneous particles.

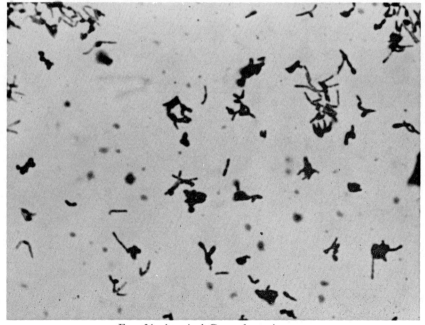

FIG. 31. Atypical *Corynebacterium* sp.

On cystine–tellurite–blood agar, atypical corynebacteria develop large black colonies with a very rough surface and strongly crenated margins, after 2–4 days growth (Fig. 32).

On blood agar, white or yellow, large, and rough colonies with crenated margins, develop within 2–4 days.

On Tinsdale plates, large, black, rough colonies without brown haloes appear.

Broth remains clear with a white or yellow veil on the surface and a granular deposit.

4. *Enzymes*

Carbohydrate: in 2–3 days, three different fermentation patterns emerge on a number of carbohydrates (Table VI). Sugars are fermented slowly like *mitis* type *C. diphtheriae*, like *C. xerosis* or irregularly. Urease, sometimes present; phosphatase, always positive, haemolysis, absent.

5. *Pathogenicity*

Non-toxigenic.

6. *Bacteriophages*

Non-lysogenic; resistant to known phages.

FIG. 32. Atypical *Corynebacterium* sp. Appearance of colonies on CTBA plates.

7. *Bacteriocins*

Present.

8. *Antigenic structure*

Not studied.

E. *Corynebacterium* variants with dwarf colonies

1. *Habitat*

In human nasopharynx sometimes causing tonsillitis with mild symptoms.

2. *Morphology*

Very short, minute rods, with pointed ends, pleomorphic, Gram-positive but over-decolourising easily. Rare metachromatic bodies (Fig. 33).

3. *Growth*

Strictly aerobic; growth occurs at 37°C in 48–72 h.
On Loeffler slants—a very thin granular culture.

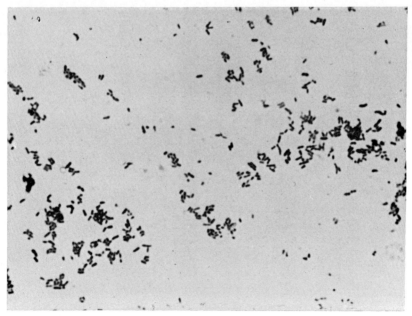

FIG. 33. *Corynebacterium* variant with dwarf colonies.

TABLE VI

Microbiological differential diagnosis among *C. diphtheriae* and other corynebacteria with human habitat

Corynebacterium species	Morphology Gram staining	Meta-chromatic bodies	Growth characters					Enzymic characters					
			Tinsdale	CTBA	BA	Loeffler	Broth	Glucose	Sucrose	Maltose	Levulose	Cystinase	Urease
C. diphtheriae	Gram-positive at ends, pleomorphic, slender curved club-shaped rods, Chinese letters	Present	Greyish black colonies surrounded by a characteristic brown halo	Black colonies, different shapes and sizes, according to the biotype	Pearl greyish, light, circular colonies	Granular white, friable culture	Turbid or clear culture with a veil on surface	+	–	+	+	+	–
C. ulcerans	Pleomorphic, Gram-positive at ends slender curved rods, and cocci of different sizes	Present	Greyish black colonies surrounded by a wide brown halo	Greyish black, smooth raised, regular margins, friable	Dense, opaque, smooth, convex colonies	Smooth, creamy yellowish colonies	Uniformly turbid with veil on surface	+	–	+	+	+	+
C. hofmannii	Gram-positive, short, regular rods, with pointed ends	Absent	Greyish black small colonies without halo	Smooth, brown colonies	Smooth, creamy sometimes yellowish colonies	Smooth, creamy, dense	Uniformly dense, turbid culture	+	–	–	–	–	+ or –

	Morphology		Colony / culture characteristics										
C. xerosis	Gram-positive medium sized barred, regularly tapered rods, palisades	Absent	Shiny black colonies without halo	Smooth, shiny black, small colonies	Small, smooth white, circular colonies	Smooth, creamy culture	Slightly uniform, turbid culture	+	+	+	+	−	+ or −
Atypical *Corynebacterium* sp.	Gram-positive, long branched, curved, club-shaped rods, Chinese letters	Present	Large black rough colonies without halo	Very dry, rough, irregular crenated margins, dark black	Dry, silverish surface, irregular edges, very friable	Rough, white or yellow culture, slow growth	Clear with veil on surface	†+	+	−	‡+	§−	+ or −
Corynebacterium variant with dwarf colonies	Gram-positive at ends, minute short, slender rods, rare corpuscles	Present	Circular, greyish, minute colonies growing slowly, slight halo	Minute, black colonies, growing slowly	Minute, silverish colonies, regular or crenated edges	Granular white, slow growth	Clear with deposit	+	−	−	+	± slowly	−

+, positive reactions.
−, negative reactions.
†, ‡, §, three different fermentative types of atypical corynebacteria.

On cystine tellurite–blood agar: minute greyish black colonies, developing very late. Their appearance is either *intermedius*-like with a depressed centre (Fig. 34) or *gravis*-like with minute, daisy headed colonies with crenated margins (Fig. 35).

On blood agar: very small, pearl-greyish colonies with a silvery white surface.

On Tinsdale plates: greyish minute colonies, only the underlying agar being light brown.

4. *Enzymes*

Carbohydrate fermentation: glucose and levulose are attacked slowly, positive reactions occurring only within 2–3 days. Cystinase: weak producer only in the aerobic portion of a stab. Urea is not hydrolysed.

5. *Antigenic structure*

Not well known. The organism could be agglutinable by diphtherial polyvalent agglutinating sera.

6. *Pathogenicity*

Non-pathogenic for animals. Toxins not produced. It is presumed to be

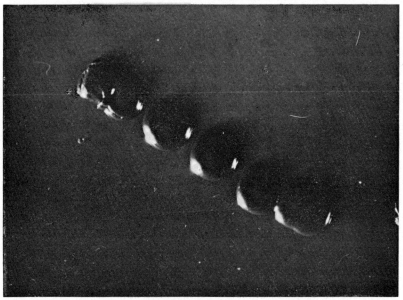

Fig. 34. Variant with dwarf colonies: *intermedius*-like on CTBA plates.

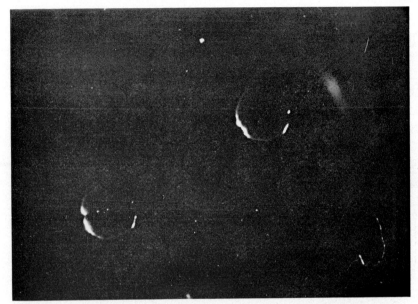

Fig. 35. Variant with dwarf colonies: *gravis*-like on CTBA plates.

a variant of *C. diphtheriae*, producing only local erythema with mild sore throat and tonsillitis in humans.

7. Neither *phage sensitivity*, lysogeny, nor bacteriocins have been demonstrated in this variant.

IX. PRACTICAL PROCEDURES FOR THE ISOLATION AND IDENTIFICATION OF *C. DIPHTHERIAE*

A. Principles

As diphtheria is a disease in which the organism is located at the site of infection (pharynx, nose, eye, ear, vagina or skin ulcer) and its toxin is rapidly spread throughout the body, the basic principle of the microbiology of its diagnosis is the need for precision and urgency.

The causative agent must be isolated, identified, characterised, and typed as soon as possible. A schematic representation of the different stages of the microbiological diagnosis in diphtheria is given in Figs. 36 and 37.

The steps in microbiological diagnosis are: swabbing, isolation of the organism in pure culture, and identification, by using all available "markers".

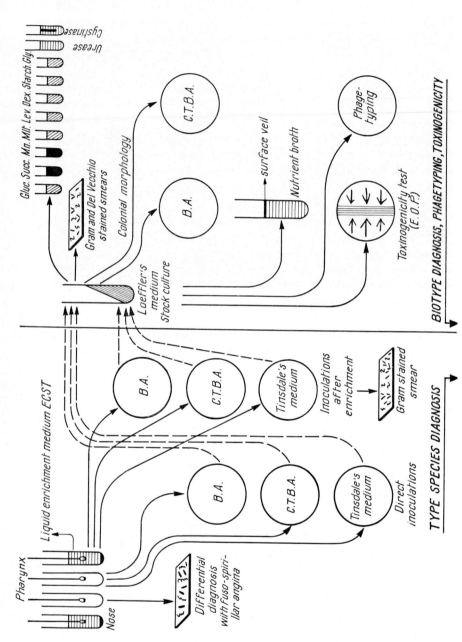

Fig. 36. Scheme for C. diphtheriae microbiological diagnosis in cases and/or suspects.

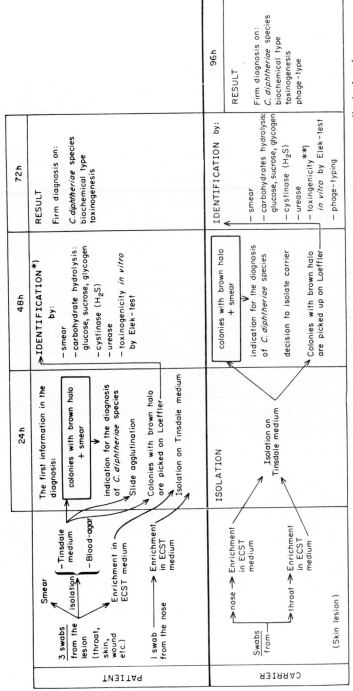

Fig. 37. Timetable of microbiological diagnosis in *C. diphtheriae*. ECST = Egg–cystine–serum–tellurite broth.

*The identification could be effected in hospitals or public health laboratories.

**Toxigenicity could also be performed *in vivo* subcutaneously in guinea pig or rabbit skin test.

B. Methods

1. *Swabbing of patients and carriers*

Swabbing should be carried out in the morning before breakfast, before any gargling with antiseptics and or treatment with antibiotics. If antibiotic treatment has been started, the laboratory should be informed.

The pathologic materials to be examined in patients are the following:

In the case of tonsillitis or faucial diphtheria—three swabs collected from the tonsillar exudates and one from the nasal secretions.

In the case of diphtheric rhinorrhoea—three swabs collected from nasal secretions and one from the pharynx or tonsils.

In the case of any other possible location (skin ulcers, conjunctivitis, otitis, burns, surgical wounds, vulvovaginitis)—three swabs should be collected from the site of infection, and two others from the throat and nose, even if the patient has no nasopharyngeal symptoms.

Reasons for the number of swabs recommended—from the first swab a smear is prepared. This will serve to establish the differential diagnosis from Plaut Vincent's angina in faucial diphtheria. However, even a skilled bacteriologist can never more than presume the presence of diphtheria bacilli in a direct smear.

In order not to lose time the second swab is streaked on selective and differential solid media. In order to avoid failure of growth, the third swab is immersed in a liquid enrichment medium and then streaked on solid selective media.

Specimens of exudates from both nose and pharynx, in extrapharyngeal diphtheria are taken to establish the relationships between these forms and cases of rhinopharyngeal diphtheria.

2. *In contacts and healthy carriers*

According to a well-designed field experiment (Saragea *et al.*, 1965) there are several basic principles that must be applied in checking contacts and/or healthy carriers, with a view to limiting an outbreak of diphtheria as quickly as possible.

Two swabs (Figs 37, 38), one from the pharyngeal and the other from the nasal exudate, are always collected from every individual assumed to be in direct or indirect contact with infected patients and from supposed healthy carriers. Swabbing is performed by thoroughly swabbing both tonsils, and both nostrils.

A large number of contacts are swabbed in order to determine the extension of the infected area.

The whole community (i.e. all the individuals to be swabbed) is simultaneously checked for carriage, in order to discover all healthy carriers and/or contacts as soon as possible, and to be able to apply simultaneously

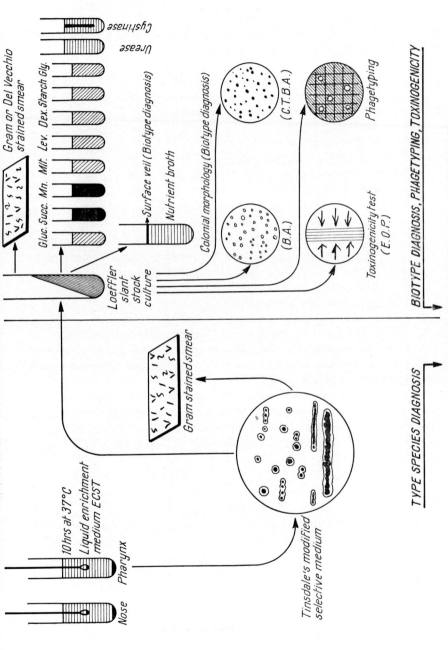

Fig. 38. Scheme for *C. diphtheriae* microbiological diagnosis in healthy carriers.

treatment and prophylaxis. If it is not possible at a given time to carry out an extensive microbiological investigation, it is advisable to delay 1 or 2 days in order to organise a complete and simultaneous check up.

In non-epidemic situations the procedures used for mapping the distribution of healthy carriers are similar, the examination of both nasal and pharyngeal exudates being obligatory.

3. Isolation of the organism in patient and carriers

(a) *In patient.* For reasons of urgency in suspected cases of faucial diphtheria, the following procedures are recommended:

On the first day

(i) The first swab is used for a direct smear.
(ii) The second swab is plated on Tinsdale and blood agar (BA) media; the swab may also be plated on CTBA. The blood agar medium is used in detecting other pathogenic species too (i.e. β-haemolytic streptococci, haemolytic staphylococci, etc.).
(iii) The third swab is immersed in the ECST enrichment medium (2–3 ml in tubes of 12/120 mm). All media are incubated overnight at 37°C; for the ECST medium, 10–12 h of incubation are sufficient.
(iv) The nasal swab is enriched in ECST medium and follows the same procedure as for swab No. 3.

On the second day

Plates are examined and specific colonies are diagnosed according to their characteristics: black coloured brown haloed on Tinsdale or buttery white greyish granular colonies on BA.

From suspected colonies on Tinsdale plates, a Gram-stained smear is effected to confirm the presence of corynebacteria bacilli, thereafter two or three colonies are picked up on a Loeffler slant and the culture is incubated overnight at 37°C. This culture will represent the stock culture for further identification.

In view of the urgency of diagnosis some isolated colonies may be picked and sugar fermentation (glucose, sucrose, glycogen), urea, toxigenicity test (Elek) as well as slide agglutination can be carried out.

If there is no growth on the plates, the enriched swab from the ECST medium is plated on Tinsdale. Isolation and identification of the developed culture follow the same steps as for the previous swab.

On the third day

The identification and typing of the stock culture from the Loeffler slant is effected by:

(i) Gram-stained smear for microbiological morphology; Methylene blue, Neisser or del Vecchio staining for metachromatic granules.

(ii) Cultural characters on solid (CTBA) and liquid (broth) media.

(iii) Enzymic characters: carbohydrate fermentations, urease and cystinase production.

(iv) Slide agglutination.

(v) Toxigenicity by *in vitro* Elek test or *in vivo* guinea pig or rabbit inoculation.

(vi) Phage typing, lysogenicity and bacteriocin typing.

In conclusion, at least 24–36 h are necessary for a rapid diagnosis in a suspected infection. In the next 48–72 h, biotype, toxigenicity and phage type can be defined.

(b) *In carrier*. Both nose and pharynx, should be swabbed for *C. diphtheriae* diagnosis.

Microbiological procedures are as follows:

On the first day

Swabs are immersed in 2–3 ml ECST medium for at least 10–12 h and incubated at 37°C.

On the second day

The enriched swabs are plated on Tinsdale medium, nose and throat each on half of the same plate, and incubated for 24 or 48 h at 37°C.

On the third day

Plates are checked for typical brown-haloed colonies and a Gram-stained smear is effected for a provisional diagnosis.

Specific colonies are picked on Loeffler slants, for stock cultures and incubated overnight at 37°C.

On the fourth day

Smears, enzymes, toxin, sero-agglutination, phage typing, lysogenicity and bacteriocin type are checked.

In conclusion, in carriers, the identification of *C. diphtheriae* can be obtained in 48 h and the biotype, toxigenicity and phage type can be done in 96 h.

4. *Identification of* C. diphtheriae

The characterisation and diagnosis of corynebacteria and the biotypes of *C. diphtheriae* species are listed in Tables VI and VII.

TABLE VII

Differential microbiological diagnosis within biotypes of *C. diphtheriae* and *C ulcerans*

Species	Growth characters on CTBA plates	Haemolysis on BA plates	Growth characters on nutrient broth	Carbohydrate fermentations								Patho-genicity
				Glu-cose	Su-crose	Man-nite	Malt-ose	Levu-lose	Dex-trine	Starch	Gly-cogen	
C. diphtheriae gravis	Greyish black daisy-headed colonies with crenated edges and radiary striations	Haemolysis + or −	Clear liquid. Veil on surface. Deposit	+	−	−	+	+	+	+	+	*tox*++† or *tox*⁻
C. diphtheriae mitis	Black, smooth, convex medium sized colonies. Shiny glossy surface, regular circular edges	Haemolysis always +	Uniformly turbid	+	−	−	+	+	−	−	−	*tox*⁺ or *tox*⁻
C. diphtheriae intermedius	Greyish black colonies, medium sized, circular edges, granular surface, inter-mediate position between *mitis* and *gravis*	No haemolysis	Slightly, turbid, granular deposit on the walls and at the bottom	+	−	−	+	+	+	−	−	*tox*⁺ or *tox*⁻
C. ulcerans type fermenta-tive *gravis*	Smooth, greyish, circular, opaque colonies	No haemolysis	(a) Slightly turbid with deposit	+	−	−	+	+	+	+	+	*tox*⁺ or *tox*⁻
type fermenta-tive *mitis*			(b) becomes clear in a few days	+	−	−	+	+	+	−	−	

+, positive reaction.

−, negative reaction.

† *tox*⁺ or *tox*⁻ = diphtherial exotoxin present or absent.

Stock cultures on Loeffler slants are tested for:

(a) *Microscopical morphology*. Gram stained smears show slender, slightly club-shaped, pleomorphic bacilli, arranged like Chinese letters; granular bodies and club forms are recognised by their intense violet staining in contrast with the *C. diphtheriae* bacillus which in Gram staining is easily decolourised (Fig. 1).

When Neisser or del Vecchio staining are used, brownish green metachromatic granules (Babes–Ernst bodies) on a yellowish green background of the bacillary body are observed.

After Methylene blue staining, reddish metachromatic granules appear in the blue bacillary bodies.

C. ulcerans shows predominantly pleomorphic cocci (very easily decolourised) (Fig. 25) and sometimes rod forms similar to those of *C. diphtheriae*.

Atypical corynebacteria show long, apparently branched club-shaped bacilli, with large Babes–Ernst bodies (Fig. 31).

The dwarf colony variant of *Corynebacterium* shows minute, short, slender rods, with rare corpuscules, Gram-positive at the ends.

Diphtheroids are intensely Gram-positive and show short rods with pointed ends, regularly arranged in palisades (*C. hofmannii*, Fig. 27), or intermediary irregular staining rods, slightly similar to *C. diphtheriae* (*C. xerosis*, Fig. 30).

(b) *Growth characters*

(i) In broth, *C. diphtheriae* develops different features of growth, accordingly to the type:

gravis: pellicle on the surface, deposit on the bottom, leaving a clear liquid.

intermedius: granular culture and deposit on the walls and at the bottom.

mitis: slightly homogeneous turbidity.

C. ulcerans resembles *gravis*, *mitis* or *intermedius* *C. diphtheriae*.

Atypical corynebacteria: white or yellow coarse pellicle on the surface, thick deposit on the bottom and clear liquid.

Dwarf colony variant of *Corynebacterium*, grows only at the bottom with no turbidity in the liquid.

Diphtheroids show a uniform heavy turbidity.

(ii) On solid media on Loeffler, *C. diphtheriae* develops with a creamy coloured, smooth or granular culture; *C. ulcerans* grows similarly, but with more luxuriant growth. The variant with dwarf colonies show a granular, slow, sparse growth.

Atypical corynebacteria develop very dry and rough growth, white or yellow in colour, after 48 h.

Diphtheroids display white or creamy smooth dense growth.

Tinsdale medium examined with the naked eye, or lens, shows for *C. diphtheriae* and *C. ulcerans* the characteristic black colonies, brownish halo surround, which might easily be differentiated from all diptheroids and atypical corynebacteria. These produce black-greyish colonies lacking the brown halo. The variant with dwarf colonies produces greyish minute colonies growing tardily (48 h) with a slight, sparse brownish halo.

On blood agar *C. diphtheriae* shows pearl greyish colonies with a slightly granular surface, the shape of the colony depending on the type (*gravis, mitis, intermedius*) (Fig. 6).

C. ulcerans is very similar to *C. diphtheriae*, with smooth yellowish dense colonies.

Atypical corynebacteria grow after 48 h with rough, dry, often large colonies, with a silverish white or yellow surface, and deeply crenated margins (Fig. 32). Usually they are difficult to emulsify and are spontaneously agglutinable.

The variant with dwarf colonies, produces minute silvery colonies with regular or crenated edges, and a very soft consistency.

Diphtheroids are shiny in surface with dense regular margins.

CTBA medium is the most useful medium for differentiating the cultural aspects of *C. diphtheriae* biotypes. At 48 h, differences between all three types are striking.

Gravis type—large grey to black colonies, granular in surface, with radial striations and crenated margins (daisy-head-like colonies) (Fig. 3).

Mitis type—intensely black, dense, middle-sized colonies, with regular margins, convex, shiny–glossy surface (Fig. 4).

Intermedius type—small or middle sized colonies with regular margins, a granular surface, flat or convex with a raised black centre (Fig. 5).

C. ulcerans develops greyish black dense smooth colonies with a shiny surface, similar to *C. diphtheriae*.

Atypical Corynebacterium develops late, after 48 h of incubation with various sized colonies, intensely black or greyish, dry, with very rough surface, deeply irregular, crenated edge and raised centre (Fig. 32).

The variant with dwarf colonies of *Corynebacterium* produces minute black to brownish colonies, with a slow growth. (Figs 34, 35)

Diphtheroids: *C. xerosis* displays shiny black colonies, similarly to *mitis*, but more smooth in consistency; *C. hofmannii* shows smooth brownish colonies (the tellurite incompletely reduced to black metallic tellurium) which become black in 5–6 days.

C. diphtheriae and *C. ulcerans* colonies on all solid media show different degrees of consistency; they tend to break into fragments (*gravis*, atypical corynebacteria), or show a soft buttery consistency. They are easily homogenised (*mitis, intermedius, C. ulcerans*).

(c) *Enzymic characters.* Carbohydrate fermentation: glucose, sucrose, maltose and laevulose are used for species differentiation. Dextrine, starch and glycogen are used to define *C. diphtheriae* biotypes (*gravis, mitis, intermedius*) (see above). Sugar fermentations are performed in test tubes on Hiss medium with Andrade's indicator.

A micromethod, Corynetest, using carbohydrate impregnated discs, was recently recommended (Ionescu-Stoian *et al.*, 1977).

Urease test is performed on Blake–Christensen's medium; this test is negative for *C. diphtheriae* and the variant with dwarf colonies and positive for *C. ulcerans*; 75–90% of diphtheroids display positive reaction.

Cystinase test is performed in stab cultures in tubes of 10/100 mm on Pisu's modified medium. The test is positive in 24 h (starting from the fifth hour of growth at 37 °C) for *C. diphtheriae* and *C. ulcerans*. A positive reaction consists of a black to brown halo along the streaked culture.

Diphtheroids, atypical corynebacteria and the variant with dwarf colonies are always negative or produce a blackening of the medium surface only.

(d) *Slide agglutination.* For the rapid confirmation of *C. diphtheriae* a polyvalent antidiphtherial serum for slide agglutination is recommended. Positive agglutination is specific for *C. diphtheriae* (87%). Spontaneously agglutinable or inagglutinable strains can also occur (13%). Therefore it should be stressed that a negative slide agglutination is not always contrary to a *C. diphtheriae* diagnosis.

Variants of *C. diphtheriae* or *C. ulcerans* with dwarf colonies may also agglutinate, but inconstantly. No other species of *Corynebacterium* agglutinate with antidiphtherial polyvalent serum.

Slide agglutination can be performed also with the isolated colonies from Tinsdale medium as a supplementary test for a rapid identification of *C. diphtheriae*.

(e) *Toxigenicity.* Toxigenicity tests present different degrees of sensitivity: the *in vitro* Elek–Ouchterlony–Frobisher test is a qualitative test showing a high sensitivity (98%); tissue cultures and rabbit skin i.d. test are quantitative tests. They can detect 1/500 and 1/1000–1/2000 DLM respectively. Guinea pig tests detect at least 1 DLM.

(i) *In vitro* tests. Elek–Ouchterlony–Frobisher test. The most commonly used test, economical, reproducible and easy to perform, is the Elek–Ouchterlony–Frobisher immunodiffusion test (EOF). It is based upon

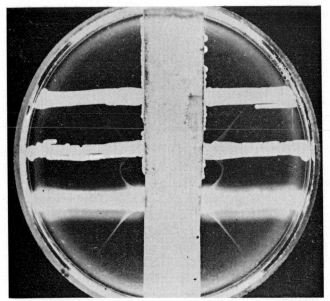

FIG. 39. *In vitro* toxinogenicity EOF test (the original technique). Diphtherial antitoxin is adsorbed on a filter paper strip.

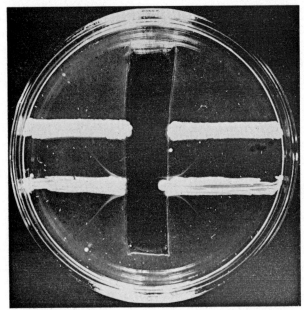

FIG. 40. *In vitro* toxinogenicity EOF test (the modified technique). Diphtherial antitoxin is introduced in a groove, cut in the agar.

the principle of double diffusion and specific toxin–antitoxin precipitation (Oudin, 1946, 1948). Elek (1948) and Ouchterlony (1948) almost simultaneously described a method whereby this reaction takes place in gel, culturing the bacteria directly on the medium in which antidiphtherial serum has been incorporated. Frobisher *et al.* (1951) recommended improved conditions for this test.

When the antitoxin is introduced on a paper strip sunk in the depth of the agar medium, precipitation lines develop within 18–48 h between the culture streak and the antitoxin lines (Fig. 39).

When the antitoxin is introduced in a groove in the medium (Maximescu and Drăgoi, 1971), the sensitivity of the test is increased (Fig. 40).

(ii) *Tissue culture test*. Sensitivity of tissue cultures to diphtherial toxin has been known for a long time. Levaditi and Muttermilch (1913a–d) as well as Burrows and Suzuki (1918) demonstrated that growth of chicken embryo fibroblasts was inhibited by diphtherial toxin. This effect was specifically neutralised by the respective antitoxin. Penso and Vicari (1957), La Placca (1957) and Souza and Evans (1957) were able to perform thorough studies on the cytotoxic activity of diphtherial toxin and also demonstrated the possibility of neutralising this activity by specific antitoxin. Calalb *et al.* (1965) compared the sensitivity of their test performed on *Macaccus rhesus* kidney cell cultures with all other tests used for determining the toxigenicity of *C. diphtheriae*, i.e. EOF plates and *in vivo* methods—guinea pig and rabbit tests—and concluded that tissue cultures were well suitable for the demonstration of diphtherial toxin.

The technique uses trypsinised monkey kidney secondary cell cultures suspended in Hanks's solution at a density of 50 000 cells/ml.

The cell suspension is distributed in 2 ml quantities, in Jena or Pyrex tubes (160/16 mm) maintained in a slanting position at 37°C for 2–3 days, until a continuous cell monolayer forms. The maintenance medium used is modified by Dubreuil and Pavilanis (1958) and contains 1% calf serum.

The broth culture of the strain under study is inoculated in 0·2 ml quantities in two to four tissue culture tubes. A tube with 1 A.U. antidiphtherial serum added is used as a control. After incubation at 37°C, the toxigenicity is appreciated by direct daily microscopic examination during the first four days following inoculation, according to the cytotoxic effect of diphtherial toxin on the tissue culture. The amount of toxin produced in an overnight broth culture shows characteristic cytopathologic effects within 24–48 h. The first cytotoxic changes show up in the arrangement of the cells. Groups of more refractory cells delimited by different sized lacunae appear. Hence holes in the cell monolayer, and cell detachment are observed. The tissue culture layer disappears almost completely within 24–48 h (Figs 41 and 42).

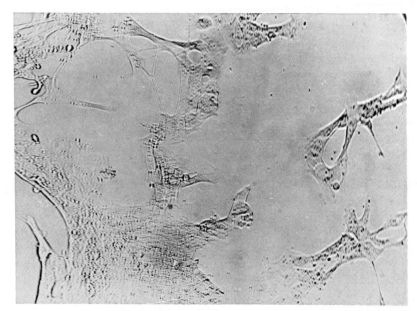

FIG. 41. Toxinogenicity test of *C. diphtheriae* on MKTC (monkey kidney tissue cultures). Activity of toxinogenic strains.

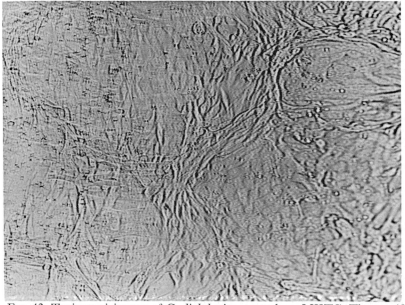

FIG. 42. Toxinogenicity test of *C. diphtheriae*, control, on MKTC. The specific diphtherial antitoxin prevents toxic damage and the monolayer remains intact.

This test using *Macaccus rhesus* cells detects minute quantities of diphtherial toxin (1/500 DLM); other tissue cultures (KB, HeLa, etc.) show variations in the degree of sensitivity according to the type of the tissue cells.

(iii) *In vivo* tests. Guinea pig test for *C. diphtheriae*: for each strain to be tested, two guinea pigs are used: one of the animals is protected with 2000 A.U. antidiphtherial serum. Two to three hours later, both animals are inoculated, subcutaneously with 3–5 ml of a 2–3 days broth culture. A toxigenic strain kills the unprotected animal in 24–96 h, while protected animals survive without any sign of toxicity. Necropsy reveals specific symptoms such as gelatinous haemorrhagic oedema at the site of inoculation, congestion of the abdominal wall; pleural and peritoneal sero-haemorrhagic exudates and congestion of the adrenals. Features of diphtherial intoxication cannot always be found together.

Stains of low toxigenicity which produce less than 1 DLM of toxin usually do not kill the animal, but can produce local necrosis; paralyses of posterior limbs or death may occur later.

Non-toxigenic strains allow the guinea pigs to survive without any pathological signs.

Guinea pig test in *C. ulcerans*: like *C. diphtheriae*, not all *C. ulcerans* strains produce a toxin which can be neutralised by diphtherial antitoxin.

The guinea pig test is carried out as for *C. diphtheriae*. When two guinea pigs of which one is protected with diphtherial antitoxin are inoculated with a broth culture of *C. ulcerans* which produces diphtherial toxin, both animals die. In the unprotected animal, death occurs in 1–2 days with all signs specific for diphtherial toxin. In the protected animal death occurs later (3–4 days). The changes are: local haemorrhagic gelatinous oedema, marked haemorrhagic congestion of all viscera, possibly haemorrhagic exudates of the pleural and peritoneal cavities (characteristic of the pathogenicity of *C. ulcerans*), but never the congestion of the adrenals. *C. ulcerans* strains, which do not produce diphtherial toxin, kill both animals in the same way as in the protected animal.

Skin test in rabbit (Fraser's test): this test is more sensitive than the previous test as it detects 1/1000–1/2000 DLM of toxin. It is not obligatory for routine work, but it may be used together with the guinea pig test, when the Elek-Ouchterlony–Frobisher test plates do not present clear-cut results.

An overnight broth culture is injected intradermally in 0·2 ml amounts on one side of a depilated back of a rabbit (2500–3000 g weight). Approximately 10–12 strains may be tested per rabbit. Before inoculation, squares are drawn with a special pencil on the depilated skin of the back of the rabbit. Three and a half hours after injection, the rabbit is injected intravenously with 2000–3000 AU diphtherial antitoxin. Half an hour after inoculation of the antitoxin, 0·2 ml amounts of the same suspension of strains is injected intradermally in the opposite half of the animal's back. Toxigenic strains produce erythematous reactions around the sites of injection within 48 h. No reaction occurs on the opposite side of the animal's back inoculated after intravenous administration of antitoxin (Fig. 43 (1), (2)).

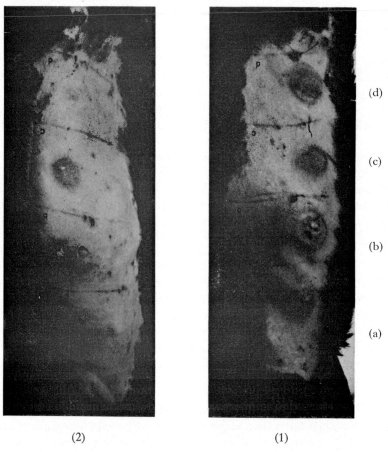

(d)

(c)

(b)

(a)

(2)   (1)

Fig. 43. *In vivo* intradermal toxigenicity test, in rabbit (according to Fraser's method). (1) Before the inoculation of diphtherial antitoxin. (a) *C. ulcerans Di-tox⁻*; (b) *C. ulcerans Di-tox⁺*; (c) *C. ulcerans Di-tox⁺*; (d) *C. diphtheriae tox⁺*. (2) Control after the inoculation of diphtheriae antitoxin. The intradermal effect of diphtherial toxin in both toxigenic *C. diphtheriae* and *C. ulcerans* is neutralised by the previously inoculated diphtherial antitoxin. (Reproduced from *J. gen. Microbiol.* 53, 125 (1968).)

This test may also be used for *C. ulcerans*, although reactions are far more intense.

Phage and bacteriocin typing are described in Sections VI and X.

X. PRACTICAL PROCEDURES OF PHAGE TYPING IN *C. DIPHTHERIAE*

A basic description of the two phage typing schemes for *C. diphtheriae* (original and additional) have already been given. The working principles and practical considerations concerning the technique will be discussed here.

A. Principles

Typing phages may be propagated in a tryptic digest broth such as Pope Linggood's broth (see Saragea and Maximescu, 1964) supplied with 0·001% L-cystine for the original typing phages and in heart broth supplied with 0·001% L-cystine for the additional typing phages.

The titre of a particular phage lysate depends upon the kind of phage propagated and the technique used. For instance, *C. diphtheriae* phages, even under the best conditions, never have routine tets dilutions (RTD) above 10^2–10^3, while *C. hofmannii* phages reach RTD titres of 10^8–10^{10} (RTD = the highest dilution still displaying confluent lysis on its propagating strain). The main problem associated with phage propagation is not the preparation of high titre lysates but maintaining the phage with its proper host range specificity. The lytic host range of every typing phage preparation should be tested against the whole set of sensitive and resistant strains from the respective phage-typing scheme. No typing phage preparation should be used unless it conforms to the accepted pattern. It is therefore advisable that for each established scheme, new phage stocks should be obtained periodically from an international reference laboratory. Diphtheria phages are relatively stable for a period of 3–4 and even 6 months. Undiluted stock phage preparations may be stored in the freezer and used as seed preparations for further propagation.

For long-term storage, diphtheria phage preparations may be lyophilised, the phages keeping their host range for at least 10 years.

B. Phage-typing procedure

An overnight stock culture is grown in 0·001% cystinated nutrient broth for $1\frac{1}{2}$ h at 37°C. A plate of brain–heart agar, supplemented by bovine serum, cystine and calcium chloride, is flooded with the culture under test. The excess fluid is removed by vacuum and the plate left to dry for 45–50 min at 37°C with the lid half open. A drop of 100 RTD of each preparation from the basic set of typing phages is then placed (at pre-determined positions) on the agar surface over the dried film of the culture. The plates are then incubated overnight at 35°C. Lytic reactions are recorded the following day and compared with the known phage type patterns.

C. Notation of results

Any degree of lytic reaction is taken into consideration. Results are marked in the following manner:

CL = confluent clear lysis
OL = confluent opaque lysis due to a film of fine growth on the area
       of the dropped phage
LCL = clear lysis with irregular margins of the lytic area
LOL = opaque lysis with irregular margins of the lytic area
SCL = semiconfluent clear lysis
SOL = semiconfluent opaque lysis

± = 5–10 lytic plaques
1+ = 10–20 lytic plaques
2+ = 20–50 lytic plaques
3+ = 50 or more plaques.

According to the size of plaques we add for normal sized plaques "n" (normal plaques) or "s" (small plaques) for instance: 1+n, 2+n or 1+s, 2+s.

The notation of lytic reactions is performed according to the position of phages in the basic set; the whole group of lytic reactions of a strain represents its lytic pattern, i.e. its phage type, which is recorded in the original phage typing scheme by a Roman numeral (Fig. 8) and in the additional phage typing scheme (Fig. 16) by capital letters.

D. Preparation of phage stocks

1. *Original phage typing scheme*

(a) *Propagating strains*. Each of the 22 typing phages has its specific propagating (indicator) strain. The first five typing phages of the basic set (nos. 1–5) are propagated on a *mitis tox⁻* strain no. 1180 (Freeman); phage no. 6 on another *mitis tox⁻* strain no. 15.872 (Romania); phages nos. 7–9 on an *intermedius tox⁻* strain no. 4465 (Romania); phages 10–13 and 16–22 on a *gravis tox⁻* strain no. 9253 (Romania); phages nos. 14 and 15 on a *gravis tox⁻* strain no. 304 (Copenhagen).

The indicator strains are stored either freeze dried or on Loeffler slants, by monthly transfer. They must be stored sealed in the dark at room temperature. *C. diphtheriae* stock cultures must never be kept in the refrigerator, since the organism dies in the cold within a few days.

(b) *Preparation of phage stocks.* With a view to starting from a single isolated plaque 0.1 ml of the diluted phage (10^{-3}–10^{-5}) together with 0·2 ml of the indicator strain (a fresh culture of $1\frac{1}{2}$ h in broth) are added to 2 ml soft agar, spread over the surface of a brain–heart agar plate and incubated overnight at 35 °C. The next day single plaques are picked with a sterile Pasteur pipette, dropped into 1–2 ml broth and stirred for 30–60 s either mechanically or by hand. The supernatant obtained after centrifugation of this preparation is dropped on the propagating strain (fresh culture) previously flooded on to the surface of brain–heart infusion agar. The plates are incubated overnight at 35 °C. The lytic areas are cut out and dropped (10–15 areas) into 30–40 ml of broth and shaken for 5–6 h overnight at 35 °C. After centrifugation at 3000 rev/min for $\frac{1}{2}$ h, the supernatant containing the propagated phage is decanted and treated with chloroform (5 drops/ml).

The preparation is left for 2 h at room temperature and gently stirred from time to time, then kept at 4 °C until the following day when the chloroform is discarded. The phage preparation is assayed for its proper host range activity and for sterility and titrated in ten-fold dilutions on its propagating strain. The RTD usually range from 10^{-2} to 10^{-4}. For phage typing 100 RTD are used. Each new phage stock is used only if it shows the accepted host range.

2. *The additional phage typing scheme*

(a) *Propagating strains* for the additional typing phages are represented by a series of 33 *C. diphtheriae* strains from different geographical areas. Their record number and their origins are listed in the second vertical row on the left of Fig. 15.

(b) *Preparation of phage stocks.* The phages are multiplied in the same way as in the previously described method.

3. *Isolation of new typing phages*

(a) *Phages of endogenous origin.* New phage lysates are obtained from the broth cultures of the supposed lysogenic strain by overnight shaking at 35 °C. The following day, an hour before centrifugation sodium citrate (0·07 M) is added to the lysates, followed by 1 h further shaking. Supernatants obtained after centrifugation of the lysates for $\frac{1}{2}$ h at 3000 rev/min. are tested by the spot method against a set of previously resistant *C. diphtheriae*. If lytic activity occurs, the propagation of the lytic areas is performed according to the methods used and described in the previous section for obtaining phage stocks. These new phage preparations may serve for further typing of non-typable local strains.

(b) *Host restrictions and modifications of the highly virulent phage* 951 *"L"* are obtained from a lysate prepared from the 951 "L" *C. diphtheriae* strain spotted on a newly chosen propagating strain from apparently non-typable strains, as described by Maximescu and Saragea (1969). An overnight shaken culture of the *C. diphtheriae* 951 "L" strain in either Pope Linggood's broth or heart broth with 0·001% L-cystine, is centrifuged 1 h after 0·07 M sodium citrate has been added. The lytic activity of the lysate is tested on a young culture (1$\frac{1}{2}$ h) flooded on a brain–heart agar plate. After overnight incubation at 35°C, even if isolated plaques occur, they should be cut out, inoculated and propagated in nutrient broth (tryptic digest or heart broth, with 0·001% L-cystine) for 24 h at 35°C in shaken cultures. The following day the preparation is centrifuged, decanted and exposed to chloroform (24 h in the refrigerator). The chloroform is discarded and the new phage preparation titrated in ten-fold dilution steps and the host-range activity tested on a large number of untypable strains isolated from the same area or community. Sometimes, as pin-point, hardly visible plaques could occur after the primary lysate has been spotted, areas without visible plaques should be propagated and tested.

E. Conclusions

In order to determine the phage type of a *C. diphtheriae* strain the two phage typing schemes are used as follows: first, strains are tested with the original set of phages; if non-typable, the additional set of phages or new phage preparations may be used.

For practical and economic purposes, a restricted number of phages are routinely used, namely: nos. 2, 3, 5, 6, 7, 8, 9, 10, 11, 12, 13, 19 and 21 from the original scheme and 19 previously mentioned (Fig. 16) from the additional scheme; for reasons of rapidity, both tests can be used simultaneously.

XI. PRACTICAL PROCEDURES OF PHAGE TYPING OF *C. ULCERANS* AND *C. OVIS*

A. Basic set of typing phages

The basic set of typing phages is represented by six *C. ulcerans* phages and a diphtherial phage restricted in a *C. ovis* strain (Fig. 26).

Phage (a) 15542/234 originates from the starch fermenting *C. ulcerans* lysogenic strain no. 15542 *Di-tox*⁺ (diphtherial toxin producer), isolated in Romania from humans.

Phage (b) 213 L/40 c was isolated from the starch fermenting *C. ulcerans* 213 L *Di-tox*⁺, reference strain (W. H. H. Jebb).

Phage (c) 16259/40 c, originates from the starch fermenting *C. ulcerans* 16259 *Di-tox*⁺, Romanian wild strain of human origin.

Phage (d) 210 L/16 L was isolated from a non-starch fermenting *C. ulcerans* 210 L *Di-tox*⁺, collection strain (W. H. H. Jebb).

Phage (g) 9304/984 was isolated from the starch-fermenting *C. ulcerans* 9304 *Di-tox*⁻, Romanian wild strain of human origin.

Phage (h) 984/984 was isolated from starch-fermenting *C. ulcerans* 984 *Di-tox*⁺, reference strain (Henriksen, *C. ulcerans* 1605).

76 c is a diphtherial phage, originating from the 76 c *C. diphtheriae* gravis *Di-tox*⁺ strain, equine, restricted in *C. ovis* 21 strain.

B. Propagation strains

C. ulcerans 234, starch-fermenting *Di-tox*⁻, Romanian wild strain, equine, propagation strain for phage (a).

C. ulcerans 40 c, starch-fermenting *Di-tox*⁻, Romanian wild strain, equine, propagation strain for phages (b) and (c).

C. ulcerans 16 L, starch-fermenting *Di-tox*⁺, reference strain (Robinson and Armstrong), propagation strain for phage (d).

C. ulcerans 984, starch-fermenting *Di-tox*⁺, reference strain (Henriksen 1605), propagation strain for phages (g) and (h).

C. ovis 21, reference strain (Dr Zaki), propagation strain for phage 76.

C. Phage typing

Phage typing in *C. ulcerans* and *C. ovis* is performed according to the technique described by Saragea and Maximescu (1964) and used in the original and additional phage typing of *C. diphtheriae* strains. Media and techniques used for multiplication of phages and for typing are the same as for *C. diphtheriae* using heart broth and brain heart agar.

XII. CHECKING LYSOGENICITY IN CORYNEBACTERIA

A. Induction of lysogenic phages in shaken cultures

1. *Preparation of lysates*

An overnight Loeffler slant culture of the strain to be tested is inoculated into a tube containing 9 ml cystinated (0·001%) tryptic digest broth (Pope Linggood) or heart broth and the liquid culture is shaken for 24 h at 35°C. The following day, 1 ml of a 0·7 M sodium citrate solution is added in order to obtain a final concentration of 0·07 M sodium citrate in the medium (Gundersen and Henriksen, 1959). The addition of sodium citrate will prevent readsorption of the liberated phage on to the bacteria.

After the addition of sodium citrate, the culture is shaken again for 1 h at 35°C, with a view to obtaining the maximum yield of phage from the lysogenic cells. The culture is then centrifuged at 3000 rev/min for 20–30 min and the decanted fluid is tested against a wide range of available sensitive *C. diphtheriae* strains. The wider the range of sensitive strains the better the chance of obtaining lysogenic phages.

Sensitive indicator strains of both schemes and even other multisensitive strains are available for checking lysogeny.

2. *Preparation of the stock phage in the sensitive (indicator) strains*

Overnight cultures of the sensitive strains on Loeffler slants are inoculated into 4–5 ml broth and incubated for 1½ h at 35°C. Brain–heart agar plates (as for phage typing) are flooded with the sensitive strains and dried for 45 min. Every supernatant is dropped on to predetermined areas recorded on the backs of the plates. The plates are incubated for 24 h at 35°C. If any degree of lysis occurs over the area of the spot, even isolated plaques, they are cut out, inoculated into a tube with 5 ml broth, and shaken overnight. After centrifugation and the addition of chloroform (for 2–3 h), the supernatant is again spotted on the sensitive strain. If propagation is successful, the suspension is titrated and checked for host range.

B. Induction of lysogenic phages by ultraviolet irradiation

An overnight fluid culture is diluted to an opacity of 0·3, shaken for 1 h, and submitted to ultraviolet irradiation. The layer should not be deeper than 2–3 mm. A distance of 76 cm and time of exposure of 135 s is used with the lamps available in our laboratory, but preliminary experiments will be needed to determine optimal conditions for other equipment. The plate is shaken continuously during irradiation. The irradiated fluid is transferred to a flask in the dark and again shaken for 3 h at 35°C before centrifugation, chloroform treatment and testing lytic activity. As control, a known lysogenic strain is always included to test the effectiveness of the ultraviolet exposure (Barksdale and Pappenheimer, 1954).

XIII. SPECIAL TECHNIQUES AND MEDIA USED IN THE BACTERIOLOGICAL DIAGNOSIS OF *C. DIPHTHERIAE*

A. Staining of metachromatic granules

1. *Preparation of alkaline Methylene blue solution (Loeffler's formula) for del Vecchio staining or for use as such*

(a) the basic solution of Methylene blue is a 10% alcoholic solution kept for 7 days in the incubator.

(b) The solution used for staining is a 10% aqueous solution prepared from (a). To this second solution 0·01% potassium hydroxide is added. After incubation for 1 week, the solution is filtered and is ready for use.

2. *Staining for metachromatic granules*

(a) *Simple staining.* The smears, dried and heat-fixed, are stained for 10 min with the alkaline Methylene blue solution and washed with tap water.

Microscopic appearance: metachromatic granules, red; the bacillary body, blue.

(b) *Double staining* by del Vecchio's method. Smears are dried and heat-fixed and then stained as follows:

Loeffler's alkaline Methylene blue, 10 min
Wash with tap water
Lugol solution, 10 min
Wash with tap water

Microscopic appearance: metachromatic granules, brown; the bacillary body, yellowish brown.

B. Media

1. *ECST enrichment fluid medium (egg–cystine–serum–tellurite)* (Calalb *et al.*, 1961)

(a) *Formula*

Normal bovine serum	150 ml
Nutrient broth with 0·2% glucose	50 ml
Potassium tellurite 2% distilled water	20 ml
L-Cystine solution 1%	2·2 ml
Egg yolks	2 each

(b) *Preparation.* Wash fresh hens eggs with soap and water and leave them for $\frac{1}{2}$ h in 70° alcohol. Place each egg in a small sterile container. With sterile scissors and forceps cut out the upper end of the egg shell and transfer the yolk aseptically to a sterile stoppered bottle, containing glass beads. The next day after having checked the sterility of the egg yolks, all the above mentioned ingredients are mixed with the sterile egg yolks, by stirring them thoroughly with the help of glass beads. The final medium should be passed through a funnel with gauze and again checked for sterility using both aerobic and anaerobic media. This medium may be stored for 1–2 months at 4°C.

2. *Tinsdale medium* (modified by Meitert and Saragea, 1967)

(a) *Tinsdale basic agar formula*

Labemco meat extract (Oxoid) or any other beef extract	1 g
Proteose peptone (Difco No. 3)	1 g
Sodium chloride	0·5 g
Bacto agar powder (or equivalent)	1·5 g
Distilled water	100 ml

Adjust to pH = 7·4 and autoclave at 120°C for 20 min. Filter through cotton, divide into suitable bottles, in known quantities and autoclave at 115°C for 30 min. The basic medium can be stored in the refrigerator at +4°C for 3–4 months.

(b) *Enrichment.* To 100 ml of the basic medium melted and cooled to 55°C is added strictly in the following order (thoroughly stirring after each ingredient):

(i)	Normal bovine or horse serum	15 ml
(ii)	Defibrinated sheep blood, formalised (per each 100 ml sheep blood: 0·125 formalin)	0·3 ml
(iii)	Sodium hydroxide N/10	6 ml
(iv)	L-Cystine 0·4% solution in N/10 HCl, pasteurised for 30 min in a waterbath at 60°C	6 ml
(v)	Potassium tellurite, 1% solution in distilled water, sterilized for 15 min at 115°C	3 ml
(vi)	Sodium thiosulphate 2·5 g/100 ml solution in sterile distilled water pasteurised at 60°C, for 30 min	1·7 ml

All solutions are stored at +4°C. The formolised sheep blood can be stored indefinitely in glass-stoppered bottles. At most five Petri plates can be prepared for 100 ml base.

In designing this medium, the suitability of different available peptones was checked. With some, the brown halo developed very late. Witte peptone, Difco's Neopeptone and Richter's Peptone gave good results in contrast to Difco Tryptose, Bactopeptone and Merck Peptone, which are therefore not recommended.

As basal agar, a plain agar prepared from fresh meat infusion was also tried with good results instead of using beef extract.

After being poured into plates, the medium must be transparent, and hay-yellow in colour. Tinsdale plates should not be used more than 48 h after preparation. During this interval plates should be stored at +4°C. They have to be dried for 5–10 min in the incubator, with the lid half-open, before inoculation.

Commercial Tinsdale base and commercial dried enrichment medium are available (Difco).

Each lot of medium whether commercial or self prepared, is checked with control reference strains for growth and production of a satisfactory strong brown halo around the C. diphtheriae colonies.

3. Blood agar medium (BA)

Nutrient agar (90 ml) melted and cooled at 45–50°C is mixed aseptically with 10 ml defibrinated sheep blood and poured in Petri dishes.

The medium may be stored at 4°C for 7 days.

On BA C. diphtheriae presents pearl-greyish colonies with granular surfaces.

C. ulcerans is very similar to C. diphtheriae with smooth yellowish, dense colonies.

Atypical corynebacteria grow with very rough colonies having white, silverish or deep yellow colonies.

4. CTBA (cystine–tellurite–blood agar) plates

(a) To 100 ml of melted nutrient agar, pH 7·4–7·6 at 45°C, add sterile:

   10 ml defibrinated sheep blood
   4 ml potassium tellurite, 1% solution
   0·1 ml L-cystine, from a 1% solution

(b) Potassium tellurite 1% solution

   1 g potassium tellurite (Merck)
   100 ml distilled water
   Dissolved by heating and autoclaved for 20 min at 105°C.

(c) L-Cystine 1% solution is obtained by first dissolving 1 g anhydrous sodium carbonate in 10 ml distilled water; after heating to boiling point, add 1 g L-cystine. Warm again and stir well while adding 90 ml distilled water. Autoclave 20 min at 105°C. The solution is kept in the dark in brown flasks at +4°C. Care must be taken not to use a solution for more than 1 month. If it changes colour from light yellow it should be discarded.

C. diphtheriae and *C. ulcerans* produce black colonies with characteristic morphology for each type: *gravis, mitis, intermedius* or *C. ulcerans*.

Atypical corynebacteria develop with black-greyish dry colonies, deeply irregular in margins (atypical daisy head).

Diphtheroids: smooth brownish to shiny black colonies.

5. *Loeffler medium (slants)*

In 100 ml peptone broth, containing 1% glucose, adjusted to pH = 7·6, add 300 ml sterile normal bovine serum. After thorough mixing, the medium is distributed either in 5–6 ml amounts in 16×160 tubes, or in 3 ml aliquots in 12×120 ml tubes; then the tubes are inspissated for 30 min at 90°C in a slanting position.

6. *Hiss medium for carbohydrate fermentation*

(a) *Preparation of basic medium.* Dissolve 7 g peptone and 1·4 g disodium phosphate (Na_2HPO_4) in 1400 ml distilled water. Boil and filter through filter paper, autoclave for 20 min at 120°C. Cool and add 250 normal bovine serum. Then steam at 100°C for 20 min; add 11 ml of sterile solution of Andrade indicator.

The basic medium distributed in bottles keeps indefinitely at +4°C. Different carbohydrate solutions may be added; 1% for all carbohydrates except starch which is used at 0·4%.

(b) *Preparation of Andrade indicator.* To 100 ml distilled water add:
   0·5 acid fuchsin
   10 ml 1 N NaOH
After autoclaving for 30 min at 115°C this keeps indefinitely at +4°C.

(c) *The carbohydrate solutions* used are: glucose, sucrose, maltose, levulose, dextrine and glycogen 10% solutions, and starch 4%.

The carbohydrate solutions are sterilized at 105°C for 20 min.

For sucrose, filtration is recommended, due to the tendency to hydrolyse during heating.

The final medium is prepared by adding sterile carbohydrates to each 90 ml basic Hiss medium, 10 ml of each carbohydrate solution. The pH should be 8·1–8·2. The final medium is distributed in 2–3 ml quantities in 12×120 mm tubes.

For practical purposes, glycogen is recommended instead of starch. At acid pH, the medium turns pink.

7. *Pisu's medium (modified by Maximescu and Nicoară, 1964) for cystinase test*

To 80 ml nutrient agar, with 0·4% agar (Difco) containing 2% peptone, pH = 7·4–7·6 melted and cooled at 45°C, add:

Bovine serum	20 ml
1% L-cystine	6 ml
10% lead acetate	1 ml

Aliquots of 2–3 ml are distributed in small tubes left to solidify vertically. Inoculation should be made by stabbing. Starting 4–6 h after inoculation, traces of black growth surrounded by a brown halo begin appearing when the organism is a cystinase producer. The reaction becomes intensified overnight. *C. diphtheriae*, *C. ulcerans* and *C. ovis* are the only high cystinase producers in this genus. The medium keeps 1 month in the refrigerator.

8. *Blake–Christensen medium for the urease test* (recommended by Kauff-mann, 1954)

Bacto-peptone Difco	1 g
NaCl	5 g
KH$_2$PO$_4$	2 g
Phenol red 1/500 solution	6 ml
Distilled water	900 ml

Dissolve by boiling all ingredients, except the Phenol red. Adjust to pH = 6·8–6·9 and filter through paper. Autoclave at 120°C for 20 min. Then add sterile 1 g glucose solution, 20 ml of a 20% urea solution which has been sterilised by filtration and 6 ml Phenol red solution 1/500 (auto-claved 20 min at 110°C). The medium should be light yellow and clear. Aliquots (2 ml) are distributed to small tubes and checked for sterility.

C. diphtheriae does not produce urease. Diphtheroids and *C. ulcerans* are rich in urease.

9. *Elek–Ouchterlony–Frobisher plates for the* in vitro *toxinogenicity test*
(a) *Base*

Meat infusion	1000 ml
Protese peptone Difco	20 g
NaCl p.a.	5 g
Bacto agar (Difco)	11 g

Dissolve all ingredients at boiling point.
Adjust to pH = 7·6–7·8 with 10% NaOH solution.
Precipitate the medium by autoclaving 20 min at 120°C.
Filter through cotton, distribute in known quantities in flasks, autoclave 20–30 min at 110°C.
Meat infusion could be replaced by beef extract (Difco) (5 g to 1000 ml distilled water).

The medium is melted, cooled at 40°C, and bovine serum added to 20% and poured in Petri dishes. A ditch (7×0.5 cm in a plate of 10 cm) is cut in the agar on the diameter of the plate, 0·5 ml diphtherial antitoxin (500 U.A./ml) is dropped in the ditch. The plate is left for 1–2 h at 37°C with open lid and then the ditch is filled with melted agar. The cultures to be tested are streaked in strips perpendicular on the ditch. Precipitation lines occur in the right angle where antitoxin meets the toxin produced by toxigenic cultures. The medium should be controlled by reference strains *C. diphtheriae tox*[+] and *tox*[−].

C. Slide agglutination for serological identification

Agglutinant polyvalent diphtherial antiserum, prepared on rabbits is used diluted 1/10–1/20. The diluent is a 0·1 M phosphate buffer solution (15·7 g $Na_2HPO_4 + 2H_2O$, 1·8 g $NaH_2PO_4 + H_2O$, 4·9 g NaCl, 1000 ml distilled water, pH = 7·6, autoclaved 20 min at 110°C.

The control is the 1/20 diluted rabbit serum, in the phosphate buffer solution.

On a slide, deposit one drop of diphtherial antiserum and one drop of control normal rabbit serum, both diluted as recommended.

Suspend in each drop a loop from the culture (of 24–48 h on Loeffler slant) to be tested. The culture should be taken from the top of the medium; do not touch the liquid from the bottom. Agglutination is produced within 1–2 min.

Results could be as follows:

Diphtherial antiserum	Normal rabbit serum	Results
+ to + +	−	*C. diphtheriae*
−	−	Non-diphtherial or inagglutinable *C. diphtheriae*
+ +	+ +	Spontaneously agglutinable

D. Media used for phage typing and lysogenicity

Phage typing and lysogenicity investigations in *C. diphtheriae*, *C. ulcerans* and *C. hofmannii* are done with fresh media prepared according to Narbutovicz (1955) and supplemented by growth factors.

In practice commercial media could replace fresh media: Difco brain–

heart infusion agar (BHIA) and brain–heart infusion broth (BHIB) (Saragea *et al.*, 1972b and unpublished data; Toshach *et al.*, 1977).

For typing procedures the solid medium recommended is BHIA Difco with a concentration of agar 1·1%.

To each 100 ml of 1·1% agar (BHIA) are added aseptically:

*Thiamine HCl 0·01%	0·1 ml
D-L-Tryptophane 0·1%	1 ml
†L-Cystine 1%	0·1 ml
CaCl$_2$, 1 M	0·1 ml
Glucose 10%	2 ml
Bovine or horse serum	10 ml
Final concentration of agar	0·97%

For fresh cultures, phage stock preparation and lysogenicity, BHIB Difco medium supplemented with 0·001% L-cystine is recommended.

Techniques for performing phage typing, bacteriocin typing and lysogenicity were described in the previous Sections.

XIV. PRINCIPLES OF EPIDEMIOLOGICAL METHODOLOGY IN THE PREVENTION AND CONTROL OF DIPHTHERIA

A. General

Dubos remarked in 1952, "medical microbiology is the study of host-parasite relationships and not that of micro-organisms alone, considered as independent living agents". This principle may be further extended to community–parasite relationships, i.e. the epidemiology of the organisms.

Accordingly the general principles recommended for the public health fight against diphtheria will be mentioned.

These principles are complex, as is evidenced by extensive practical and theoretical research of the authors of this Chapter in Romania in a vast field experiment, lasting about 20 years and extending throughout the country.

The epidemiological features of diphtheria, as well as socioeconomic factors, determine the strategy and tactics of public health services in the prevention and control of this disease. This activity involves a scientific programme of surveillance and control.

In Romania, in 1955, in spite of compulsory diphtherial immunisation,

---

* Amino acid solutions are filtered through Millipore membranes.
† L-Cystine is prepared as described in the previous Section.

a fairly high diphtheria morbidity (only 60% Schick negative reactors) and a high frequency of toxigenic *C. diphtheriae* (86%) could still be observed. Under these conditions two principles were followed to stop the epidemic spreading. These were

Surveillance and control of antidiphtheria immunity.

Surveillance and control of the organism and its mechanisms of spread.

This epidemiological surveillance work helped the Medical Authorities of the country to determine the extent of the problem and its place in the order of priority among other problems related to communicable and non-communicable diseases and to work out short- and long-term strategies for its control.

B. Surveillance and control of antidiphtheria immunity

(1) Control was achieved by manufacturing highly immunogenic diphtherial prophylactics and associated vaccines (purified and aluminium phosphate adsorbed diphtherial toxoid, diphthero-tetanus toxoid and diphthero-tetanus-pertussis vaccine).

(2) Planning, together with our Health Authorities, a vast programme of antidiphtheria immunisation including all sensitive age groups up to 18 years of age.

(3) Randomised field control of the application of this programme and of antidiphtheria acquired immunity by means of Schick conversion rates (90% Schick negative reactors) and by serum antitoxin titrations in different age groups of the population.

(4) Active intervention by vaccination when necessary, in order to avoid gaps in the general immunity level.

To recognise spreading when it occurred, it was important to have good bacteriological detection schemes and typing methods.

1. *Principles of surveillance*

The surveillance and control of the pathogenic agent were carried out by systematic investigations regarding the incidence and frequency of wild organisms throughout the country.

To this end, the Diphtheria Laboratory of the Cantacuzino Institute, Bucharest, was organised as a National Reference Centre for diphtheria microbiological, epidemiological and immunological surveys. Close and permanent relations were established between this centre and all regional public health laboratory services in the country, and theoretical and practical advice were given when necessary.

(a) Regarding surveillance of the pathogenic agent, the chief task was to obtain representative samples from all epidemic situations, from the sporadic cases, and their contacts, as well as from healthy carriers of non-epidemic communities from all over the country.

(b) Thus the distribution of biotypes and phage types of *C. diphtheriae* in each geographical region of the country is well known, and the penetration of a new phage type can easily be detected.

2. Results

Valuable general conclusions have been drawn from past results. For instance, a striking parallelism could be observed between the incidence of the *gravis* toxigenic biotype and diphtheria morbidity during the whole epidemic period. Thus, the occurrence of *gravis* toxigenic strains always represented an epidemic indicator and an alarm signal for epidemiologists, necessitating immediate prophylactic measures. Phage typing provided the origin and spread of a certain type in a region or in the country, the origin and extent of an epidemic, the filiation of the cases, and relationships between the outbreaks even when distant in space and time. It also helped us to establish the right epidemiological relationships and prognosis, to arrest the spread of infection and to elucidate the role and right attitude to be adopted today towards non-toxigenic healthy carriers.

C. Scheme of intervention in a diphtheria outbreak

Three categories of methods are applied, i.e. epidemiological, microbiological and immunological (Fig. 44).

1. *Epidemiological methods*

(a) Outline of a common plan developed together with or solely by the local Health Authorities involving epidemiologists, bacteriologists, pediatricians, physicians and medical officers.

(b) Identification and hospitalisation of the cases.

(c) Epidemiological inquiries.

(d) Delimitation of the area of epidemic spread by investigating all contacts.

(e) Clinical and bacteriological examination of contacts.

(f) Study of the recorded vaccinations of the whole community.

(g) Diagnosis, isolation and treatment of contacts (treatment by antibiotics).

(h) Follow-up and control of the area after specific intervals of time: 3, 6, 12 months.

2. *Microbiological methods*

(a) Swabbing of all cases, suspects, or contacts established by the

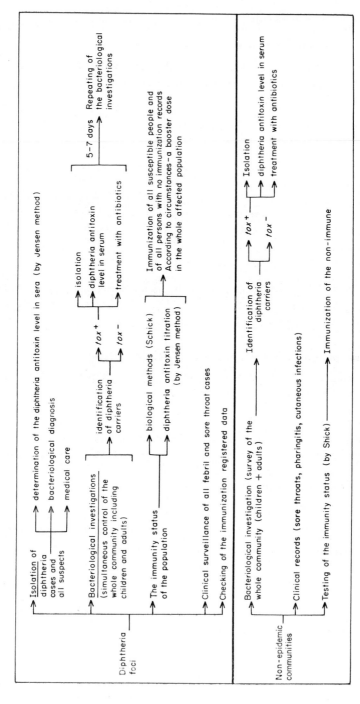

Fig. 44. Outline of a plan of investigation in diphtherial foci and in non-epidemic communities.

epidemiological inquiry (compulsory: nasal and pharyngeal exudates): households, family, school and neighbourhood contacts.

(b) If one of the family members goes to another school or community, the latter will also be checked.

(c) Swabs are collected in the morning before breakfast and before any gargling and/or antibiotic treatment.

(d) Swabbing of the whole community is recommended as early as possible after occurrence of the first cases on a large number of individuals in the epidemic period, and simultaneously of the whole number of persons established by the epidemiological inquiry.

3. *Immunological methods*

(a) Checking of diphtherial antitoxin level in patients (less 0·03 AU) to confirm diphtheria case (Stănică *et al.*, 1971).

(b) After the occurrence of the first cases, one should immediately check the level of immunity of the susceptible age groups, below 18 years of age, and adults in close contact with children, by means of the Schick test or Jensen test.

(c) Vaccination of all Schick positive reactors, even of adults (teachers, parents, etc.).

D. General recommendations for the organisation of a reference diphtheria laboratory

1. *At any moment adequate quantities of the following material must be available*

(a) Sterile swabs introduced in sterile small tubes (12 × 120 mm).

(b) Sterile glassware, plates, plastic plates, pipettes, etc.

2. *Equipment*

(a) An optical microscope.

(b) A dissecting microscope or stereoscopic microscope.

(c) Platinum loops.

3. *Media*

(a) Enrichment medium: ECST.

(b) Tinsdale's base.

(c) Agar base for BA (blood agar) and CTBA (cystine–tellurite–blood agar).

(d) Plain nutrient broth and 0·2% glucose broth.

(e) Tryptic digest broth (commercial).

(f) Agar base for EOF plates.

(g) Pisu's modified medium.
(h) Hiss's base.
(i) Blake–Christensen's liquid medium or slants.
(j) Loeffler slants.
(k) Brain–heart agar base (commercial Difco or self-prepared).

4. *Biological materials and solutions*

(a) Normal bovine or horse serum.
(b) Horse diphtherial antitoxin either plain or purified (500–1000 AU/ml).
(c) Sheep or rabbit erythrocytes.
(d) L-Cystine, 1% solution.
(e) Sodium citrate, 0·7 M solution.
(f) Calcium chloride, molar solution.
(g) Potassium tellurite, 1% solution (for CTBA and Tinsdale).
(h) Potassium tellurite 2% solution (for ECST).
(i) Defibrinated sheep erythrocytes formolised (for each 100 ml erythrocytes, 0·125 ml formalin) (for Tinsdale).
(j) NaOH N/10 (for Tinsdale).
(k) L-Cystine 0·4 N/10 HCl (for Tinsdale).
(l) Sodium thiosulphate solution, 2·5 g% (for Tinsdale).
(m) Carbohydrate solutions: glucose, sucrose, mannite, maltose, levulose, dextrin and glycogen 10% solutions; starch, 4% solution.
(n) Lead acetate, 10% solution (for Pisu's medium).
(o) Phenol red sen, 1/500 solution (for Blake–Christensen medium).
(p) Thiamine solution, 0·01%.
(r) Tryptophan solution, 0·5%.
(s) Alkaline Methylene blue solution (Loeffler's stain).
(t) Gram stains.

5. *Animals*

Guinea pigs (250 g weight).
Rabbits (2500–3000 g weight) with white skin.

6. *General remarks*

(a) It is strongly recommended that each new batch of medium (even if commercial media) should be tested before being introduced into use with reference strains. Thus:
(b) Reference strains: biotypes *gravis*, *mitis*, *intermedius*, toxigenic and non-toxigenic, *C. ulcerans*, *C. hofmannii*, *C. xerosis* must always be available, in all the laboratories.

(c) Reference strains have to be stored either by monthly transfers on sealed Loeffler slants and kept at room temperature in the dark, or lyophilised.

7. Biological materials for phage typing

It is advisable to perform phage typing only in specialised reference laboratories. Recommended are the sets of 22 original and 33 additional phages, preferably supplied by the Cantacuzino Institute, Bucharest, Romania, stocked or produced in each reference laboratory (from stocked phage strains available from the Cantacuzino Institute).

8. Biological materials for lysogenicity

A large set of indicator sensitive strains from every biotype is recommended. These sensitive strains may either be provided by the Cantacuzino Institute or be selected in each country from local strains.

9. Biological materials for bacteriocin typing

A set of 20 bacteriocinogenic corynebacteria are today available in the Cantacuzino Institute. In England, Dr Zamiri (Sheffield) and in Canada, Sheila Toshach (Edmonton Alberta University) may provide reference strains.

However, it is advisable in the present state of bacteriocin typing to also select additional local bacteriocinogenic strains.

XV. GENERAL CONCLUSIONS

The effectiveness of the national epidemiological programme of surveillance and control of diphtheria in Romania has led to a spectacular drop in diphtheria morbidity during the last 20 years (from 6 per 100 000 in 1955 to zero case in 1976, 1977).

The level of antidiphtherial immunity has increased from 60% in 1955, to 96% Schick negative reactors in 1977, and concomitantly there has been a decrease in toxigenicity among circulating strains of *C. diphtheriae* (from 86% in 1955–1966 to 5% in 1977).

REFERENCES

Anderson, J. S., Happold, F. C., McLeod, J. W. and Thompson, J. G. (1931). *J. Path. Bact.*, **34**, 667–681.

Anderson, J. S., Cooper, K. E., Happold, F. C. and McLeod, J. W. (1933). *J. Path. Bact.*, **36**, 169–182.

Babes, V. (1886). *Bull. Soc. Anat.* (*Paris*), **61**, 72.

Banach, T. M. and Hawirko, R. Z. (1966). *J. Bact.*, **92**, 1304–1310.

Barber, C., Calalb, G., Meitert, E., Saragea A., Stănică, E. and Stoian, C. (1963) *Arch. roum. Path. exp.*, **22**, 357–367.

Barber, C., Meitert, E. and Saragea, A. (1965). *Pathologia Microbiol.*, **28**, 274–286.
Barber, C., Meitert, E., Bica-Popii, V. and Saragea, A. (1966a). *Pathologia Microbiol.*, **29**, 377–386.
Barber, C., Lazăr, I. and Meitert, E. (1966b). *Pathologia Microbiol.*, **29**, 84–94.
Barksdale, L. (1955). *C. r. hebd. Séanc. Acad. Sci.*, **240**, 1831–1833.
Barksdale, L. (1970). *Bact. Rev.*, **34**, 378–422.
Barksdale, L. and Pappenheimer, A. M., Jr. (1954). *J. Bact.*, **67**, 220–232.
Barksdale, L., Garmise, L. and Horibata, K. (1960), *Ann. N.Y. Acad. Sci.*, **88**, 1093–1108.
Barksdale, L., Garmise, L. and Rivera, R. (1961). *J. Bact.*, **81**, 527–540.
Barksdale, L. and Arden, S. B. (1974). *Ann. Rev. Microbiol.*, **28**, 265–299.
Barratt, M. M. (1933). *J. Path. Bact.*, **36**, 369–397.
Belsey, M. A., Sinclair, M., Roder, M. R. and Leblanc, D. R. (1969). *New Engl. J. Med.*, **280**, 135–141.
Berezkina, G. N. and Kravcenko, N. N. (1970). *Zh. Mikrobiol. Epidem. Immun.*, **47**, 111–114.
Bergey, O. (1974). "Bergey's Manual of Determinative Bacteriology" (Eds R. E. Buchanan and N. E. Gibbons) 8th edn, pp. 602–603. Williams and Wilkins, Baltimore.
Bezjack, V. (1958). *Acta med. yugos.*, **12**, 270–281.
Blumberg, B. S. and Warren, L. (1961). *Biochim. biophys. Acta*, **50**, 90–101.
Bonciu, C., Petrovici, M., Meitert, E., Saragea, A. and Calalb, G. (1965). *Arch. roum. Path. exp.*, **24**, 41–54.
Bonciu, C., Bona, C., Petrovici, M., Calalb, G., Meitert, E. and Saragea, A. (1966). *Arch. roum. Path. exp.*, **25**, 27–44.
Bordet, J. and Ciucă, M. (1921a), *C. r. hebd. Séanc. Soc. Biol.*, **84**, 747–748.
Bordet, J. and Ciucă, M. (1921b), *C. r. hebd. Séanc. Soc. Biol.*, **84**, 748–750.
Branham, S., Hiatt, C. W., Cooper, A. D. and Riggs, B. (1959), *J. Immun.*, **82**. 397–408.
Bray, J. (1944), *J. Path. Bact.*, **56**, 497.
Brooks, G. F., Benett, V. J. and Feldman, R. (1974). *J. infect. Dis.*, **129**, 172–178.
Burrows, M. T. and Suzuki, J. (1918). *J. Immun.*, **3**, 219–232.
Calalb, G., Saragea, A., Maximescu, P., Cioroianu, N., Popescu, A., Popa, S. and Mihailescu, A. (1961), *Arch. roum. Path. exp.*, **20**, 95–101.
Calalb, G., Saragea, A., Meitert, E. and Cosman, M. (1963). *Arch. roum. Path. exp.*, **22**, 123–130.
Calalb, G., Stănică, E., Maximescu, P., Stoian, C. and Stoian, I. (1965). *Arch. roum. Path. exp.*, **24**, 71–76.
Calalb, G., Stănică, E., Stoian, C. and Meitert, E. (1966). *Arch. roum. Path. exp.*, **25**, 409–414.
Calalb, G., Stoian, C. and Mațepiuc-Stănică, M. (1968). *Microbiologia, Parazit. Epidem.*, **13**, 441–446.
Carne, H. R. (1968). *Nature, Lond.*, **217**, 1066–1067.
Christensen, E. P. (1957). *Acta path. microbiol. scand.*, **41**, 67–78.
Christie, A. B. (1969). "Infectious Diseases Epidemiology and Clinical Practice". Churchill Livingstone, Edinburgh.
Ciucă, M., Calalb, G., Saragea, A. and Maximescu, P. (1960). *Arch. roum. Path. exp.*, **19**, 485–491.
Collier, R. J. (1967). *J. molec. Biol.*, **25**, 83–98.
Collier, R. J. and Kandel, J. J. (1971). *J. biol. Chem.*, **246**, 1496–1503.

Collier, R. J. and Pappenheimer, A. M., Jr. (1964). *J. exp. Med.*, **120**, 1019–1039.

Cook, G. T. and Jebb, W. H. H. (1952). *J. clin. Path.*, **5**, 161–164.

Conradi, H. and Troch, P. (1912). *Münch. med. Wschr.*, **59**, 1652–1653.

Cummins, C. S. (1954). *Br. J. exp. Path.*, **35**, 166–180.

Cummins, C. S. (1965). *Am. Rev. resp. Dis.*, **92**, 63–72.

Cummins, C. S. and Harris, H. (1956). *J. gen. microbiol.*, **14**, 583–600.

Davis, B. D., Dulbecco, R., Eisen, H. N., Ginsberg, H. S. and Wood, B. W. (1973). "Microbiology—Including Immunology and Molecular Genetics", 2nd edn, p. 682. Harper and Row, Hagerstown, Maryland, New York, Evanston, San Francisco, London.

D'Herelle, F. D. (1917). *C. r. hebd. Séanc. Acad. Sci.*, **165**, 373.

Doegge, T. C. and Walker, R. J. (1962). *South Afr. med. J.*, **55**, 144–149.

Drăgoi, T., Saragea, A., Michel, J. and Iancu, L. (1979). *Bact. Virol. Parasit. epidem.* (in press).

Drew, R. M. and Miller, J. H. (1951). *J. Bact.*, **62**, 549–559.

Dubos, R. J. (1952). "Bacterial and Mycotic Infections of Man", 2nd edn. Pitman Medical Publishing Co. Ltd., London; J. B. Lippincott Company, Philadelphia.

Dubreuil, R. and Pavilanis, V. (1958). *Can. J. Microbiol.*, **4**, 543–550.

Durand, P. (1918). *C. r. hebd. Séanc. Soc. Biol.*, **81**, 1011–1013.

Durand, P. (1920a), *C. r. hebd. Séanc. Soc. Biol.*, **83**, 611–613.

Durand, P. (1920b), *C. r. hebd. Séanc. Soc. Biol.*, **83**, 613–615.

Eagleton, A. J. and Baxter, E. M. (1923). *J. Hyg., Camb.*, **22**, 107–122.

Ehrlich, P. (1897). *Klin. Jb.*, **6**, 299.

Ejowa, G. G. (1967). *Zh. Mikrobiol. Epidem. Immunobiol.*, **44**, 93–98.

Elek, S. D. (1948). *Br. med. J.*, **1**, 493–496.

Emilianov, P. I., Musonova, A. G. and Lavnik, O. N. (1968). *Zh. Mikrobiol. Epidem. Immunobiol.*, **45**, 143.

Endemann, D. (1960). "4 Colloquium uber Fragen der Lysotypie", p. 173. Herausgegeben vom Zentrallaboratorium für Lysotypie beim Bezirks-Hygiene-Institut, Wernigerode/Harz.

Ernst, P. (1888). *Z. Hyg. Infekt. Krankh.*, **4**, 25–46.

Ewing, J. O. (1933). *J. Path. Bact.*, **37**, 345–351.

Fahey, J. E. (1952). *Can. J. publ. Hlth*, **43**, 167–170.

Favorova, L. A. and Kostjukova, N. N. (1966). *Zh. Mikrobiol. Epidem. Immunobiol.*, **43**, 55–61.

Ferris, A. A. (1950). *J. Path. Bact.*, **62**, 165–174.

Freeman, V. J. (1951). *J. Bact.*, **61**, 675–688.

Freeman, V. J. and Minzel, G. H. (1950). *Am. J. Hyg.* **51**, 305–309.

Freeman, V. J. and Morse, I. U. (1952). *J. Bact.*, **63**, 407–414.

Frobisher, M., Jr., Adams, M. L. and Kuhns, W. J. (1945). *Proc. Soc. exp. Biol. Med.*, **58**, 330–334.

Frobisher, M., Jr., King, E. O. and Parsons, E. I. A. (1951). *Am. J. clin. Path.*, **21**, 282.

Gabrilovitch, I. M. (1966). *Zh. Mikrobiol. Epidem. Immunobiol.*, **43**, 53–55.

Gibson, L. F. and Colman, G. (1973). *J. Hyg. Camb.*, **71**, 679–689.

Gibson, L. F., Cooper, G. N., Saragea, A. and Maximescu, P. (1970). *Med. J. Aust.* **1**, 412–417.

Gilbert, R. and Stewart, F. C. (1926). *J. Lab. clin. Med.*, **12**, 756–761.

Gill, D. M. and Dinius, L. L. (1971). *J. biol. Chem.*, **246**, 1485–1491.

Gill, D. M., Pappenheimer, A. M., Jr., Brown, R. and Kurnick, J. T. (1969). *J. exp. Med.*, **129**, 1–21.

Gillies, R. R. (1964). *J. Hyg., Camb.*, **62**, 1–9.

Goor, R. S. and Pappenheimer, A. M., Jr. (1967). *J. exp. Med.*, **126**, 899–912.

Groman, N. B. (1955). *J. Bact.*, **69**, 9–15.

Gundersen, W. B. and Henriksen, S. D. (1959). *Acta path. microbiol. scand.*, **47**, 65–74.

Hajderi, H. I., Koci, I. and Luloci, I. (1967). *Arch. roum. Path. exp.*, **26**, 912–918.

Hamon, Y. (1965). *Pathol. Biol.*, **13**, 806–824.

Havens, L. C. (1920). *J. infect. Dis.*, **26**, 388–401.

Heide, K., Haupt, H. (1962). *Naturwissenschaften*, **49**, 15–16.

Henriksen, S. D. (1955). *Acta path. microbiol. scand.*, **37**, 65–70.

Henry, J. E. (1920). *J. Am. med. Ass.*, **75**, 1715.

Hewitt, L. F. (1947). *Br. J. exp. Path.*, **28**, 338–346.

Hofmann, G. (von) (1888). *Wien. med. Wochschr.*, **38**, 65–108.

Holmes, R. K. and Barksdale, L. (1969), *J. Virol.*, **3**, 586–598.

Holmes, R. K. and Barksdale, L. (1970). *J. Virol.*, **5**, 783–794.

Honjo, T., Nishizuka, Y., Hayaishi, O. and Kato, I. (1968). *J. biol. Chem.*, **243**, 3553–3555.

Huang, C. H. (1942a). *Am. J. Hyg.*, **35**, 317–324.

Huang, C. H. (1942b). *Am. J. Hyg.*, **35**, 325–336.

Ionescu-Stoian, F., Maximescu, P., Popa, A., Michel, I. and Vieru, E. (1977). *Arch. roum. Path. exp.*, **36**, 157–172.

Jebb, W. H. H. (1948). *J. Path. Bact.*, **60**, 403–412.

Jebb, W. H. H. and Martin, T. D. (1965). *J. clin. Path.*, **18**, 757–758.

Jensen, Cl. (1933). *Acta path. microbiol. scand. Suppl.*, XIV, 1–211.

Johnstone, K. J. and McLeod, J. W. (1949). *Publ. Hlth Rep.*, **64**, 1181–1187.

Kato, M. (1970). *J. Bact.*, **101**, 709–716.

Kato, I. and Pappenheimer, A. M., Jr. (1960). *J. exp. med.*, **112**, 329–349.

Kauffmann, F. (1954). "Enterobacteriaceae", p. 317. Ejnar Munksgaard Publisher, Copenhagen.

Keogh, E. V., Simmons, R. T. and Anderson, G. (1938). *J. Path. Bact.*, **46**, 565–570.

Klebs, E. (1883). *Verh. Cong. Inn. Med. Wiesbaden*, **2**, 139–154.

Klett, A. (1900). *Z. Hyg. Infekt. Krankh.*, **33**, 137–160.

Kostjukova, N. N., Kadîrova, Kh. V. and Ezepciuk, Yu. V. (1970). *Zh. Mikrobiol. Epidem. Immunobiol.*, **4**, 59–64.

Kostjukova, N. N., Kojecinikova, L. A., Gekasian, G. G., Ejowa, G. G., Trifonova, V. I., Kudriatev, N. K. and Cabilov, I. R. (1971). *Zh. Mikrobiol. Epidem. Immunobiol.*, **48**, 60–66.

Kröger, E. and Thofern, E. (1952a). *Z. Hyg.*, **134**, 474–487.

Kröger, E. and Thofern, E. (1952b). *Z. Hyg.*, **135**, 254–274.

Krîlova, M. D. (1969). *Zh. Mikrobiol. Epidem. Immunobiol.*, **46**, 11–15.

Krîlova, M. D. (1972). *Zh. Mikrobiol. Epidem. Immunobiol.*, **49**, 27–33.

Krîlova, M. D., Markina, S. S., Kuznina, Yu. G., Neiman, Z. J., Bacikova, V. A., Hiskarova, E. D. and Serghienko, A. G. (1969), *Zh. Mikrobiol. Epidem. Immunobiol.*, **46**, 21–25.

Krîlova, M. D., Reihstat, G. N., Ostrovkaia, Z. S., Tkacenco, A. M., Kuznețova, N. S., Agafonova, L. I. and Pozina, V. S. (1973). *Zh. Mikrobiol. Epidem. Immunobiol.*, **50**, 93–98.

La Placca, M. (1957). *Riv. Ist. sieroter. Ital.*, **32**, 350.

Langer, H. (1916). *Zentbl. Bakt. Abt. I. Orig.*, **78**, 117 (quoted by Robinson, D. and Peeney, A. L., 1936).

Lautrop, H. (1950). *Acta path. microbiol. scand.*, **27**, 443–447.

Lautrop, H. (1955). *Acta path. microbiol. scand.*, **36**, 274–288.

Lehmann, K. B. and Neumann, R. (1896). "Atlas und Grundriss der Bakteriologie und Lehrbuch des Speziellen bakteriologischen Diagnostik", 1st edn, p. 350. J. F. Lehmann, München.

Levaditi, C. and Muttermilch, S. (1913a). *C. r. hebd. Séanc. Soc. Biol.*, **74**, 379–382.

Levaditi, C. and Muttermilch, S. (1913b). *C. r. hebd. Séanc. Soc. Biol.*, **74**, 614–616.

Levaditi, C. and Muttermilch, S. (1913c). *C. r. hebd. Séanc. Soc. Biol.*, **74**, 1305–1308.

Levaditi, C. and Muttermilch, S. (1913d). *C. r. hebd. Séanc. Soc. Biol.*, **74**, 1379–1382.

Loeffler, F. (1884). *Mitt. Klin. Gesundh., Berlin*, **2**, 421–499.

Mair, N. (1928). *J. Path. Bact.*, **31**, 136–137.

Marcuse, N. K., Gilbert, O. and Grand, M. (1973). *J. Am. med. Ass.*, **224**, 305–310.

Matsuda, M. and Barksdale, L. (1967). *J. Bact.*, **93**, 722–730.

Matsuda, M., Kanei, C. and Yoneda, M. (1972). *Biken J.*, **15**, 111–114.

Mauss, E. A. and Keown, M. J. (1946). *Science, N.Y.*, **104**, 252–253.

Maximescu, P. (1967). *Microbiologia, Parazitol., Epidemiol.*, **12**, 355–359.

Maximescu, P. (1968). *J. gen. Microbiol.*, **53**, 125–133.

Maximescu, P. and Drăgoi, T. (1971). *Arch. roum. Path. exp.*, **30**, 493–496.

Maximescu, P. and Nicoară, I. (1964). *Microbiologia, Parazitol., Epidemiol.*, **9**, 551–552.

Maximescu, P. and Saragea, A. (1969). *Arch. roum. Path. exp.*, **28**, 1059–1068.

Maximescu, P., Pop, A., Oprişan, A. and Potorac, E. (1968). *Arch. roum. Path. exp.*, **27**, 733–750.

Maximescu, P., Saragea, A. and Drăgoi, T. (1972). *Arch. roum. Path. exp.*, **31**, 357–366.

Maximescu, P., Oprişan, A., Pop, A. and Potorac, E. (1974a). *J. gen. Microbiol.*, **82**, 49–56.

Maximescu, P., Pop, A., Oprişan, A. and Potorac, E. (1974b). *J. Hyg. Epidemiol. Microbiol. Immunol.*, **18**, 324–328.

McCloskey, R. V., Saragea, A. and Maximescu, P. (1972), *J. infect. Dis.*, **126**, 196–199.

McLeod, J. W. (1943), *Bact. Rev.*, **7**, 1–41.

Meitert, E. (1969a). *Arch. roum. Path. exp.*, **28**, 1082–1085.

Meitert, E. (1969b). *Arch. roum. Path. exp.*, **28**, 1086–1097.

Meitert, E. (1972). "Annual Scientific Meeting of Cantacuzino Institute", X, pp. 154–155.

Meitert, E. (1973). "Annual Scientific Meeting, Cantacuzino Institute", XI, pp. 101–102.

Meitert, E. (1975). "IV Symp. sur les bactériophages et les phénomènes de bactériophagie, Bucharest, 27–28 September 1974. Résumés. *Arch. roum. Path. exp.*, **34**, 195.

Meitert, E. and Bica-Popii, V. (1971). "Annual Scientific Meeting of Cantacuzino Institute," IX, 18–19 February, pp. 100–102.

Meitert, E. and Bica-Popii, V. (1972). *Arch. roum. Path. exp.*, **31**, 475–480.

Meitert, E. and Bica-Popii, V. (1975). "IV Symp. sur les bactériophages et les phénomènes de bactériophagie", Bucharest, 27–28 September 1974. Résumés. *Arch. roum. Path. exp.* (1975), **34**, 194–195.

Meitert, E. and Saragea, A. (1967). *Microbiologia, Parazitol., Ep idemiol.*, **12**, 361–368.

Meitert, E., Saragea, A., Petrovici, A. and Ionescu, M. (1972a). *Arch. roum. Path. exp.*, **31**, 31–37.

Meitert, E., Saragea, A., Petrovici, M. and Filloti, A. (1972b). *Arch. roum. Path. exp.*, **31**, 373–378.

Morijama, T. and Barksdale, L. (1967). *J. Bact.*, **94**, 1565–1581.

Morton, H. E. and Anderson, T. F. (1941). *Proc. Soc. exp. Biol. Med.*, **46**, 272–276.

Mueller, J. H. and Miller, P. A. (1941). *J. Immunol.*, **40**, 21–32.

Murata, R., Akama, K., Hirose, S., Kameyama, S., Nakano, T. and Yamamoto, A. (1959). *Jap. J. med. sci. Biol.*, **12**, 319–330.

Murray, J. F. (1935). *J. Path. Bact.*, **41**, 439–445.

Narbutovicz, B. (1955). *Acta Microbiol. Pol.*, **4**, 245.

National Communicable Disease Centre (1968). Surveillance Reports, no. 10. Summary 1969, p. 31.

Nicholas, J. (1896). *C. r. hebd. Séanc. Soc. Biol.*, **68**, 1025–1027.

Niggemeyer, H. (1955). *Ann. Pediatr. (Basel)*, **183**, 1–?.

Nomura, M. (1967). *Ann. Rev. Microbiol.*, **21**, 257–284.

Oehring, H. (1963a). *Arch. Hyg. Bakt. (Berlin)*, **147**, 432–439.

Oehring, H. (1963b). *J. Bact.*, **86**, 266–273.

Ortalli, A. V., Princivalle, M. and Zampieri, A. (1956). "Proceedings of the Ninth National Congress on Microbiolgoy", Palermo, p. 214.

Oudin, J. (1946). *C. r. hebd. Séanc. Acad. Sci.*, **222**, 115–116.

Oudin, J. (1948). *Ann. Inst. Pasteur*, **75**, 30–51.

Ouchterlony, O. (1948). *Acta pathol. microbiol. scand.*, **25**, 186–191.

Panaitescu, M., Maximescu, P., Michel, I. and Potorac, E. (1977).

Pappenheimer, A. M., Jr. and Gill, D. M. (1973) *Science, N.Y.*, **182**, 353–358.

Pappenheimer, A. M., Jr. and Johnson, S. J. (1936). *Br. J. exp. Path.*, **17**, 335–341.

Pappenheimer, A. M., Jr., Uchida, T. and Harper, A. A. (1972). *Immunochemistry*, **9**, 891–906.

Park, W. H. and Williams, A. W. (1896). *J. exp. Med.*, **1**, 164–185.

Penso, G. and Vicari, G. (1957). *Rendiconti Ist. super. di Sanita*, **20**, 655–659.

Petrie, G. F. and McClean, D. (1934). *J. Path. Bact.*, **39**, 635–663.

Petrovici, A., Meitert, E., Bica-Popii, V., Saragea, A., Popescu, M. and Dunăreanu, G. (1974). *Arch. roum. Path. exp.*, **33**, 57–64.

Petrovici, A., Saragae, A., Maximescu, P. and Constantinescu, S. (1969). *Arch. roum. Path. exp.*, **28**, 821–830.

Pisu, I. and Guarnacci, M. (1971). *Ann. Igiena*, **51**, 247–252.

Pope, C. G. and Stevens, M. F. (1958). *Br. J. exp. Path.*, **39**, 139–149.

Püschel, H. (1936). *Klin. Wschr.*, **15**, 375–378.

Ramon, G. (1923). *C. r. hebd. Séanc. Acad. Sci.*, **177**, 1338–1380.

Ramon, G. (1928). *Ann. Inst. Pasteur*, **42**, 959–1009.

Rische, H. and Endemann, D. (1962). *Arch. roum. Path. exp.*, **21**, 337–341.

Robinson, D. T. and Peenay, A. L. (1936). *J. Path. Bact.*, **63**, 403–418.

Rountree, Ph. and Carne, H. R. (1967). *J. Path. Bact.*, **94**, 19–27.

Rurka, G. A. R. T. (1974). *Can. J. med. Technol.* **36** 310–313.

Saragea, A., Maximescu, P., Marion, M. and Olinescu, E. (1962). *Arch. roum. Path. exp.*, **21**, 503–510.

Saragea, A. (1963).

Saragea, A. (1965). *C. diphtheriae* Phage Typing. Doctoral Thesis, Institute of Medicine and Pharmacy, Bucharest.

Saragea, A. and Maximescu, P. (1964). *Arch. roum. Path. exp.*, **23**, 817–838.

Saragea, A. and Maximescu, P. (1966). *Bull. Wld. Hlth. Org.*, **35**, 681–689.

Saragea, A. (1969). *Bull. Inst. Pasteur*, **11**, 2515–2525.

Saragea, A. and Maximescu, P. (1969). *Arch. roum. Path. exp.*, **28**, 1053–1059.

Saragea, A., Petreanu, R., Maximescu, P. and Meitert, E. (1965). *Arch. roum. Path. exp.*, **24**, 211–229.

Saragea, A., Maximescu, P., Bîrzu, I., Gândac, G., Tiereanu, N., Popescu Pretor, I., Bărbulescu, E. and Rădulescu, S. (1966a). *Arch. roum. Path. exp.*, **25**, 83–94.

Saragea, A., Maximescu, P. and Meitert, E. (1966b). *Zentbl. Bakt. Abt. I Orig.*, **200**, 441–448.

Saragea, A., Meitert, E. and Maximescu, P. (1967a). *Microbiologia, Parazitol., Epidemiol.*, **12**, 369–376.

Saragea, A., Maximescu, P. and Meitert, E. (1967b). *Arch. roum. Path. exp.*, **26**, 919–934.

Saragea, A., Maximescu, P., Meitert, E., Dumitrescu, S., Dumitrescu, Gh., Luca, Fl., Ionescu, T., Badiu, L., Dăscălescu, C., Alexandrescu, N. and Luştrea, V. (1968). *Microbiologia, Parazitol., Epidemiol.*, **13**, 137–147.

Saragea, A., Carraz, M. and Guillermet, Fr. (1972a). *Rev. Inst. Pasteur (Lyon)*, **5**, 203–211.

Saragea, A., Maximescu, P. and Toshach, S. (1972b). (unpublished).

Saragea, A., Michel, I., Meitert, E. and Petrovici, Al. (1975). *Arch. roum. Path. exp.*, **34**, 299–306.

Saragea, A., Maximescu, P. and Meitert, E. (1973). *Corynebacterium diphtheriae.* In "Lysotypie und andere spezielle epidemiologische Laboratoriumsmethoden", pp. 425–436, VEB Gustav Fischer Verlag, Jena.

Saxholm, R. (1951). *J. Path. Bact.*, **63**, 303–311

Schick, B. (1913). *Münch. med. Wschr.*, **60**, 2608–2610.

Scott, H. H. (1934). "Milk-borne Diphtheria. Some Notable Epidemics". Edward Arnold, London.

Smith, J. (1923). *J. Hyg.*, **22**, 1–5.

Solovieva, Yu. B. and Ananieva, E. P. (1945). *Zh. Mikrobiol. Epidem. Immunobiol.*, **22**, 41–45.

Souza, G. P. and Evans, D. G. (1957). *Br. J. exp. Path.*, **38**, 644–649.

Stănică, E., Maximescu, P., Stoian, C., Pop, A., Oprişan, R. and Potorac, E. (1968). *Arch. roum. Path. exp.*, **27**, 555–560.

Stănică, E., Stoian, C., Holban, E. and Calalb, G. (1971). *Arch. roum. Path. exp.*, **30**, 281–288.

Strauss, N. and Hendee, E. D. (1959). *J. exp. Med.*, **109**, 144–163.

Suslova, V. S. (1964). *Zh. Microbiol. Epidem. Immunobiol.*, **41**, 13–16.

Tarnowski, C. (1942). *Acta path. microb. scand.*, **19**, 300.

Tashpulatova, N. V. and Krîlova, M. D. (1967). *Zh. Mikrobiol. Epidem. Immunobiol.*, **44**, 78–80.

Tasman, A. and Lansberg, H. P. (1957), *Bull. Wld. Hlth. Org.*, **16**, 939–973.

Thibaut, J. and Frédéricq, P. (1956a). *C. r. hebd. Séanc. Soc. Biol.*, **150**, 1039.

Thibaut, J. and Frédéricq, P. (1956b). *C. r. hebd. Séanc. Soc. Biol.*, **150**, 1512–1514.

Tinsdale, G. F. W. (1947), *J. Path. Bact.*, **59**, 461–466.
Toshach, S. (1950). *Can. J. pub. Hlth*, **41**, 332–336.
Toshach, S., Valentine, A. and Sigurdson, S. (1977). *J. infect. Dis.*, **136**, 655–660.
Twort F. W. (1915). *Lancet*, **ii** 1271.
Tucker, F. L., Walper, J. W., Appleman, M. D. and Donahue, J. (1962). *J. Bact.*, **83**, 1313–1314.
Uchida, T. and Yoneda, M. (1967). *Biochim. biophys. Acta*, **145**, 210–213.
Uchida, T., Gill, D. M. and Pappenheimer, A. M., Jr. (1971). *Nature New Biol.*, **35**, 8–11.
Van Heyningen, W. E. and Arseculeratne, S. N. (1964). *Ann. Rev. Microbiol.*, **18**, 195–216.
Van Wazer, J. R. (1958). "Phosphorus and its Compounds." Interscience Publishers Inc., New York.
Welsch, M. and Thibaut, T. J. (1948). *Rev. Belge Path.*, **19**, 53–57.
Wildführ, G. (1949). *Zentbl. Bakt. Abt. I Orig.*, **154**, 14–17.
Wilson, G. S. and Miles, A. A. (1946). "Topley and Wilson's Principles of Bacteriology and Immunity," 4th edn, p. 536. Edward Arnold Ltd, London.
Wilson, T. S. and Toshach, S. (1957). *Can. J. Surg.*, **1**, 57–63.
Wilson, G. S. and Miles, A. A. (1966). "Topley and Wilson's Principles of Bacteriology and Immunity," 5th edn, p. 406. Edward Arnold Ltd, London.
Wong, S. C. (1938), *Proc. Soc. exp. Biol. Med.*, **38**, 107–110.
Wong, S. C. (1940). *Proc. Soc. exp. Biol. Med.*, **45**, 850–852.
Wong, S. C. and T'ung, T. (1938), *Proc. Soc. exp. Biol. Med.*, **39**, 422–423.
Wong, S. C. and T'ung, T. (1939a). *Proc. Soc. exp. Biol. Med.*, **41**, 160–162.
Wong, S. C. and T'ung, T. (1939b). *Proc. Soc. exp. Biol. Med.*, **42**, 824–828.
Wong, S. C. and T'ung, T. (1940). *Proc. Soc. exp. Biol. Med.*, **43**, 749–753.
Wright, H. A. and Christison, M. H. (1935). *J. Path. Bact.*, **41**, 447–467.
Wright, H. D., Sonne, H. R. and Tucker, F. R. (1941). *J. Path. Bact.*, **52**, 111.
Zalma, V. M., Older, J. J. and Brooks, J. F. (1970). *J. Am. Med. Ass.*, **211**, 2125–2129.
Zamiri, I., McEntegart, M. and Saragea, A. (1972), *J. Hyg., Camb.*, **70**, 619–625.
Zamiri, I. and McEntegart, M. G. (1973). *J. med. Microbiol.*, **6**, 21.

CHAPTER V

Bacteriophage Typing of *Shigella*

TOM BERGAN

Department of Microbiology, Institute of Pharmacy, and Department of
Microbiology, Aker Hospital, University of Oslo, Oslo, Norway

I. GENERAL INTRODUCTION

Shigellosis constitutes a very significant problem on a world scale. Historically, dysentery has decided the fate of battles, due to epidemics within army contingents, but today—at least in regions with a developed sewage and food control system—it is particularly associated with closed communities with deficient hygiene or close contact such as between inmates of mental institutions, children's homes, prisons, refugee or deportation camps. Often, explosive epidemics are caused by contaminated milk, or milk products like butter, cheese, ice cream as well as lettuce, or contaminated water (Ziesché and Rische, 1973). In many countries, e.g. in Scandinavia, shigellosis is now infrequent and usually not endemic, but reflects imported cases, *Shigella sonnei* dominates with only rare isolates of *S. flexneri*. In the U.S.A., *S. sonnei* constitutes some 85% and *S. flexneri* about 14% of the shigellae (Pruneda and Farmer, 1977). *S. sonnei* is also the most important species in Japan and large parts of Africa (Ziesché and Rische, 1973). In non-industrialised countries, *S. dysenteriae* is more frequent than in industrialised nations, where *S. sonnei* and *S. flexneri* are the dominating species.

A factor that would tend to select for *S. flexneri* and *S. sonnei* is the circumstance that these species survive longer outside the body, in e.g. water, milk, and other foods (Nass, 1965; Steltzner and Urbach, 1969; Ziesché and Rische, 1973).

Elucidating the source of a *Shigella* infection may present a considerable problem. Phage typing of the isolates may be an essential tool in both endemic and epidemic situations. Typing may be the only way to determine the source of infection of persons who have been exposed to several possible sources. In our days of rapid transport, it may be important that the type of an infecting strain be recognised as identical to one that dominates in a particular geographical area through which the patient has recently travelled.

Shigella has been typed by a number of procedures such as serological typing, biotyping, and bacteriocin typing, in addition to phage typing. Serotyping is not sufficiently distinctive in an epidemiological context, but is invaluable for species diagnosis in this genus (*vide infra*).

Phage typing is of particular significance, because it enables subtyping of each serotype and therefore is satisfactory for efficient epidemiological tracing. For *S. dysenteriae* and *S. boydii*, for which phage typing would be of considerable importance in large parts of the world, only few reports have appeared (Bercovici *et al.*, 1972; Goldhar and Eylan, 1974; Goldhar *et al.*, 1974). In contrast, the number of different phage typing sets for *S. sonnei* and *S. flexneri* have proliferated. Many typing sets have been used only in one or a few laboratories, so that comparison becomes difficult. Sets from one region have often been unsuitable for bacterial isolates from another geographical area, because of a large locally dominating percentage of untyped or non-classifiable types. Consequently, the desirability of international standardisation and co-ordination led in 1967 to the establishment of an International Working Group for *Shigella* Phage Typing. For *S. flexneri*, the Ludwik Hirszfeld Institute of Immunology and Experimental Therapy, Polish Academy of Science, Wroclaw, Poland, serves as reference laboratory with S. Ślopek as the director. For *S. sonnei*, the Department of Bacteriology, National Bacteriological Laboratory, Stockholm, Sweden, with L. O. Kallings as director, serves in the same capacity. Collaborating laboratories are located in nine countries (Bucarest, Budapest, Moscow, Paris, Prague, Sofia, Stockholm, Wernigerode, and Wroclaw). The standardisation reached through the efforts of the working group will be reflected in the methodology of this Chapter as it has been presented at joint meetings and published reports of international symposia.

This Chapter will emphasise the current status of phage typing and deal mainly with *S. flexneri* and *S. sonnei*, which are the two systems that have received the most attention.

II. TAXONOMY OF *SHIGELLA*

The genus *Shigella* is named after the Japanese bacteriologist K. Shiga who is usually credited with being the first to establish the aetiology of dysentery (Shiga, 1898). The same microbe appears, however, to have been studied ten years earlier by Chantemesse and Widal (1888) who only presented a cursory description of it, but succeeded in inducing ulcers of the colon in animals. The cultures of the two groups of investigators were subsequently determined to have the same bacterial properties (Vaillard and Dopter, 1903).

Much of the very early work centred on *S. dysenteriae*. In 1900, Flexner isolated *S. flexneri* from American military personnel in the Philippines (Flexner, 1900a, b). In 1915, Sonne in Denmark described in detail what later became *S. sonnei*. In 1907, this had been called the dysentery bacillus type E by Kruse *et al.* (1907). A number of additional types were described and the situation was confusing until the Norwegian Thjøtta (1919) first proposed a nomenclatural scheme based on serology:

Type I: Shiga's bacillus
Type II: Flexner's bacillus and isolates reacting serologically with it
Type III: Sonne's type III bacillus.

This scheme was later substantiated by an extensive biotyping scheme elaborated in Denmark by Bojlén (1934) who extended Thjøtta's classification with a type IV, which corresponds to Schmitz's bacillus and is now *S. dysenteriae* type 2.

In 1949, Ewing proposed the present taxonomic scheme for the genus *Shigella* dividing it into, *S. dysenteriae*, *S. flexneri*, *S. boydii*, and *S. sonnei*. Six serotypes, characterised by Boyd between 1928 and 1938, became *S. boydii* serotypes 1–6 (Ewing *et al.*, 1971).

Soviet bacteriologists have used another classification scheme according to which (a) *S. grigorev-shiga* and *S. stuzeri-schmitzii* correspond to *S. dysenteriae*, (b) *S. flexneri* and *S. newcastle* equal the international *S. flexneri*, and (c) *S. boyd-novgorod* correspond to *S. boydii*.

Shigella isolates are Gram-negative, catalase-negative, non-motile, non-encapsulated, aerobic, facultatively anaerobic rods. They cannot use citrate or malonate as sole source of carbon or produce H_2S, gas from glucose (except aerogenic strains of *S. flexneri* type O:6), acid from mucinate, arginine hydrolase (except *S. sonnei*, a few strains), or lysine decarboxylase, but *S. sonnei* as a rule has ornithine decarboxylase. No growth occurs in the presence of KCN (Carpenter, 1974).

Although shigellae usually do not produce acid from lactose, some *S. sonnei* possess β-galactosidase (exhibit a positive ortho-nitrophenyl-β-D-galactoside test, ONPG). A few strains are even clearly lactose positive to the same extent as the most common *Escherichia coli*. A few *S. sonnei* strains are indole positive. Differentiating biochemical–cultural characteristics appear in Table I.

III. SEROLOGY OF *SHIGELLA*

A. Major serogroups of *Shigella*

The antigenic composition of shigellae is important for variation in susceptibility to phages, since their receptor sites are surface structures. Serology also forms the basis of *Shigella* classification.

Shigella has O-antigens and envelope antigens of the K-variety. Group and type specific O-antigens are present on the colony S-form of cells. The more infrequent K-antigens inhibit agglutinability in O-sera. K-antigens are removed by boiling in water for 1 h. R-form colonies have core R-antigens which are common for all shigellae.

Electron microscopic studies (Vaneeva *et al.*, 1976) have shown that S-form cells have a relatively smooth surface. They differentiated between phase II cells and R-cells. The former had a smooth surface with cell wall fragments split off from it, whereas the latter had a coarse, folded surface with numerous fragments split off most cells.

TABLE I

Species differentiation within the genus *Shigella*[a]

Characteristics	Group A S. dysenteriae	Group B S. flexneri	Group C S. boydii	Group D S. sonnei
No. of O-antigens	10	8 (+subserogroups)	15[1]	1
Dulcitol	−(10) (+serotype 5)	−(0·2)[2]	d (7)[5]	−(0)
Lactose	−(0)	−(0)	−(1)	(+) (2)
ONPG[b]	−(50)[c]	−(0·8)	−(11)	+(95)
Mannitol	−(0)[3]	+(93)[4]	+(98)	+(99)
Saccharose	−(0)	−(1)	−(0)	(+)[6]
Xylose	−(4)	−(2)	d (11)	−(1)
Indole	d (44)	d (50)	d (29)	−(0)
Ornithine	−(0)	−(0)	−(3)	+(99)
Arginine	−(2)	−(8)	−(18)	−(0·5)

[a] Numbers in parentheses indicate percentage with positive reaction (Edwards *et al.*, 1971). Such figures may show geographical differences. + indicates 90% or more positive within two days, − indicates 90% or more no reaction, d indicates different reactions (positive reactions in 11–89%). Uniformly negative genus characteristics: oxidase, adonitol, citrate, urease, gelatinase, production of acetylmethylcarbinol, by some decarboxylation, H₂S, KCN, phenylalanine deaminase. All strains of serogroups A–C do not produce gas.

[1] Two provisional serogroups in addition (Ślopek, 1968).
[2] + some serotype 6 strains; type 6, 80% + ; type X, 86% ; type Y, 100%.
[3] Rare mannitol + strains in subgroup 3 (Report, 1958).
[4] − some serotype 6 and serotype 4a strains (Carpenter, 1974; Ślopek, 1968; Edwards and Ewing, 1972; Ewing, 1949; Ewing *et al.*, 1971).
[5] Negative for all types, but percentage positive in serosubgroups (serotype) 3, 75% ; type 4, 28% ; type 6 and 10, 100% ; type 11, 34% ; type 12, 14%.
[6] 0·1% positive after only one day 88% delayed positive.

[b] ONPG = ortho-nitrophenyl-β-D-galactopyranoside reaction.
[c] All *S. dysenteriae* type 1 are ONPG positive, all type 5, 8, 10 strains negative, other species types are intermediate.

S–R variation is frequently accompanied by a loss of O-antigen and emergence of the R-antigen. Serological relationships exist between *Shigella* and other enterobacteria, particularly *Salmonella* and *Escherichia* (Ørskov *et al.*, 1977).

The O-specific polysaccharide of *S. dysenteriae* type 3 is an acidic branched hexosaminoglycan. The repeating unit is a pentasaccharide composed of D-galactosamine, D-glucopyranose, and an unidentified acidic component with galactofuranoside–glucosidic linkage (Dmitriev *et al.*, 1973, 1975a, b).

Type 5 *S. dysenteriae* has lipopolysaccharide containing carboxyethyl-rhamnose and carboxyethylglucose (Kochetkov, 1976).

In *S. flexneri* types 1b and Y, the repeating unit consists of glucose, N-acetyl-glucosamine, and two rhamnose residues. In the type 1b, N-acetyl-glucosamine and two rhamnose residues are found (Beer and Selt-mann, 1973; Lindberg *et al.*, 1972). All *S. flexneri* serotypes except type 6, have O-specific polysaccharides with the same repeating unit, N-acetyl-glucosamine and two rhamnose residues. The individual serotypes, in addition, have side chains of glucose or O-acetylation (Seltmann, 1972). As in *Salmonella*, *S. flexneri* cell walls have lipid-A, an R-specific core and an S-portion which is the basis for the group and the type antigens in *Shigella* (e.g. Seltmann, 1972; Seltman and Beer, 1972). The *S. flexneri* core contains galactose, glucosamine, glucose, heptose, and 3-deoxy-octulosonic acid (KDO) (Hannecart-Pokorni *et al.*, 1976). The lipopolysac-charide also contains 3,6-dideoxyhexose, galactosamine, mannose, rhamnose, and L-quinovosamine. By chemical composition, eight chemotypes have been differentiated, i.e. fewer different categories than are differentiable serologically.

The O-antigens are divided into separate serogroups labelled A, B, C, and D, each corresponding to one of the four *Shigella* nomenspecies. Each serogroup is subdivided by distinctive antigens referred to as serotype, subserogroup, or main antigen, designated by arabic numerals (Table II).

Legend to Table II

References: Carpenter (1974), Ewing and Carpenter (1966), Ślopek (1968), Kauffmann (1954).

[a] *S.b.* = *S. boydii*
S.d. = *S. dysenteriae*
S.f. = *S. flexneri*
S.s. = *S. sonnei*

[b] Shiga's bacterium. Produces the exotoxin Shiga toxin. Formerly called *S. shigae*, *S. shiga krusei*, or *Shiga–Kruse bacillus*.

[c] Previously named *S. ambigua* or *S. schmitzii.*, *B. ambiguus*.

[d] Previously named *S. arabinotarda A*. *S. dysenteriae* types 3–7 belonged to previous Large–Sachs group.

[e] Previously named *S. arabinotarda B*. Synonyms: *S. paradysenteriae* and *Bacterium paradysenteriae*.

[f] Flexner's bacterium.

[g] Previously named *S. newcastle*.

[h] Previously named *S. etousa*.

[i] *S. ceylonensis*, Sonne–Duval bacillus, and Kruse type E = *Bacterium paradysenteriae* belong to *S. sonnei*.

[j] Should contain mannitol negative cultures that contain main type 4 antigen-like entities variously referred to as *S. rabaulensis*, *S. rio*, and *S. saigonensis* (Kauffmann, 1954).

TABLE II

Antigenic scheme of *Shigella*

Subgroup, species	Serotype	Subserotype	Antigenic formula	Cross-reactions[a]
Subgroup A				
S. dysenteriae	1[b]			
	2[c]			*S.b.* 1, *S.b.* 15
	3[d]			*S.b.* Prov 3615–53
	4[e]			
	5			
	6			
	7			
	8			*S.b.* 15
	9			
	10			*S.b.* 1, *S.b.* 4
Subgroup B				
S. flexneri	1	1a[f]	I:2, 4	*S.b.* 2, *S.b.* 11, *S.b.* 12
		1b	I:"S":6:2, 4	*S.b.* 2, *S.b.* 12
	2	2a	II: 3, 4	*S.b.* 12
		2b	II:7, 8	*S.b.* 11, *S.b.* 12
	3	3a	III:6, 7, 8	*S.b.* 12, *S.b.* 13
		3b	III:6, 3, 4	*S.b.* 12, *S.b.* 13
		3c	III:6	
	4[j]	4a	IV:"B" 3, 4	*S.b.* 5
		4b	(IV):"B":6:3, 4	
	5		V:7, 8	
	6[g]		VI:(2), 4	*S.b.* 5, *S.b.* 12
	X		−:7, 8	
	Y		−:3, 4	*S.b.* 2, *S.b.* 5, *S.b.* 9, *S.b.* 11
Subgroup C	1			*S.d.* 2, *S.d.* 10
S. boydii	2			
	3			
	4			*S.f.* 5
	5			
	6			*S.s.* (phase II)
	7[h]			
	8			
	9			*S.f.* 2, *S.f.* 4, *S.f.* 5
	10			
	11			
	12			*S.f.* 6
	13			
	14			
	15			*S.d.* 2, *S.d.* 8
Provisory	3615–53			
	2710–54			*S.d.* 3
Subgroup D,	1	Phase I		
S. sonnei[i]		Phase II		*S.b.* 6

Serogroup A, designated *S. dysenteriae*, contains ten serotypes including Shiga's and Schmitz's bacilli as types 1 and 2. There is no common antigen applicable to the whole group, but each type has a distinctive antigen. There are a few minor cross-reactions within *S. dysenteriae* and between it and members of the genus *Escherichia*. Serogroup A strains do not ferment mannitol. Only serotype 1 produces the protein exotoxin associated with dysentery (Shiga's toxin), the main target of which is small blood vessels.

Cross-reactions have been described between *S. dysenteriae* serotype 1 and *E. coli* (Alkalescens-Dispar) O:1; *S. dysenteriae* type 2 and S. *boydii* 1 and 15; *S. dysenteriae* 8a, 8b and *S. dysenteriae* 8a, 8c; *S. dysenteriae* 8a, 8b and *S. boydii* 15; *S. dysenteriae* 10 and *S. dysenteriae* 2; *S. boydii* 1 and *E. coli* (Alkalescens-Dispar) O:1 (Edwards and Ewing, 1972).

Serogroup B, *S. flexneri*, consists of 13 serological entities with a number of common antigens. Types X and Y have only the common group antigen. Types 1–11 which lose their type antigen become types X or Y, but often in a culture it is possible to find some colonies in X- or Y-suspect cultures which have retained type antigen (Edwards and Ewing, 1972). Cross-reactions to *S. boydii* and within *S. flexneri* are many as shown in Table 66 of Edwards and Ewing (1972). *S. flexneri* is mannitol fermenting.

Serogroup C, *S. boydii*, consists of strains which are referred to 15 established and two provisional distinct types with only few intraspecies cross reactions, except types 10 and 11 which have strong cross-reactions, but type 1 cross-reacts with types 4 and 11; type 4 with 1, 9, and 11; type 5 with 9; type 9 with 4, 5, and 13; type 10 with 11; type 11 with 1, 4, and 10; and type 13 with 4 and 9.

Non-mannitol fermenting strains which do not possess the common group antigens of *S. flexneri* are referred to this species. The species encompasses a few strains which utilize mannitol, but possess the *S. boydii* antigens.

There are cross-reactions with *S. dysenteriae*, *S. boydii* and *E. coli* (Alkalescens-Dispar) (see Table 71 of Edwards and Ewing (1972) and Ślopek and Metzger (1958)).

Serogroup D, for which the name *S. sonnei* is used, is serologically more homogeneous than the other groups. *S. sonnei* has only one serotype, but this varies between two states, phase I (S-form) and phase II (R-form) (Edwards and Ewing, 1972; Rauss and Kontrohr, 1964) (Table III). Edwards and Ewing (1972) prefer to reserve the term phase for variation in flagellar antigens and use the expressions form I and form II. *S. sonnei* is the only species within this genus which regularly is ONGP positive and it ferments mannitol.

S. sonnei simultaneously with the change in phase also exhibits a change from smooth colonies (S-form) in phase I and rough colonies (R-form) in

TABLE III

Agglutination titres of *S. sonnei* in unabsorbed sera of phase I and phase II (Ślopek, 1968)

	Sera against	
Cells	Phase I	Phase II
Phase I	10240	0
Phase II	0	5120

phase II. The grape vine leaf reminiscent R-form colonies are so typical that they should make one suspect the diagnosis of *S. sonnei*. The S-form colonies possess O-antigens, the R-form R-antigens.

The antigenic variation which accompanies colony variation was first described by Thjøtta in 1917 (Thjøtta, 1918), i.e. three years before Arkwright's well-known description of the S–R-phenomenon (1920). Thjøtta and Waaler (1932) demonstrated that the R-forms were sensitive to normal serum so that a mixture of S- and R-forms can be converted to a pure S-strain by cultivation in the presence of normal guinea-pig serum. Also other species of the genus *Shigella* easily dissociate from S- to R-forms, but not with the same ease as *S. sonnei*, as described by Waaler (1935).

The extent to which phase I converts to phase II is medium dependent. Development of phase II of *S. sonnei* is repressed on SS-agar, whereas old broth cultures and Endo-agar enhance the emergence of phase II.

The composition of the O-antigen polysaccharide has been determined for a number of serotypes (Table IV). There are cross-reactions with *E. coli* (Edwards and Ewing, 1972).

B. Antigenic variability

Antigenic variability has been documented most extensively within *S. flexneri*. The main variation is the loss of subgroup (type) antigens and emergence of the main group antigen. For instance, the *S. flexneri* subgroups X and Y are derived from other distinct and separate subgroups (e.g. X from the types 2b, 3, or 5, and Y from 2a and 4a). The groups may exist *in vivo* or appear after subcultures *in vitro* (Ślopek, 1968).

C. O-antigen cross-reactions between *Shigella* and other species

Shigella exhibits cross-reactions with several *E. coli* antigens. Ślopek (1968) and Edwards and Ewing (1972) have presented detailed lists of the reactions between types within *Shigella* and *Escherichia*. This is of practi-

TABLE IV

Composition of group antigen polysaccharides in *Shigella*

Serogroup		D-Galactose	D-Glucose	Hexos-amine	D-Mannose	L-Rhamnose
S. dysen-	1	+	+	+		+
teriae	2, 3, 4, 6, 7	+	+	+		
	5	+	+	+	+	
S. flexneri	1a, 2a		+	+		+
	3a, 4a, 5a, x, 4, 6	+	+	+		+
S. boydii	1	+/−	+	+		+/−
	2, 4, 7	+	+	+		+
	3	+	+	+		
	5, 12	+	+	+	+	+
	6	+	+	+	+	
	8	+/−	+	+		
	9		+	+		+
	10	+/−	+	+	+	
	11		+	+	+/−	+
S. sonnei[a]	I	+	+	+		
	II	+	+	+/−		

[a] *S. sonnei* may or may not have xylose.

+ = present; − = absent. (Modified from Ślopek, 1968.)

cal importance, particularly in relation to the anaerogenic *E. coli*, which are often slow lactose and sucrose utilisers, and between which the cross-reactions to *Shigella* are particularly well developed. In these situations, the delineation between *Shigella* and *E. coli* may be difficult. It is then advisable to agglutinate with *E. coli* "Alkalescens-Dispar" specific sera to facilitate classification. An almost imperceptibly continuous series of intermediates connects the two taxons. Biochemically, the definition for the genus *Shigella* will include a few isolates of *E. coli*. The separation between the entities is aided by the presence of lysine decarboxylase, citrate utilis-ation (Christensen's medium), and utilisation of acetate and mucate which are typical of *E. coli*.

D. Thermolabile antigens

In addition to the thermostable somatic antigens, thermolabile fimbrial antigens have been identified within the *S. flexneri* serotypes 1a, 2a, 2b, 3a, 4a, 4b, 5, X, and Y (Ślopek, 1973). The ability to develop fimbriae and corresponding antigens has been transferred by conjugation from *E. coli* to *S. flexneri* (Mulczyk and Lachowicz, 1968).

K-antigens which occasionally cause inagglutinability in O-sera of live cells have been demonstrated in *S. dysenteriae* O:1–10, *S. boydii* O:1:15, *S. flexneri* O:2a and 6, and *S. sonnei* (Edwards and Ewing, 1972; Ślopek, 1973).

E. Serology and biotype

The stability of the biotyping scheme developed by Bojlén (1934) for *Shigella* has been verified by Hammarström (1949), Szturm-Rubinsten (1957, 1964, 1965, 1972), Ślopek (1973), and Ziesché and Rische (1973). Each serotype is divided into a few distinct stable biotypes which may be employed in epidemiological typing. Biotyping has been used alone (Szturm-Rubinsten, 1964), with phage typing (Hammarström, 1949) or with colicin typing (Ziesché and Rische, 1973). As an example, Table V shows the *S. flexneri* biotypes.

TABLE V

Biotypes of *Shigella flexneri*

			Serotype								
1	2	3	4	5				6			
								Biotype			
Characteristic						1	2	3	4	5	6
Gas (from glucose)	−	−	−	−	−	−	−	−	+	+	+
Dulcitol	−	−	−	−	−	−	−	+	−	−	+
Glycerol	−	−	−	−	−	−	−	+	−	−	+
Mannitol	+	+	+	+	+	−	+	+	−	+	+

Reference: Ślopek (1973).

Characteristic	Serotypes 1–5	Serotype 6
Dulcitol	−(0)[a]	d (9)
Lactose	−(0)	−(0)
ONPG[b]	−(0·8)	−(0)
Mannitol	+(94)	d (83)
Saccharose	−(2)	−(0)
Xylose	−(2)	−(0)
Indole	d (62)	−(0)
Ornithose	−(0)	−(0)
Arginine	−(0)	d (49)
Gas production (glucose)	−(0)	d (18)
Raffinose	d (53)	−(0)
Glycerol	−(0)	d (60)

Reference: Ewing *et al.* (1971).

[a] +, positive reaction of 90% or more of isolates; −, negative reactions of 90% or more of isolates; d, different reactions, positive reactions in 11–89% of isolates.

[b] ONPG = ortho-nitrophenyl-β-D-galactopyranoside reaction.

IV. BACTERIUM–PHAGE INTERACTIONS

A. Source of *Shigella* bacteriophages

A large number of *Shigella* isolates are lysogenic (Hammarström, 1947, 1949; Quynh, 1967, 1969; Rische, 1968a, b). Among *S. sonnei*, the frequency of lysogenic strains has been analysed in a number of studies (Table IV). Gromkova (1966) found that 12 of 30 strains were lysogenic, some even polylysogenic. *S. flexneri* may be less lysogenic than *S. sonnei*. Thus, Milch *et al.* (1968) using 11 indicators found only 9% lysogeny among 272 strains. Lysogeny has also been determined in *S. dysenteriae* and *S. boydii* (Bercovici *et al.*, 1972; Goldhar and Eylan, 1974). *Shigella* phages have also been isolated from residual waters (Bercovici *et al.*, 1972).

B. Modification and restriction of phage DNA

Host modification and restriction of phages are well-known phenomena in enterobacteriae. The phenomena are important in the phage typing

TABLE VI

Frequency of lysogeny observed in *S. sonnei*

Reference	Year	No. of strains studied	Procedure of determination	Lysogenic strains (% in parenthesis)	Comments
Gromkova	1966	30	UV and spontaneous	12(40)	
Emelyanov	1966	120	Spontaneous	43(36)	Twenty-four of the strains were virulent
Quynh	1967	15	Spontaneous	12(80)	Donor strains of same lysotype used four indicator strains
Stepenkovskaja[a]	1968	1500	UV	(45·2)	
Quynh	1969	97	Spontaneous	29(30)	Used five indicator strains
Quynh	1969	112	Spontaneous	35(31)	
Ziesché and Rische	1973			(60)	Unpublished data of Rische and Horn
Chanishvili and Chanishvili	1975			84	

[a] Cited from Ziesché and Rische (1973) (in which not referenced).

techniques. Modification and restriction implies that certain phages, R-factors, or colicin factors, when introduced into a bacterial cell reduce or abolish its sensitivity to phages which previously were lytic on the strain (László and Rimanóczy, 1976; Strobel and Nomura, 1966). Modification and restriction is mediated by specific enzymes which recognise certain short, recurring base pair sequences of DNA and cause either a modification or a restriction of them (Lewin, 1977). Modification in *E. coli* is mediated through methylation of the base recognised by the particular enzyme. Restriction results from cleavage of both strands of the DNA duplex at the enzyme recognition site. This is followed by degradation of the cleaved DNA by other nucleases that are actually non-specific. The important point is that any DNA in the cell, whether it is in principle the endogenote or an exogenote, like a plasmid or phage, may be modified if it possesses the decisive base sequence, the recognition sequence, at which point modification or restriction takes place. Essentially, this system distinguishes the endogenote DNA and that of the exogenote. The above information is derived from our particularly detailed information for *E. coli* (Eskridge *et al.*, 1967; Horn and Taubeneck, 1970; Meselson and Yuan, 1968) and *Salmonella* (László and Rimanóczy, 1976). Horn and Taubeneck (1970), Tschäpe and Rische (1970a), and Rohde (1975) have demonstrated restriction also in *S. sonnei* and László and Rimanóczy (1976) in *S. flexneri*. Host modification has also been studied in this species (Horn and Tschäpe, 1969). In *S. flexneri*, R-factors that induce host restriction have been divided into nine groups according to the changes they cause in phage susceptibility of the recipients (László and Rimanóczy, 1976; László, 1973a, b).

Restriction must be distinguished from loss of phage receptor site specificity.

The fact that R-factors may implement changes in phage type has serious implications for the interpretation of phage typing results, since R-factors typically spread epidemically (Tschäpe and Rische, 1970a, b). R-factors are very common in *Shigella*, where 30–84% of the isolates have such exogenote DNA (Chanishvili and Chanishvili, 1975; Schön and Mandliková, 1975; Tschäpe and Rische, 1970b).

Host modification has been found to cause serological change of $\frac{3}{4}$ of the phages after propagation in shigellae.

C. Lysogenic conversion of bacterial antigens

Uptake of exogenote DNA may lead to changes in serotype and susceptibility to bacteriophages. This has been demonstrated in *S. flexneri* after conjugation with *E. coli* Hfr (lac+) (Ślopek *et al.*, 1972) or Hfr (C, rha) (Jankowski *et al.*, 1974). Transduction mediated by phage PE5 has been noted in *S. flexneri* (Financsek and Kétyi, 1976). Most of the time the recombinant acquired a different group or type antigen and lost antigens originally present. Other examples of antigenic changes after transduction have been mediated by typing phages in a number of studies (Gemski *et al.*, 1975: László *et al.*, 1973a, b; Milch *et al.*, 1968). Parallel changes in the chemical composition of O-type polysaccharide have been observed by Financsek *et al.* (1975).

D. Phage receptor chemistry

In *S. dysenteriae*, a protein carrier of the O-antigen or common component present in the lipopolysaccharide has been identified as a probable phage receptor substance (Goldhar *et al.*, 1973, 1974, 1975, 1976).

Burnet and McKie (1930) reported that most phages act similarly on smooth and rough colonies of the same strains of *S. flexneri*. On the other hand the phages selected by Hammarström (1949) were mostly lytic against phase II strains, R-phages, and only some had equal or reduced propensity against phase I strains (R- and S-phages). Pruneda and Farmer (1977) reported that smooth and rough variants of the same strains had different lysotypes. They developed their proposed phage typing set for R-colonies.

V. PROPERTIES OF *SHIGELLA* PHAGES

A. Morphology of phages

The phages of *Shigella* possess polyhedric heads and tails of varying appearance according to phage strain. The phages of the international typing set of *S. flexneri* and *S. sonnei* have been subdivided into seven morphology types (Krzywy *et al.*, 1972). For *S. flexneri*, six types have been identified; for *S. sonnei*, seven morphological types have been described (Krzywy *et al.*, 1970, 1971). Inhomogeneity of phage strains has been demonstrated for one *S. flexneri* and four out of 16 *S. sonnei* typing phage preparations (Krzywy *et al.*, 1972). The phages are morphologically similar to those observed in other species of the *Enterobacteriaceae*, *Brucella*, and *Bacillus mycoides* (Krzywy *et al.*, 1970, 1971; Quynh, 1968).

The capsids are either icosahedrons or octahedrons. Their uncontracted tails are long or short, cylindrical or conical. Some have flexible tails. Rare phage strains have non-contractable, conical tails. Fibrils, collars, and base plates have been identified in most phages.

B. Serology of phages

Classification of bacteriophages according to serology, has demonstrated seven groups on the basis of neutralisation assays in *S. flexneri* phages (László *et al.*, 1973a, b; Gromkova, 1967). Among the phages of Hammarström (1947), Ślopek *et al.* (1961) have demonstrated four immunotypes.

There was no relationship between antigenic type of the phages and their lytic spectra.

C. Phage sensitivity to physical agents

Phage sensitivity to physical agents generally varies with the pH and type of suspension medium. Dilute phage suspensions of *S. sonnei* are rapidly inactivated, whereas concentrated suspensions with 10^9–10^{11} phages/ml keep for years at $+4°C$. A suitable pH for survival appears to be 7·2 (Ziesché and Rische, 1973).

VI. METHOD OF PHAGE TYPING

A. Development of lysotyping

Polyvalent *Shigella* phages were used early diagnostically to differentiate *Shigella* from other enterobacteria (Marcuse, 1931; Massa, 1931; Sartorius, 1932; Sartorius and Reploh, 1931; Sonnenschein, 1925a, b; Thomen and Frobisher, 1945). Although the phages are mainly restricted in their action to shigellae, they may also lyse some *E. coli, Enterobacter aerogenoides, E. cloacae, Salmonella, Arizona, S. dysenteriae, S. flexneri*, and *Proteus* (Grunow, 1965c). *Shigella* phages have not lysed *Citrobacter* or *Klebsiella*. Conversely, *Shigella* has been resistant to even avid phages from *Alcalescens, Proteus, Salmonella*, and *Aeromonas shigelloides* (Rische *et al.*, 1969).

Another interesting diagnostic aspect is the isolation of phage from faeces to indicate *Shigella* colonisation of the gut (Hammarström, 1949; Grunow and Köhler, 1965; Wheeler and Burgdorf, 1941).

Phage typing sets have evolved gradually since the early work of Burnet and McKie (1930) who were the first to demonstrate the fact that phages produced different lysotypes in *Shigella*. Burnet in 1933 developed a typing set for *S. flexneri*.

Since phages are mostly lytic only for one serogroup, phage typing sets are specialised for each species in the genus. Methodology is roughly the same for each species, but there are small, probably coincidental, differences in recommended techniques for each set. Accordingly, even though they may work for all sets, the techniques for each one will be described separately.

B. Phage typing of *S. dysenteriae*

The phages of *S. dysenteriae* have different lytic spectra. The only formal typing set developed specially for this species is one elaborated mainly for serotypes 1 and 2 (Meitert *et al.*, 1968). This, though, also differentiates strains of *S. flexneri*. There is epidemiological need for a further refined method for all serotypes of this species.

Goldhar and Eylan (1974) have demonstrated the chemical composition of the cellular phage receptor sites.

C. Phage typing of *S. flexneri*

1. *Previous developments*

A review of phage typing methods for *S. flexneri* has been presented by Rische (1968b). In 1968, three typing sets were published (Meitert *et al.*, 1968; Milch *et al.*, 1968; Ślopek *et al.*, 1968a, b, 1969). The set of Meitert *et al.*, was designed for only five *S. flexneri* serotypes and two *S. dysenteriae* types (*vide supra*), and consisted of 12 phages. Analogously, Istrati *et al.* (1962) have presented a set of 13 phages for only *S. flexneri* 2a. Other typing sets have been published by Metzger *et al.* (1958), Istrati (1960a, b), Ślopek and Mulczyk (1961), Mulczyk and Ślopek (1961), Istrati *et al.* (1964), and László and Milch (1975).

2. *International typing set*

A number of phages from the above sets have been combined in a new panel constituting the proposed international typing set for *S. flexneri* (Ślopek, 1968). The selected phages and their homologous host-propagating strains are shown in Table VII. Plaque morphology and viron morphology are shown in Fig. 1.

3. *Media (Ślopek, 1973)*

(a) *Medium for phage propagation and growth of bacterial inocula before typing*

Peptone (Difco)	10 g
NaCl	3 g
Na_2HPO_4	2 g
Agar (Difco)	12 g

Distilled water to 1000 ml
pH 7·2 after sterilisation

For broth, this medium is used without agar.

(b) *Phage typing medium*

Peptone (Difco)	5 g
Meat extract	3 g
Lactose	10 g
Bromocresol purple	0·025 g
Agar (Difco)	20 g

Distilled water to 1000 ml
pH 6·8–7·0 after sterilisation

TABLE VII

Properties of typing phages of international typing set of *S. flexneri*

Phage strain designations	Morphological type	Phage sero- gical type	Host propa- gating strain (serotype)	Original source
F1	A-I-1		1a (Weil)	S. Ślopek, hen faeces
F2	B-I-1	F	4a (Weil)	S. Ślopek, hen faeces
F4	B-I-1	B	2b (Weil)	S. Ślopek, hen faeces
F5	B-I-1	D	2b (Weil)	S. Ślopek, hen faeces
F6	A-I-1	D	1b (Weil)	S. Ślopek, hen faeces
F7	C-III-1			S. Ślopek, hen faeces
F9	B-I-1	B	2a (Weil)	S. Ślopek, hen faeces
F10	C-III-1	A	6　(435)	E. Hammarström (origin- ally called X), hen faeces
a	B-I 2	A	4　(Istrati)	G. Istrati
α	C-III-1		4　(118)	G. Istrati
D8	C-III-1	B	1b (Milch)	H. Milch, lysogenic bacteria
D2a	C-III-1	*	4g (106)	H. Milch, lysogenic bacteria
D2b	B-I-1	*	3a (109)	H. Milch, lysogenic bacteria
D2c	C-II-1	*	1b (449)	H. Milch, lysogenic bacteria

The F1, F5, F6, F4, and F9 bacteriophages were isolated in 1942 by Ślopek and adapted to *S. flexneri* serotypes 1a, 5, 2a, 2b, and 4 respectively.

The F2 bacteriophage was isolated in 1955 by Mulczyk and Kucharewicz and adapted to *S. flexneri* serotype 4a.

The F7 bacteriophage was isolated in 1957 by Mulczyk and adapted to *S. flexneri* serotype 2c.

* The parent preparation from which the D2a, D2G, and D2c were isolated, belonged to phage serotype F.

References: Krzywy and Ślopek (1974); Mulczyk and Ślopek (1961); Ślopek (1973); Ślopek and Mulczyk (1961).

(c) *Maintenance of bacterial strains.* Three parts egg (white and yolk) and one part physiological saline. Sterilised by heating to 85°C three days in a row.

(d) *Subcultivation.* Any medium suitable for growing enterobacteria may be used to prepare precultures before typing.

4. *Propagation of phages*

The phages are propagated on their homologous host strains as shown in Table VII. To 9 ml broth is added 1 ml of a 3-h liquid culture of the host propagating strain and 0·5 ml of phage lysate. After 4 h incubation

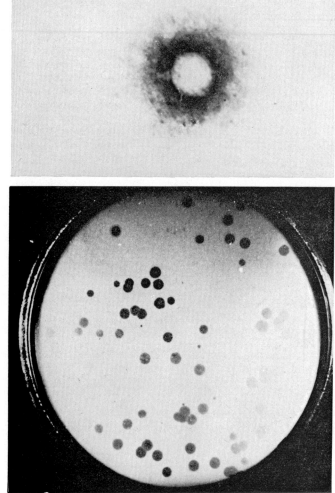

FIG. 1. Plaque morphology and electron microscopy of typing phages of the proposed international typing set for *S. flexneri*. Preparation of microscopic preparations described by Krzywy and Ślopek (1974). Classified by morphological type.

Phage F1 for *S. flexneri* (×240,000). The virion head is of a regular icosahedron shape, the tail conical. Type A-I-1.

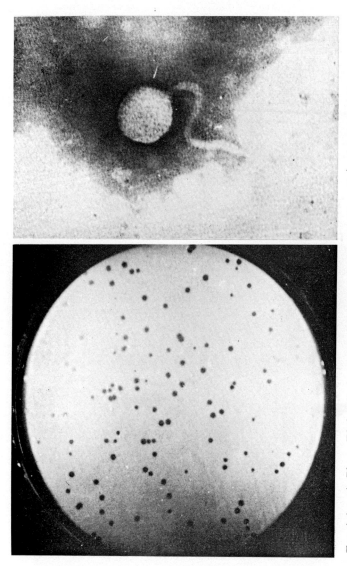

Fig. 1 (cont.). Phage F2 for *S. flexneri* (× 240,000). Tail longer than diameter of head. Type B-I-1.

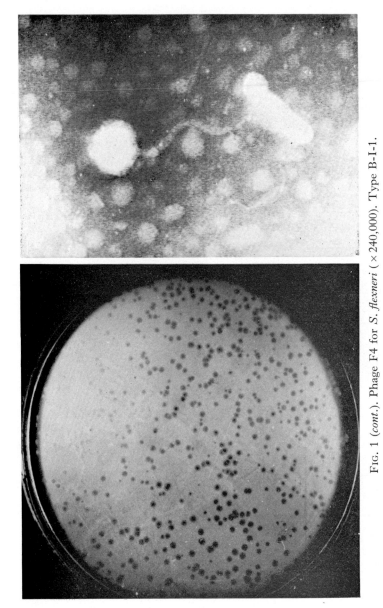

Fig. 1 (*cont.*). Phage F4 for *S. flexneri* (×240,000). Type B-I-1.

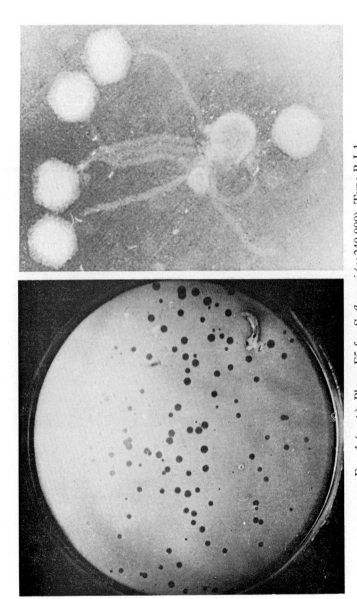

Fig. 1 (cont.). Phage F5 for *S. flexneri* (×240,000). Type B-I-1.

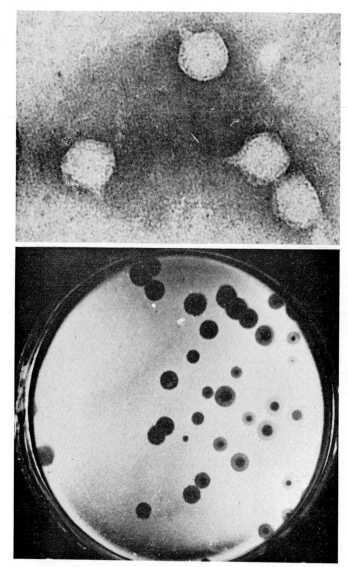

Fig. 1 (*cont.*). Phage F6 for *S. flexneri* (×240,000). Type A-I-1.

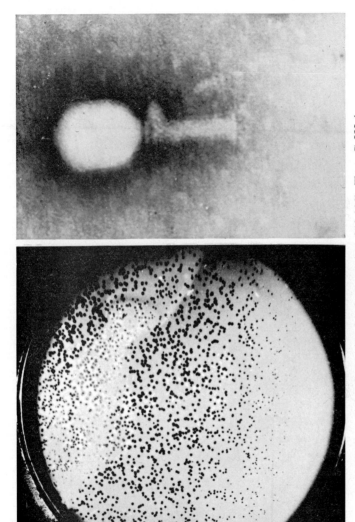

Fig. 1 (*cont.*). Phage F7 for *S. flexneri* (×240,000). Type C-III-1.

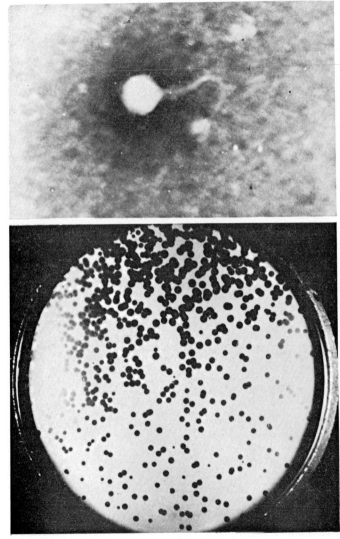

Fig. 1 (*cont.*). Phage F9 for *S. flexneri* (×240,000). Head shaped as regular icosahedron. Type B-I-1.

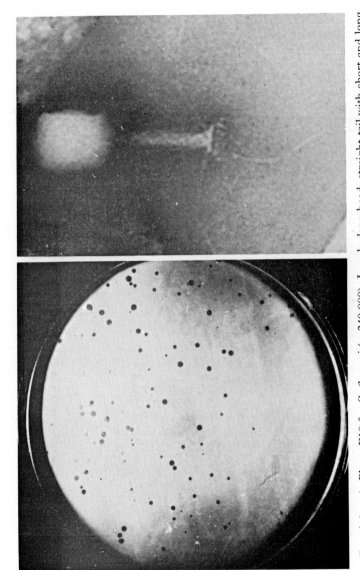

Fig. 1 (*cont.*). Phage F10 for *S. flexneri* (×240,000). Icosahedron head, straight tail with short and long tail fibres. Type C-III-1.

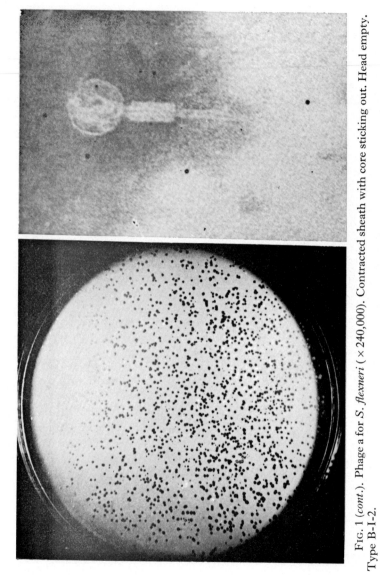

Fig. 1 (*cont.*). Phage a for *S. flexneri* (×240,000). Contracted sheath with core sticking out. Head empty. Type B-I-2.

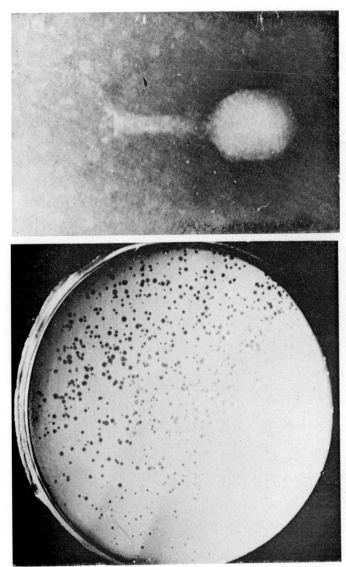

FIG. 1 (*cont.*). Phage α for *S. flexneri* (×240,000). Clearly distinct neck and sheath with basal plate of virion tail. Type C-III-1.

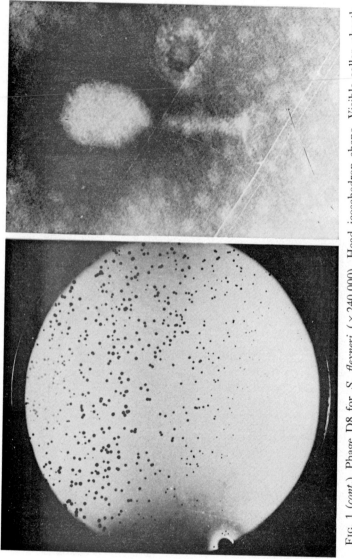

Fig. 1 (*cont.*). Phage D8 for *S. flexneri* (×240,000). Head icosahedron shape. Visible collar, sheath and basal plate. Long fibrils. Type C-III-1.

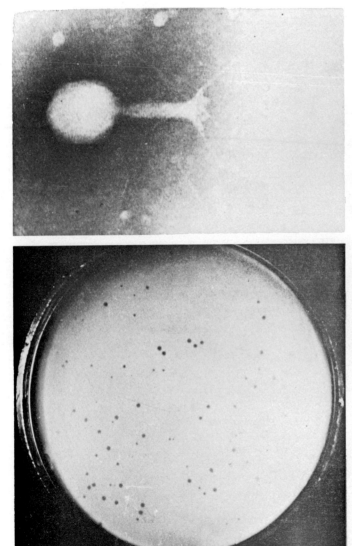

FIG. 1 (*cont.*). Phage D2a for *S. flexneri* (×240,000). Icosahedric head. Clearly visible collar, sheath, basal plate and long fibrils. Type C-III-1.

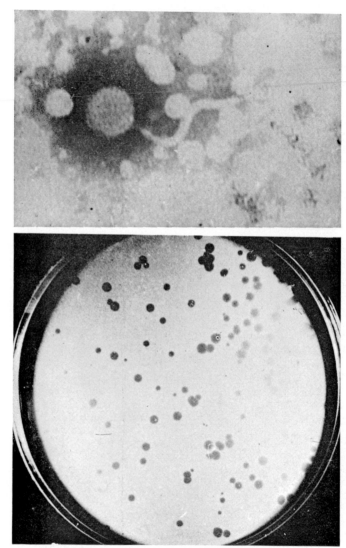

FIG. 1 (*cont.*). Phage D2b for *S. flexneri* (×240,000). Long flexible tail, distinct collar. Type B-I-1.

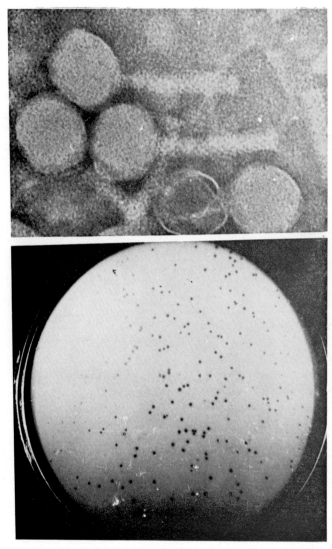

FIG. 1 (*cont.*). Phage D2c for *S. flexneri* ($\times 240{,}000$). Filled heads and one empty head. Collars visible. Straight tail. Type C-II-1.

at 37°C, most phage cultures have become transparent. At that time, the culture is filtered through a membrane filter of 0·45 μm and kept as a stock suspension. In preparations with secondary growth the titres are lower.

5. Determination of strength and lytic spectra

The phage stock is diluted 10^{-1}, 10^{-2}, etc., up to 10^{-9}. It is important to use new pipettes for each dilution to avoid carry-over. The plate is inoculated by flooding with a fresh 3-h bacterial culture, pipetting off the excess, and drying at room temperature for 30 min. The phage suspensions are applied to the plates, incubated at 37°C for 7 h and then placed at 4°C until the reactions are recorded on the following day. The highest dilution which still has sufficient phage to induce confluent lysis on the host propagating strain is recorded. This is termed the critical dilution (CD). The results are recorded as follows:

> CL = confluent lysis
> SCL = semiconfluent lysis
> OL = opaque lysis
> + + + = > 80 isolated plaques
> + + = 40–80 isolated plaques
> + = 20–40 isolated plaques
> − = no plaques.

The lytic spectrum of the typing set is shown in Table VIII.

6. Typing procedure

Both bacterial incula and phage suspensions must be prepared with the utmost consideration for meticulous repetition of details. Plates with the strains to be typed are inoculated as described in Section VI.C, 5 of this Chapter.

The phages are used at their routine test dilutions (RTD), which for *S. flexneri* is 3–5-fold more concentrated than CD (Table IX). As soon as the plates have dried, 10 μl volumes of each phage suspension (at RTD) are applied to the agar surface with a multi-inoculator. Many inoculators have been constructed. The model designed by Lidwell (1959) (Fig. 2), which may be procured from Biddulph and Co. (Manchester) Ltd, Commercial Street, Knott Mill, Manchester 15, England, is particularly suitable.

It is essential that the phage drops have dried completely before the plates are incubated, otherwise the phage number will increase in the drop to render apparent inocula above RTD. After 20–30 min at room temperature without lids, the plates are immediately transferred to 37°C for 7 h

TABLE VIII

Lytic spectra of proposed international phage typing set for *S. flexneri* (Ślopek, 1973)

No. and symbol of the phage	Bacteriophages (RTD)													
	1	2	3	4	5	6	7	8	9	10	11	12	13	14
1 F1	CL	CL	–	–	CL	SCL	–	–	+	–	CL	CL	–	CL
2 F2	CL	CL	CL	CL	CL	+	CL	CL	CL	–	SCL	SCL	CL	CL
3 F4	CL	CL	CL	CL	CL	CL	CL	CL	CL	CL	CL	SCL	CL	CL
4 F5	CL	CL	CL	CL	CL	CL	CL	+	CL	CL	CL	SCL	CL	CL
5 F6	CL	CL	–	–	+	CL	–	+	CL	+	CL	CL	–	+
6 F7	–	–	CL	CL	+	CL	CL	–	CL	CL	CL	SCL	CL	+
7 F9	–	–	CL	CL	–	CL	CL	–	CL	CL	CL	CL	–	–
8 F10	–	–	–	–	–	–	–	CL	–	–	–	–	CL	SCL
9 α	CL	SCL	CL	CL	CL	CL	CL	CL	CL	SCL	CL	–	CL	–
10 a	–	–	–	–	–	–	–	–	–	+	–	–	–	SCL
11 D8	CL	CL	–	–	CL	CL	–	–	CL	SCL	CL	SCL	CL	–
12 D2a	–	CL	–	–	–	+	–	–	+	+	CL	CL	–	–
13 D2b	–	CL	CL	CL	–	–	CL	CL	–	SCL	–	–	CL	SCL
14 D2c	CL	CL	–	–	CL	CL	–	–	–	–	–	–	–	CL

For explanation of symbols see Section VI.C, 5.

TABLE IX

Critical dilutions (CD) and routine test dilutions (RTD) of
S. flexneri typing phages

Phage designation	Ten-fold dilution of freshly prepared phage suspension commonly yielding		Relative dilution of CD suspension to give RTD
	CD	RTD	
F1	10^{-6}	10^{-2}	$\times 10^{-4}$
F2	10^{-7}	10^{-3}	$\times 10^{-4}$
F4	10^{-6}	10^{-2}	$\times 10^{-4}$
F5	10^{-5}	10^{-1}	$\times 10^{-4}$
F6	10^{-6}	10^{-3}	$\times 10^{-3}$
F7	10^{-7}	10^{-3}	$\times 10^{-4}$
F9	10^{-6}	10^{-2}	$\times 10^{-4}$
F10	10^{-6}	10^{-2}	$\times 10^{-4}$
α	10^{-6}	10^{-2}	$\times 10^{-4}$
a			Not diluted
D8	10^{-7}	10^{-3}	$\times 10^{-4}$
D2a	10^{-6}	10^{-3}	$\times 10^{-3}$
D2b	10^{-7}	10^{-4}	$\times 10^{-3}$
D2c	10^{-7}	10^{-3}	$\times 10^{-4}$

Reference: Ślopek (1973).

when the plates are taken out and stored at 4°C until they are read the next morning.

7. Recording of results

The results are recorded as phage types, basically according to the 54 patterns published by Ślopek (1973) (Table X). Additional lysotypes can be seen. László et al. (1973b) thus observed 89 different phage types. In evaluating new typing patterns, it is important to distinguish between coincidental deviations from established types and new patterns which are stable and reproducible. Each pattern is represented by a standard strain which, with carefully prepared and adjusted phage suspensions, exhibits a characteristic reaction pattern with properly quantitated (RTD) phage suspensions (Table X).

8. Phage type frequencies

The frequency of the phage types vary considerably. Some are only represented by a few, rare strains, whereas others constitute dominating types. In one study of 1360 S. flexneri strains, two of 54 types represented

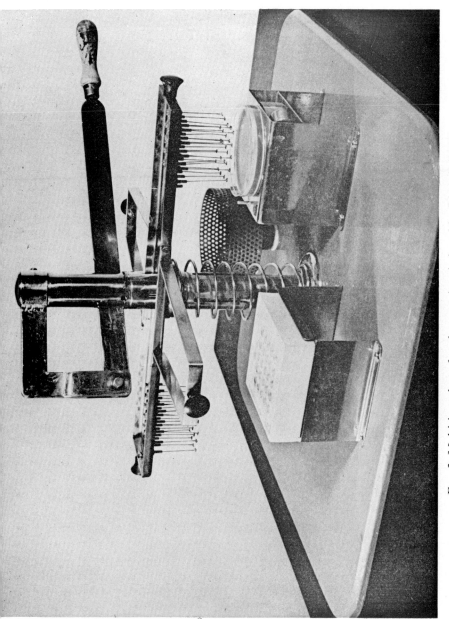

Fig. 2. Multi-inoculator for phage typing designed by Lidwell (1959).

TABLE X

Lytic patterns of *S. flexneri* phage types and designation of standard bacterial strain exhibiting corresponding type (Ślopek, 1973)

Phage type	1 F1	2 F2	3 F4	4 F5	5 F6	6 F7	7 F9	8 F10	9 α	11 D8	12 D2a	13 D2b	14 D2c	Standard strain[a]
1	CL	CL	CL	CL	CL	CL	CL	CL	CL	CL	CL	CL	CL	1 (1a)
2	CL	CL	CL	CL	SCL	SCL	CL	–	CL	CL	SCL	CL	CL	177 (2a)
3	CL	CL	CL	SCL	CL	–	CL	–	CL	CL	SCL	CL	CL	1113 (4a)
4	CL	CL	CL	–	CL	++	–	–	SCL	CL	++	CL	CL	628 (1b)
5	CL	++	++	–	CL	++	++	–	SCL	CL	++	++	–	630 (1b)
6	CL	CL	±	±	CL	CL	+++	–	SCL	CL	SCL	+	CL	679 (4a)
7	CL	CL	–	–	CL	CL	–	CL	CL	CL	SCL	–	SCL	465 (4a)
8	SCL	CL	–	–	CL	CL	–	–	CL	CL	SCL	–	CL	582 (X)
9	CL	CL	–	–	CL	CL	–	–	CL	CL	SCL	–	–	492 (1b)
10	CL	CL	–	–	CL	SCL	–	–	CL	CL	CL	–	CL	91 (1b)
11	CL	–	–	–	CL	CL	–	SCL	–	–	–	–	CL	449 (1b)
12	CL	–	CL	CL	CL	SCL	CL	CL	CL	CL	SCL	CL	CL	115 (2b)
13	CL	–	CL	CL	CL	++	CL	SCL	CL	CL	SCL	CL	CL	51 (2a)
14	CL	–	CL	CL	CL	+++	CL	–	–	CL	++	CL	–	670 (2b)
15	CL	–	CL	CL	CL	+	CL	–	CL	CL	SCL	CL	CL	556 (2a)
16	CL	–	CL	CL	CL	SCL	CL	–	CL	CL	SCL	CL	CL	184 (2a)
17	SCL	–	CL	CL	CL	SCL	CL	–	CL	+	–	CL	–	985 (2a)
18	CL	–	CL	CL	CL	SCL	CL	–	CL	±	–	CL	CL	1032 (2a)
19	CL	–	CL	–	SCL	SCL	–	CL	–	SCL	++	CL	CL	1408 (2b)
20	CL	–	CL	–	CL	CL	–	CL	CL	CL	SCL	+++	CL	942 (3a)
21	CL	–	CL	–	SCL	SCL	–	–	CL	CL	SCL	CL	CL	548 (2a)
22	SCL	–	SCL	SCL	SCL	SCL	CL	CL	CL	SCL	SCL	SCL	–	1180 (2a)
23	CL	–	–	CL	CL	–	CL	CL	CL	CL	SCL	–	–	1455 (5a)
24	CL	±	–	CL	CL	–	CL	–	–	SCL	++	–	–	168 (2a)

															Strain[a]
25	CL	−	SCL	±	±	CL	SCL	++	CL	CL	SCL	CL	−		1306 (2a)
26	CL	−	CL	−	−	CL	CL	−	CL	CL	CL	CL	+		924 (3a)
27	CL	−	++	−	−	CL	+	−	CL	CL	++	CL	+		12 (1b)
28	CL	−	SCL	−	+	CL	SCL	−	CL	CL	++	CL	CL		1224 (2a)
29	CL	−	CL	−	−	CL	CL	−	CL	−	SCL	CL	−		17 (2a)
30	SCL	−	SCL	−	−	−	SCL	−	SCL	CL	−	SCL	CL		1226 (2a)
31	SCL	−	SCL	−	−	−	SCL	−	SCL	CL	SCL	SCL	−		19 (4a)
32	−	SCL	−	CL	CL	CL	−	CL	SCL	CL	−	CL	CL		174 (2a)
33	−	SCL	−	CL	CL	CL	−	CL	SCL	CL	++	CL	CL		273 (3a)
34	−	+++	−	CL	CL	CL	−	CL	CL	CL	++	CL	+		262 (3a)
35	−	SCL	−	CL	CL	CL	−	CL	CL	CL	±	CL	CL		121 (X)
36	−	SCL	−	CL	CL	CL	−	CL	CL	CL	−	−	−		109 (3a)
37	−	SCL	−	CL	CL	−	−	−	CL	−	−	−	−		534 (1a)
38	−	CL	+	+	±	CL	±	−	CL	±	+	−	CL		681 (4)
39	−	CL	+	+	+	CL	+	SCL	+	CL	+	−	−		615 (1b)
40	−	CL	−	−	CL	CL	CL	−	CL	CL	SCL	SCL	CL		841 (4)
41	−	CL	−	±	−	SCL	±	SCL	−	CL	−	−	−		886 (4a)
42	−	CL	−	−	SCL	−	−	−	SCL	−	−	−	SCL		152 (4b)
43	±	−	−	CL	CL	CL	CL	−	−	CL	CL	SCL	−		347 (3a)
44	−	−	−	CL	CL	CL	CL	CL	CL	CL	CL	SCL	+		301 (3a)
45	−	−	−	CL	CL	SCL	CL	−	CL	SCL	CL	CL	++		958 (3a)
46	±	−	+	CL	CL	SCL	CL	−	+	SCL	±	CL	CL		994 (3a)
47	±	−	+	CL	CL	SCL	CL	−	CL	SCL	CL	CL	SCL		567 (3a)
48	−	−	−	CL	CL	SCL	−	CL	CL	SCL	CL	CL	CL		959 (3a)
49	−	−	−	−	±	SCL	−	−	CL	SCL	CL	−	−		31 (3a)
50	−	−	−	−	±	SCL	−	−	CL	SCL	CL	−	−		1329 (3a)
51	−	−	−	−	−	−	−	−	CL	−	−	−	−		364 (6)
52	−	−	−	−	−	−	−	−	±	SCL	++	++	−		1191 (3a)
53	−	−	−	−	−	−	−	−	CL	−	−	−	−		435 (6)
54	−	−	−	−	−	−	−	CL	−	CL	−	−	−		184 (X)

[a] Numbers in parentheses indicate serotype of standard control strain of *S. flexneri*. For explanation of symbols see Section VI.C, 5.

as much as 54% of the bacterial isolates (type 15: 18·7%; and type 43: 34·8%) (Ślopek, 1973). Another eight types contributed 34·6% of the strains. The results will vary according to local variation of dominating types. The phage type frequencies of Ślopek (1973) are compared in Table XI with the results of László et al. (1973a, b).

An example of the distribution of phage types between bacterial serotypes is shown in Table XII.

D. Phage typing of S. boydii

Bercovici et al. (1967, 1972) have been concerned with phage typing of S. boydii. Bercovici et al. (1972) ultimately used 20 phages which were either

(1) from residual waters and adapted to reference strains of S. boydii,
(2) from Salmonella typhimurium, but also active on S. boydii,
(3) from E. coli adapted to S. boydii, or
(4) from S. dysenteriae, serotype 2 and adapted to S. boydii.

The phages were designated 1, 2, 3, 4, 6, 7, 8, 9, 10, 13, 14, 18, 68, 100, 204, 210, 338, 390, and 394.

The preparation of plates followed the procedure outlined for S. flexneri (Section VI.C, 6, this Chapter). Three-hour cultures (37°C in 0·8% Nutrient Broth (Difco)) were used to inoculated plates with Nutrient Broth (Difco)) and 2% agar.

The plates were read after 16–18 h at 37°C and lytic patterns recorded according to the key in Table XIII. Among the 35 different lysotypes, the dominating one contributed some 35%, and four other types were responsible for some 5% of the isolates.

When standardized laboratory techniques are used, S. boydii phage types appear stable, as evidenced by serial isolates from the same individuals and by the patterns of strains derived from two particular epidemic outbreaks (Bercovici et al., 1972). The patterns also have remained stable after storage of the bacteria in vitro for six to seven months.

E. Phage typing of S. sonnei

1. Previous developments

Due to the serological homogeneity of S. sonnei, phage typing is important for the subdivision of this species. Earlier phage typing methods for this species have been reviewed by Rische (1968).

The first published lysotyping scheme for S. sonnei was developed by Hammarström (1947, 1949) during the years 1943–1947. At that time, S.

TABLE XI

Distribution of 4606 *S. flexneri* serotypes on different phage types
(László *et al.*, 1973a, b)

Serotype	No. of strains	Frequent phage types	No. of strains	No. of rare phage types	Rare phage types
1a	28	62	10	10	1, 10, 16, 17, 22, 67, 72, 75,
		59	4		81, 89, NT
1b	344	62	136	29	5, 7, 9, 10, 12, 16, 17, 20, 22,
		52	56		23, 27, 28, 41, 42, 49, 53, 56,
		73	34		60, 61, 67, 68, 70, 72, 75, 76,
		59	23		81, 84, 85, 89, Degr., NT
2a	977	22	202	45	1, 2, 3, 4, 5, 6, 7, 9, 10, 14, 15,
		71	147		17, 18, 19, 21, 24, 25, 28, 29,
		16	125		34, 38, 40, 41, 42, 44, 45, 46,
		20	84		47, 48, 50, 53, 54, 56, 58, 59,
		69	72		60, 62, 67, 68, 73, 76, 81, 84,
		23	53		85, 87, Degr., NT
2b	12	68	5	6	2, 14, 16, 20, 22, 31
3a	1800	42	415	48	2, 4, 7, 9, 10, 13, 14, 16, 18, 19,
		5	407		20, 21, 22, 23, 25, 27, 28, 30,
		19	363		32, 33, 34, 36, 38, 39, 41, 44,
		37	47		45, 47, 50, 52, 54, 57, 58, 59,
		43	42		60, 61, 62, 63, 64, 68, 69, 81,
		31	39		83, 84, 85, 86, 89, Degr., NT
		67	38		
		26	37		
		40	35		
3b	12	14	4	5	3, 5, 16, 18, 55
3c	5	—	—	5	12, 19, 55, 60, 69
4a	689	81	495	26	2, 5, 6, 7, 10, 15, 28, 31, 37, 42,
		78	48		51, 59, 61, 62, 63, 67, 68, 69,
		89	32		73, 80, 82, 83, 84, 85, 86, 87,
					Degr., NT
4b	37	81	29	5	18, 19, 62, 72, 78
5	2	—	—	2	16, 84
6	407	85	358	18	5, 9, 10, 17, 20, 21, 23, 27, 41,
					42, 46, 62, 70, 73, 81, 83, 84,
					89, Degr., NT
var. X	259	32	49	32	2, 4, 7, 9, 10, 12, 13, 17, 18, 19,
		5	31		20, 23, 24, 28, 29, 31, 37, 38,
		3	30		42, 44, 55, 60, 61, 62, 63, 66,
		84	21		67, 69, 81, 83, 85, 89, Degr., NT
var. Y	25	—	—	12	2, 5, 10, 14, 23, 28, 32, 62, 68,
					81, 84, 85, NT
1b–3b	9	—	—	6	6, 14, 20, 23, 49, 81, NT

NT = non-typable.
Degr. = degraded reaction.

TABLE XII

Distribution of *S. flexneri* phage types among different serotypes
(Ślopek, 1973)

Phage type	1a	1b	2a	2b	3a	3b	4	4a	4b	5a	5b	6	X	Y	Number of strains
1	3	1	1		8			1					47	6	67
2		13													13
3								1							1
4		1													1
5		1													1
6		1						1							2
7	1	4	1					1							7
8		2											1		3
9	14	65	1					9							89
10		6						7	1						14
11		2													2
12			14	11	62						3		1		91
13			1		3										4
14				1											1
15	1		245		6	1								1	254
16			14												14
17			5												5
18			1												1
19				1											1
20					4										4
21			5												5
22			1												1
23	3									1				1	5
24			1												1
25			2												2
26					15					1					16
27		1	1												2
28	1	1	3						3						8
29					1										1
30			1												1
31								1						1	2
32			4		16								7		27
33					10								3	1	14
34					1										1
35		1	1		19								1	1	23
36					1										1
37	1														1
38							1	7							8
39		1			1										2
40							2								2
41							45	3	6						54
42								4						1	5
43			8	2	462									1	463
44					1										1

TABLE XII (*continued*)

Phage type	Serotype 1a	1b	2a	2b	3a	3b	4	4a	4b	5a	5b	6	X	Y	Number of strains
45				2											2
46		2		2											4
47				4											4
48				20											20
49				2											2
50				3											3
51												5			5
52				1											1
53												88			88
54														1	1

sonnei was more common in Scandinavia than today (Fig. 3). Hammarström (1949) evaluated the method very carefully and subdivided 1834 strains (from 142 outbreaks and additionally hundreds of individual isolates) into 64 lysotypes. The list of types has since been extended by Kallings *et al.* (1968) to 94 types. Among these, 75% are very rare and some have been recorded but a few times.

Other phage typing sets have also been developed such as that of Tee (1955). This typed all 829 strains tested, but three-quarters of the strains belonged to the same phage type, and lysotype instability was observed. No further report using this set is known.

Gromkova and Trifonova (1967) with 13 locally isolated phages differentiated 20 lysotypes among R-forms, and with ten typing phages observed 12 lysotypes among S-forms. The typing efficiency increased upon raising the strength of the phage suspension from RTD to $10 \cdot 000 \times$ RTD. Then, $1 \cdot 8\%$ of the R-forms and 8% of the S-forms were nontypable. Sixty-five per cent of the strains belonged to one single phage type. There was no relationship between lysotype and biotype.

Typing sets have been designed specially for the less phage-sensitive S-forms by Chernova (1971) and Korsakova and Sabrodina (1954). It is difficult from the literature to adequately assess the utility of these typing sets. Ślopek *et al.* (1968) proposed a *S. sonnei* typing set of 15 phages which differentiated 38 lysotypes. This consisted of the Hammarström set supplemented by three phages.

A larger number of phages were later tested by Ślopek *et al.* (1973) who combined a few Hammarström phages and some isolates of their own to construct two typing sets. These typed all 2219 bacterial isolates tested. Pruneda and Farmer (1977) compared a number of previous sets and, after

TABLE XIII

Lytic patterns of *S. boydii* (Bercovici *et al.*, 1972)

Phages in RTD	$\frac{1}{50}$	$\frac{1}{10}$	$\frac{1}{10000}$	$\frac{1}{10000}$	$\frac{1}{1000}$	$\frac{1}{10000}$	$\frac{1}{500}$	$\frac{1}{100}$	$\frac{1}{1000}$
Phage type	1	204	4	14	3	338	2	390	100
1 London a	2+m	≪SCL	CL	<SCL	≪SCL	–	CL	CL	<SCL
1 Paris b	–	≪SCL	SCL	<SCL	≪SCL	–	SCL	SCL	<SCL
T. 7040 c	CL	≪SCL	SCL	<SCL	<SCL	SCL	CL	SCL	SCL
T. 6701 d	CL	≪SCL	SCL	SCL	<SCL	SCL	–	SCL	SCL
T. 5947 b e	CL	<SCL	SCL	SCL	SCL	≪SCL	–	SCL	–
T. 5948 a f	CL	<SCL	<SCL	SCL	SCL	3+m	CL	SCL	–
T. 3912 g	CL	<SCL	CL	SCL	SCL	SCL	–	CL	SCL
T. 5592 h	CL	<SCL	CL	SCL	SCL	SCL	CL	CL	SCL
T. 5556 i	–	≪SCL	<SCL	<SCL	<SCL	–	CL	SCL	–
T. 6137 j	–	≪SCL	≪SCL	–	–	–	CL	–	SCL
T. 6140 l	–	SCL	SCL	<SCL	<SCL	–	SCL	2+s	–
T. 1166 m	–	<SCL	SCL	–	–	–	CL	<SCL	<SCL
2 London a	CL	≪SCL	≪SCL	SCL	2+m	<SCL	SCL	SCL	2+s
2 Paris a	CL	≪SCL	<SCL	SCL	±m	<SCL	SCL	SCL	2+s
7177 b	SCL	<SCL	<SCL	SCL	<SCL	<SCL	–	SCL	<SCL
5814 c	CL	≪SCL	≪SCL	SCL	–	<SCL	CL	SCL	2+s
T. 3992 d	–	–	–	≪SCL	–	–	–	<SCL	–
3 London a	CL	<SCL	<SCL	≪SCL	≪SCL	<SCL	CL	SCL	SCL
3 Paris b	CL	<SCL	SCL	<SCL	<SCL	<SCL	CL	CL	<SCL
4 London a	CL	≪SCL	SCL	<SCL	<SCL	SCL	CL	CL	–m
4 Paris b	CL	≪SCL	SCL	<SCL	≪SCL	SCL	CL	CL	<SCL
x. 3338 b c	CL	≪SCL	SCL	SCL	SCL	SCL	CL	CL	SCL
T. 3513 d	CL	≪SCL	SCL	SCL	SCL	CL	CL	SCL	≪SCL
O. 2218 e	CL	–	<SCL	<SCL	SCL	SCL	–	CL	<SCL
5 London a	CL	SCL	–	≪SCL	+m	SCL	SCL	<SCL	–
5 Paris b	SCL	<SCL	≪SCL	≪SCL	<SCL	<SCL	–	SCL	–
6 London a	SCL	<SCL	SCL	<SCL	≪SCL	SCL	SCL	SCL	SCL
6 Paris b	SCL	<SCL	SCL	<SCL	≪SCL	SCL	SCL	SCL	SCL
7 London a	SCL	<SCL	≪SCL	–	<SCL	≪SCL	<SCL	SCL	–
7 Paris b	SCL	<SCL	≪SCL	–	SCL	≪SCL	≪SCL	SCL	+s
x. 9422 c	CL	<SCL	≪SCL	<SCL	SCL	<SCL	–	CL	±s
5184 d	CL	<SCL	<SCL	SCL	<SCL	SCL	–	SCL	+s
T. 5638 e	CL	≪SCL	<SCL	SCL	<SCL	<SCL	SCL	SCL	≪SCL
O. 2722 f	–	<SCL	–	–	–	–	–	–	–s
8 London a	–	≪SCL	–	SCL	<SCL	–	–	SCL	<SCL
8 Paris b	3+m	≪SCL	–	SCL	<SCL	–	–	SCL	≪SCL
9 London a	CL	≪SCL	<SCL	–	<SCL	≪SCL	–	SCL	–
9 Paris a	CL	≪SCL	≪SCL	–	<SCL	<SCL	–	SCL	–

$\dfrac{1}{100}$	$\dfrac{1}{300}$	$\dfrac{1}{100}$	$\dfrac{1}{300}$	$\dfrac{1}{10}$	$\dfrac{1}{500}$	$\dfrac{1}{300}$	$\dfrac{1}{30}$	$\dfrac{1}{10}$	$\dfrac{1}{300}$	$\dfrac{1}{300}$
7	8	394	18	68	6	9	210	368	10	13
CL	CL	CL	–	–	<CL	–	–	–	–	–
2+m	–	–	–	–	–	–	–	–	–	–
SCL	CL	SCL	CL	CL	SCL	CL	2+s	SCL	–	–
SCL	SCL	SCL	CL	CL	SCL	SCL	<SCL	≪SCL	–	–
CL	CL	CL	CL	CL	CL	CL	<SCL	≪SCL	–	–
SCL	SCL	SCL	CL	CL	SCL	SCL	–	–	–	–
–	–	<SCL	CL	CL	–	CL	SCL	<SCL	–	–
CL	SCL	SCL	SCL	SCL	SCL	SCL	–	–m	–	–
CL	CL	CL	–	–	CL	–	–	–	–	–
SCL	SCL	SCL	–	–	SCL	–	–	–	–	–
–	–	<SCL	–	–	–	–	–	–	–	–
–	–	SCL	–	–	–	–	–	–	–	–
–	SCL	–	≪SCL	–	–	–	–	2+s	–	–
–	SCL	–	–m	–	–	–	–	OL	–	–
≪SCL	CL	–s	–	–	–	–	–	OL	–	–
3+m	SCL	<SCL	–	–	≪SCL	–	–	OL	–	–
–	–	SCL	–	–	–	–	–	–	–	–
CL	SCL	SCL	CL	CL	SCL	2+m	–	+m	–	–
CL	SCL	SCL	CL	CL	SCL	CL	<SCL	CL	–	–
–	–	–	–	–	–	–	–	–	–	–
–	2+m	–	CL	CL	–	CL	–	SCL	–	–
<SCL	–	<SCL	<SCL	–	–	–	–	–	–	–
–	–	–	CL	–	–	–	–	–	–	–
–	–	–	CL	–	–	–	–	–	–	–
–	–	–	–	–	–	–	–	–m	SCL	<SCL
–	–	–	–	–	–	–	–	–	SCL	SCL
SCL	–	SCL	SCL	SCL	–	SCL	SCL	–	–	–
SCL	–	–	2+m	–	–	–	+m	–	–	–
–	–	–	–	–	–	–	–	+m	–	SCL
–	–	–	–	–	–	–	–	–	–	–
–	–	–	2+m	CL	–	–	SCL	–	≪SCL	CL
–	–	–	2+m	CL	–	–	SCL	–	≪SCL	–
–	–	–	–	–	–	–	–	SCL	–	–
–	–	–	–	–	–	–	–	–	–	–
SCL	SCL	<SCL	–	–	<SCL	–	–	–	–	–
<SCL	–	≪SCL	–	–	–	–	–	–	–	–
–	–	–	–	–	–	–	–	SCL	–	–
–	–	–	–	–	–	–	–	CL	–	–

TABLE XIII (*continued*)

Phages in RTD	$\frac{1}{50}$	$\frac{1}{10}$	$\frac{1}{10000}$	$\frac{1}{10000}$	$\frac{1}{10000}$	$\frac{1}{10000}$	$\frac{1}{500}$	$\frac{1}{100}$	$\frac{1}{1000}$
Phage type	1	204	4	14	3	338	2	390	100
10 London a	CL	<SCL	CL	SCL	<SCL	SCL	CL	−	<SCL
10 Paris b	CL	<SCL	CL	SCL	<SCL	SCL	CL	≪SCL	≪SCL
T. 672 c	CL	<SCL	CL	SCL	SCL	SCL	CL	SCL	<SCL
T. 3501 d	SCL	<SCL	SCL	SCL	SCL	SCL	CL	SCL	<SCL
T. 3399 e	CL	≪SCL	SCL	<SCL	<SCL	SCL	−	SCL	SCL
T. 1669 f	CL	<SCL	−	−	−	<SCL	−	−	−
T. 3368 g	CL	<SCL	SCL	<SCL	<SCL	SCL	CL	SCL	SCL
T. 3644 h	SCL	<CL	SCL	−	−	SCL	CL	−	−
11 London a	SCL	<SCL	SCL	SCL	<SCL	SCL	SCL	SCL	−
11 Paris a	CL	≪SCL	SCL	SCL	≪SCL	SCL	SCL	SCL	−s
12 London a	CL	<SCL	CL	SCL	≪SCL	CL	CL	SCL	−
12 Paris a	CL	<SCL	CL	SCL	≪SCL	CL	CL	SCL	−
13 London a	CL	−	−	−	−	−	−	−	−
13 Paris a	CL	−	−	−	−	−	−	−	−
14 London a	CL	≪SCL	CL	−	≪SCL	SCL	<SCL	≪SCL	<SCL
14 Paris a	CL	<SCL	SCL	−	≪SCL	SCL	<SCL	≪SCL	<SCL
15 London a	SCL	−	<SCL	−	−	2+m	2+m	−m	−
15 Paris b	2+m	≪SCL	−	−	−	−	−	−	−

For explanation of symbols see Section VI. C, 5, page 208.
2+ = + +
3+ = + + +
s, m = plaques of diminishing sizes.
<, ≪ = degrees of lysis somewhat below those indicated by abbreviations.

computer analysis, proposed a new set consisting of 12 phages which divided 265 bacterial strains into 87 patterns. Further sets have been published by Trifonova and Bratoeva (1975) and by Chanishvili and Chanishvili (1975), who noted that particular phage types had a geographical distribution.

The phage typing set of Hammarström has had the most considerable impact on lysotyping of *S. sonnei*. Originally, the set comprised 12 phages designated I–XII. Alone or in conjunction with other phages, the Hammarström phages have been employed in a number of laboratories around the world (Aldová and Suchá, 1973; Giese, 1967; Grunow, 1965a–e; Helmholz, 1966; Junghans, 1958; Kolbe, 1969; Kucharewicz, 1959; László and Kerekes, 1969; Ludford, 1953; Mayr-Harting, 1952; Meitert *et al.*, 1969; van Oye *et al.*, 1968; Rantasalo and Uotila, 1961; Rische *et al.*,

$\frac{1}{100}$	$\frac{1}{300}$	$\frac{1}{100}$	$\frac{1}{300}$	$\frac{4}{10}$	$\frac{1}{500}$	$\frac{1}{300}$	$\frac{1}{30}$	$\frac{1}{10}$	$\frac{1}{300}$	$\frac{1}{300}$
7	8	394	18	68	6	9	210	368	10	13
SCL	÷	SCL	—	—	—	—	—	−m	—	—
SCL	÷	SCL	—	—	—	—	—	≪SCL	—	—
SCL	—	<SCL	CL	CL	—	CL	÷	SCL	—	—
—	—	—	CL	SCL	—	<SCL	≪SCL	2+s	—	—
CL	—	SCL	CL	CL	—	CL	÷	2+m	—	—
CL	—	SCL	SCL	CL	—	CL	—	<SCL	—	—
SCL	−m	SCL	SCL	SCL	—	SCL	—	÷	—	—
≪SCL	—	2+m	—	—	—	—	—	—	—	—
—	—	—	—	—	—	—	—	<SCL	—	—
—	—	—	—	—	—	—	—	<SCL	—	—
SCL	<SCL	SCL	÷	CL	SCL	—	—	—	—	—
SCL	<SCL	SCL	÷	CL	SCL	—	—	—	—	—
—	—	—	—	—	—	—	—	—	—	—
—	—	—	—	—	—	—	—	—	—	—
<SCL	SCL	<SCL	—	—	<SCL	—	—	—	—	—
<SCL	SCL	<SCL	—	—	SCL	—	—	—	—	—
—	—	—	—	—	—	—	—	—	—	SCL

1965; Szturm-Rubinsten, 1957, 1964, 1965, 1972). It also is the basis for current attempts to refine the phage typing procedure for *S. sonnei*.

2. *International typing set*

The phages of Hammarström constitute the core of the proposed international phage typing set for *S. sonnei* (Kallings and Sjöberg, 1975). A basic set is selected from the phages: I, II, III, IV, IVA, V, VI, VII,VIII, VIIIA, IX, X, XI, XII, F12, and γ66. In Table XIV, the morphology, phage serotype, host-propagating strain, relative lytic activity, and source of the 16 phages are indicated. Figures 4 and 5 show plaques and viron morphology of the phages in the proposed set. Several phages are not specific for *S. sonnei* but also attack *Salmonella*, *E. coli*, and *Arizona* (Junghans, 1961). Of particular interest is the fact that phage XII lyses the

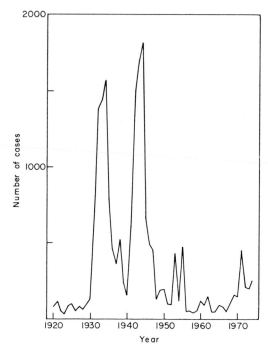

FIG. 3. Incidence of *Shigella* in Sweden.

majority of the phase II strains, although a few phase I strains are also attacked by the phages. Hammarström (1949) mainly used phage XII as a control showing that the bacteria are in phase II, i.e. suitable for typing. Phage XII was not originally intended for inclusion in the phage type patterns.

3. *Media*

(a) *Broth for pregrowth of bacterial inocula*

Nutrient Broth (Difco)   8 g
Distilled water to   1000 ml
pH adjusted to 7·4 after sterilisation

(b) *Agar for typing, determination of RTD, etc.*

Nutrient Broth (Difco)   8 g
Agar (Difco)   15 g
Distilled water to   1000 ml
pH adjusted to 7·4 after sterilisation

TABLE XIV

Properties of typing phages of international typing set of *S. sonnei*

Phage strain designa- tions	Morpho- logical type	Phage sero- type	Host propagating strain	Frequency of lysis	Source
I	B-I 2	I	Lysotype 4 type strain	30	E. Hammarström
II	C-III-1	II	Lysotype 4 type strain	82	E. Hammarström
III	C-I-2, C-II-1	II	Lysotype 4 type strain	45	E. Hammarström
IV	C-II-1, C-III-1	II	Lysotype 5 type strain	78	E. Hammarström
IVA	C-III-1, C-III-1	II	Lysotype 4 type strain	91	E. Hammarström
V	C-II-1	II	Lysotype 5 type strain	94	E. Hammarström
VI	C-II-1	II	Lysotype 4 type strain	76	E. Hammarström
VII	B-II-1	II	Lysotype 4 type strain	62	E. Hammarström
VIII	C-II-1	II	Lysotype 4 type strain	61	E. Hammarström
VIIIA	C-II-1		Lysotype 4 type strain	80	E. Hammarström
IX	B-II-1	III	Lysotype 7 type strain	79	E. Hammarström
X	C-III-1	II	Lysotype 4 type strain	75	E. Hammarström
XI	B-III-1	II	Lysotype 3 type strain	92	E. Hammarström
XII	B-II-1	IV	Lysotype 4 type strain	79	E. Hammarström
F12	A-I-1		9773	86	S. Ślopek
γ66	C-I-1		H 56		H. Rische

[a] Derived from set of Hammarström (1947).
i = morphologically inhomogeneous.
5386 Swedish isolates.
The Hammarström (1949) phages are wild strains for which the origin of each is not known. We know only that they were selected from a collection of 37 phages derived from human (19) and animal (14) faeces, sewage (5), and a lysogenic bacterial strain (1).
References: Krzywy and Ślopek (1974).

Poured in 9 cm diameter plates, 25 ml in each. The plates are left at room temperature overnight till use, preferably on the day after production.

(c) *Storage of bacteria.* The media as described for *S. flexneri* (*vide supra*) or deep agar (1% agar) are suitable.

(d) *Subcultivation media.* For starter cultures before typing, endo-agar is recommended to ensure that the colonies are in phase II.

(e) *Soft agar for quantitation.* As (b), typing agar, with only 0·6% agar.

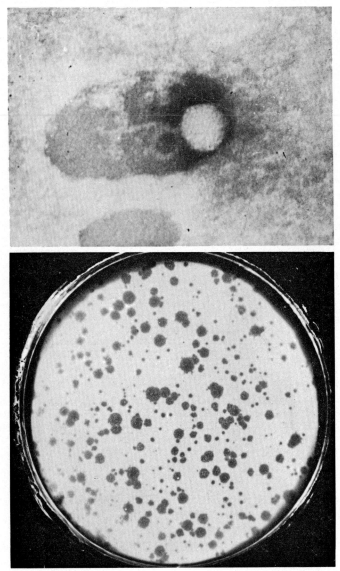

Fig. 4. Plaque morphology and electron microscopy of typing phages of the prcposed international typing set for *S. sonnei*. Preparation of microscopic preparations described by Krzywy and Ślopek (1974). Classified by morphological types.

Phage I for *S. sonnei* (×240,000). Icosahedric head. Tail characteristically coiled. Type B-I-2.

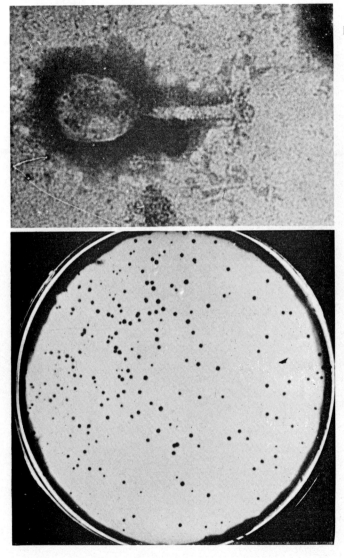

FIG. 4 (*cont.*). Phage II for *S. sonnei* (×240,000). Long fibrils and contracted sheath seen. Type C-III-1.

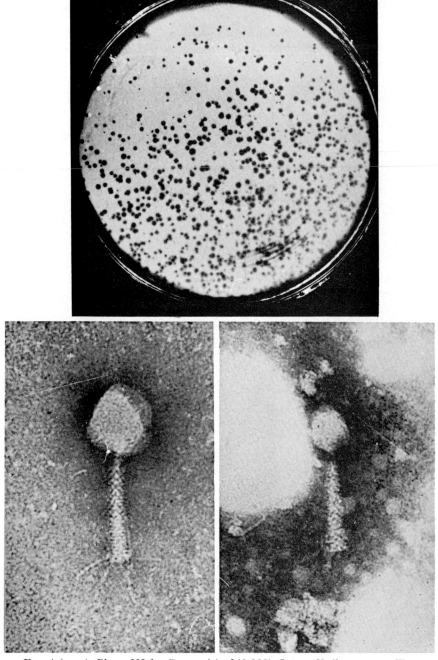

F_IG_. 4 (*cont.*). Phage III for *S. sonnei* (× 240,000). Loose fibril structure. Types C-I-2 and C-II-1.

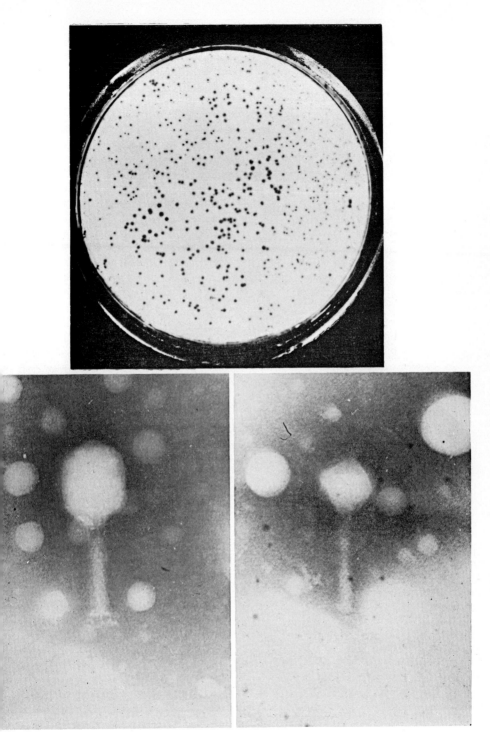

FIG. 4 (*cont.*). Phage IV for *S. sonnei* (× 240,000). Two head shapes, Left figure shows basal plates. Types C-II-1 and C-III-1.

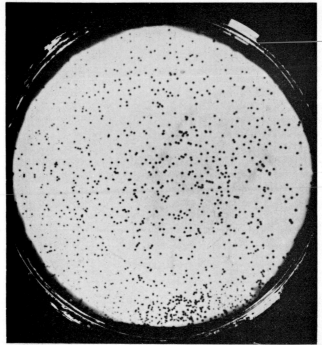

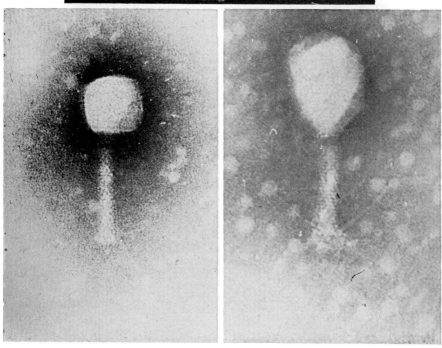

Fig. 4 (*cont.*). Phage IVA for *S. sonnei* (× 240,000). Octahedral head, Long fibrils. Clearly distinct collar. Types C-II-1 and C-III-1.

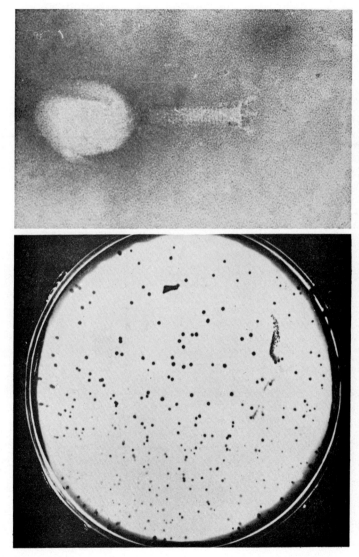

FIG. 4 (*cont.*). Phage V for *S. sonnei* (×240,000). Distinct basal plate. Type C-III-1.

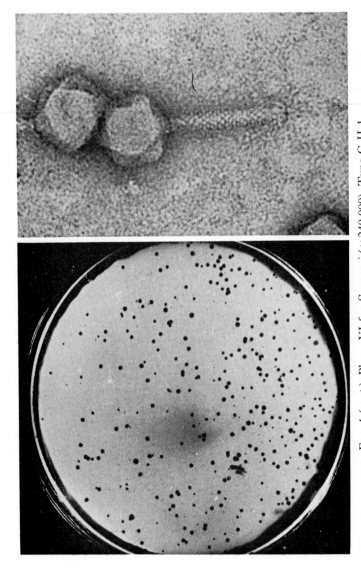

FIG. 4 (*cont.*). Phage VI for *S. sonnei* (×240,000). Type C-II-1.

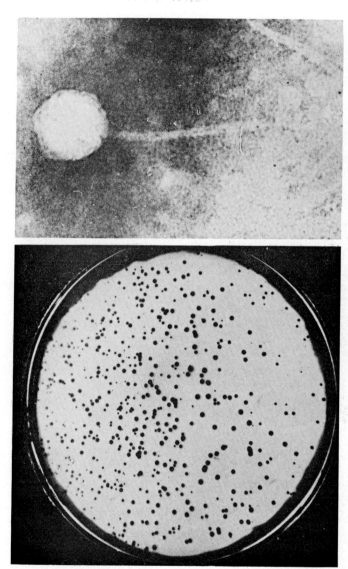

FIG. 4 (*cont.*). Phage VII for *S. sonnei* (×240,000). Octahedral head. Long, thin tail. Canal of tail visible. Type B-II-1.

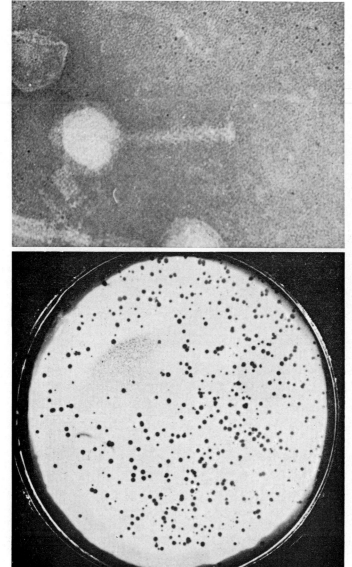

Fig. 4 (cont.). Phage VIII for *S. sonnei* (×240,000). Icosahedral head, short collar and long fibrils. Type C-II-1.

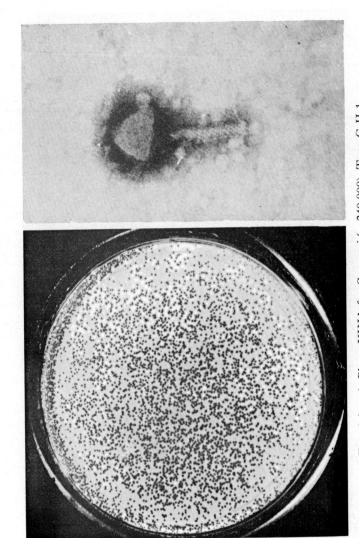

FIG. 4 (*cont.*). Phage VIIIA for *S. sonnei* (×240,000). Type C-II-1.

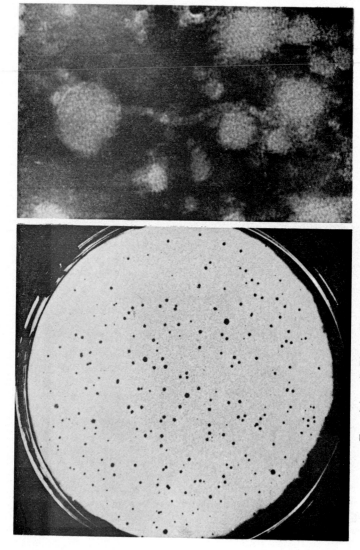

FIG. 4 *(cont.)*. Phage IX for *S. sonnei* (×240,000). Type B-II-1.

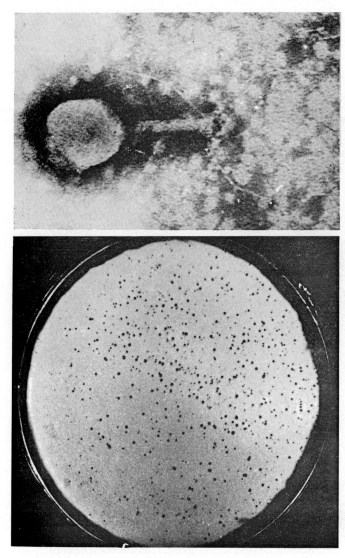

FIG. 4 (*cont.*). Phage X for *S. sonnei* (×240,000). Collar, sheath, basal plate and long fibrils visible. Type C-III-1.

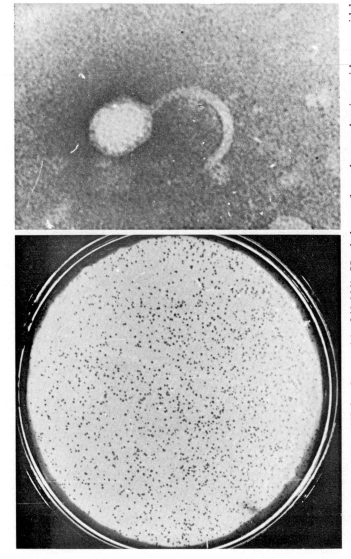

Fig. 4 (*cont.*). Phage XI for *S. sonnei* (×240,000). Head shape of an elongated prism with pyramidal ends. Curved tail. Type B-III-1.

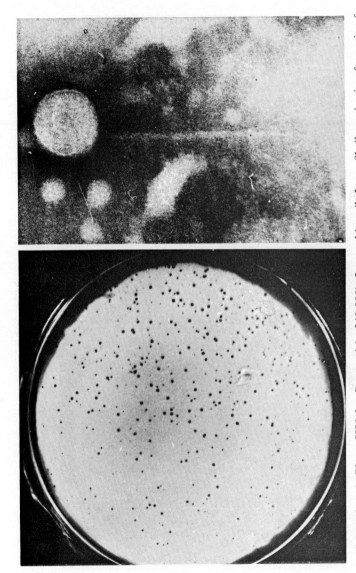

FIG. 4 (cont.). Phage XII for S. sonnei (×240,000). Long, thin tail. Long fibrils projecting from tip of tail. Central core sticking out of sheath. Type B-II-1.

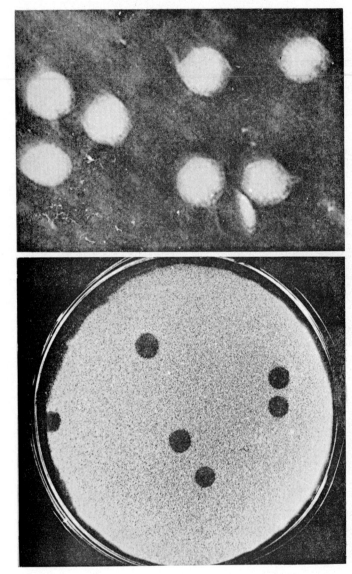

FIG. 4 (*cont.*). Phage F12 in typing set for *S. sonnei*. Originally from *S. flexneri* (×240,000). Type A-I-1.

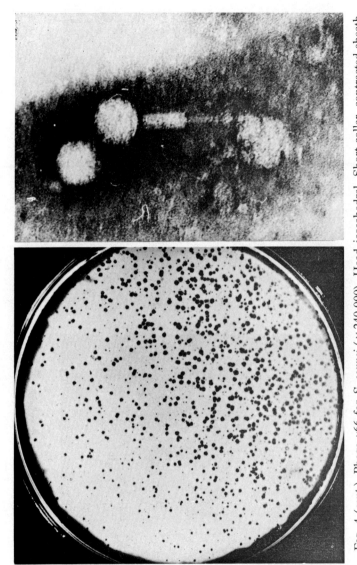

FIG. 4 (*cont.*). Phage γ66 for *S. sonnei* (×240,000). Head icosahedral. Short collar, contracted sheath. Canal in core seen. Type C-I-1.

FIG. 5. Plaque morphology of *S. sonnei* phages propagated at the reference laboratory for phage typing of *Shigella sonnei*. *S. sonnei*, phage I.

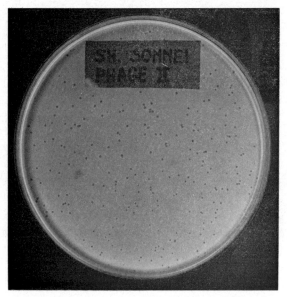

FIG. 5 (*cont.*). *Shigella sonnei*, phage II.

Fig. 5 (*cont.*). *Shigella sonnei*, phage III.

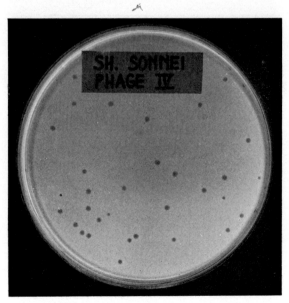

Fig. 5 (*cont.*). *Shigella sonnei* phage IV.

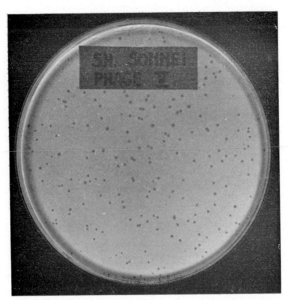

Fɪɢ. 5 (cont.). *Shigella sonnei*, phage V.

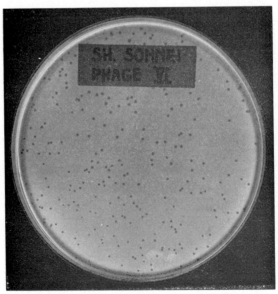

Fɪɢ. 5 (cont.). *Shigella sonnei*, phage VI.

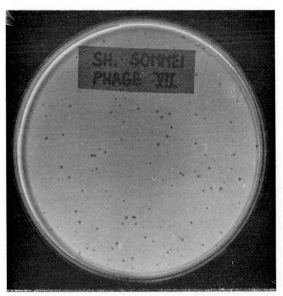

Fig. 5 (*cont.*). *Shigella sonnei*, phage VII)

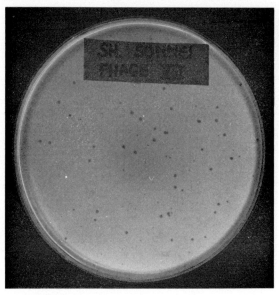

Fig. 5 (*cont.*). *Shigella sonnei*, phage VIII.

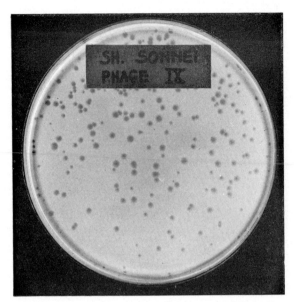

Fig. 5 (*cont.*). *Shigella sonnei*, phage IX.

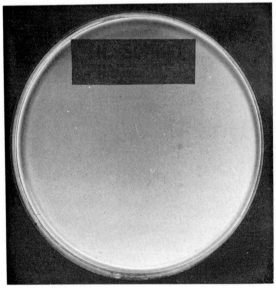

Fig. 5 (*cont.*). *Shigella sonnei*, phage X.

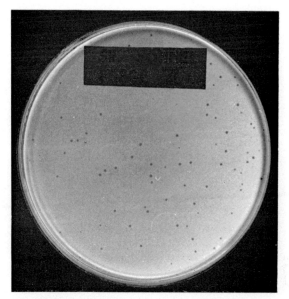

FIG. 5 (*cont.*). *Shigella sonnei*, phage XI.

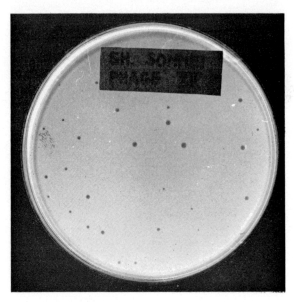

FIG. 5 (*cont.*). *Shigella sonnei*, phage XII.

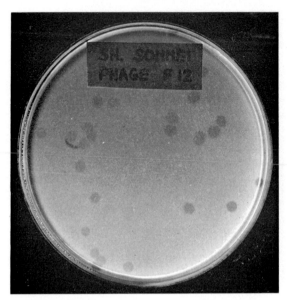

FIG. 5 (*cont.*). *Shigella sonnei*, phage F 12.

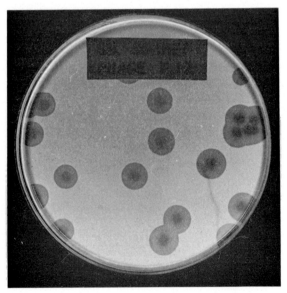

FIG. 5 (*cont.*). *Shigella sonnei*, phage F 12.

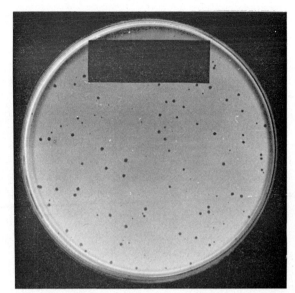

FIG. 5 (*cont.*). *Shigella sonnei*, phage γ 66.

4. Propagation of phages

(a) *Passage in broth*. On day 1, one volume of bacteria in log-phase is mixed with one volume of phage dilution (multiplicity of infection 1/200) and incubated in a shaker at 37°C for 10 min. Thereafter 18 ml of pre-warmed broth are added. After 4 h, the culture is lysed and the lysate is incubated at 4°C overnight. The following day it is centrifuged at 10,000 × *g* for 20 min at 4°C. The supernatant is filtered through a 0·22 μm Millipore or equivalent filter. The filtrate represents the phage stock.

(b) *Passage on agar*. On day 1, 5 ml of log phase bacteria and 5 ml of diluted phages at a multiplicity of infection of 1/200 are mixed and incubated in a shaker incubator at 37°C for 10 min when 90 ml nutrient agar with 0·6% agar at 45°C are added. Portions of 8 ml are plated by spreading on top of twelve 14 cm diameter plates containing 1·5% agar. The plates are incubated at 37°C overnight.

The next day the plates are flooded by 10 ml of broth. The top layer is scraped off, harvested by suction and subjected to a thorough shaking until fairly homogeneous. Subsequently, the material is centrifuged and filtered as above for broth passage.

5. Determination of strength and lytic spectra

For quantitation, 10-fold dilutions (with change of pipettes) are made in nutrient broth, and assayed by agar overnight plates. To 0·1 ml of each phage dilution are added (i) 0·25 ml of the homologous propagating strains (4-h broth culture agitated at 37°C), and (ii) 2 ml of melted soft agar at 45°C (heat block is suitable).

The ingredients are quickly mixed (air bubbles avoided) and the whole volume spread on the surface of a 9 cm diameter nutrient agar plate. The plates are incubated overnight at 37°C.

The RTD is determined on agar prepared by flooding with overnight undiluted broth cultures (Ziesché and Rische use 3-h broth cultures) for inoculation. After removal of excess liquid by suction, the plates are dried at 37°C without lids for 15–30 min. Ten-fold dilutions of the phages (10^{-1}–10^{-8}) are spotted on the homologous host-propagating strains by the help of the same nichrome coiled eye wire as is used in the multi-inoculator typing apparatus.

Following phage application, the plates are dried at room temperature and incubated at 37°C in the upright position for 5 h. RTD is adjusted by making intermediate dilution steps between the points where decreasing activity against the homologous strain is noticed. These intermediate dilutions (e.g. 1×10^{-3}; 0.75×10^{-3}; 0.5×10^{-3}; 0.25×10^{-3}; 1×10^{-4}) are tested against a limited set of type strains with varying sensitivity, *S. sonnei* lysotype standard strains 3, 4, 5, 7, 22, 28, 53, 58, 61, 67, 68, and 96.

Before new preparations are used in routine typing, their lytic spectra are checked by typing on all the 96 characterized types. The appearance of the reactions on each strain is important. The reactions of some key strains are shown in Table XV and Fig. 6. The maintenance (stability) of RTD is checked at monthly intervals by repeat typing on the limited set of control bacteria.

Reactions are scored as for *S. flexneri*. Common results with fresh preparations of phages and the relationship between RTD and CD are shown in Table XVI.

The number of phages per millilitre necessary to give RTD varies with plaque size. Hammarström (1949) demonstrated how plaque size compared with the upper and lower number of plaques per reaction (Table XVII).

The RTD is not the same as the CD, although the two are closely related. The number of plaque-forming units per millilitre varies with the type and even production batch of the same brand of medium (Pruneda and Farmer, 1977).

6. Typing procedure

From agar plates (endo-plates are said to be better for selecting right phase (Ziesché and Rische, 1973)), material from the edge of single, isolated phase II colonies is transferred to two tubes of 4 ml nutrient broth. If a strain shows evidence of differences in colony morphology, one broth

TABLE XV

Patterns of control strains of *S. sonnei* phage typing set

Test strain	I	II	III	IV	V	VI	VII	VIII	IX	X	XI	XII
2	CL	CL	CL	SCL	SCL	SCL	SCL	CL	CL	+	CL	CL
3	–	CL	SCL	SCL	SCL	SCL	SCL	CL	CL	+	CL	CL
5	–	–	CL	SCL	SCL	SCL	SCL	CL	SCL	+	CL	SCL
6	CL	CL	–	CL	SCL	SCL	CL	CL	CL	+	CL	CL
7	–	CL	–	CL	SCL	CL	CL	SCL	CL	+	CL	CL
9	–	SCL	CL	SCL	–	SCL	–	CL	SCL	–	CL	SCL
13	–	–	–	SCL	SCL	SCL	SCL	CL	SCL	+	CL	SCL
15	–	–	CL	SCL	–	SCL	CL	CL	CL	+	CL	CL
17A	–	CL	–	–	SCL	CL	SCL	CL	CL	+	CL	CL
19	–	CL	–	CL	++	–	–	CL	–	+^{m}	SCL	SCL
20	–	CL	CL	CL	+	CL	SCL	CL	CL	+	CL	–
23	–	CL	–	CL	SCL	SCL	+	CL	CL	+	CL	CL
25	–	CL+	CL	CL	SCL	–	+	SCL	CL	+	CL	SCL
28	–	+	–	SCL	±	SCL	SCL	CL	CL	–	CL	CL
30	SCL	CL	–	–	SCL	–	–	–	–	SCL	SCL	–
32	–	CL	CL	SCL	++	SCL	–	CL	–	+	CL	CL
33	CL	CL	–	SCL	SCL	++	SCL	CL	–	+	SCL	SCL
35	CL	CL	–	+	SCL	SCL	–	SCL	SCL	+^{m}	SCL	CL
36	–	SCL	–	–	SCL	CL	SCL	CL	CL	++	CL	–
37	–	–	–	–	–	–	CL	CL	CL	+	SCL	–
42	–	–	–	–	–	–	CL	–	CL	–	CL	–
46	–	–	–	–	–	–	–	CL	–	SCL	SCL	SCL
55	–	CL	–	–	SCL	–	–	–	–	SCL	CL	CL
66	–	+	–	–	SCL	CL	–	–	–	SCL	CL	SCL
77	–	CL	–	–	–	CL	–	CL	–	–	CL	–
86	–	CL	–	SCL	SCL	–	–	–	–	+	–	–
89	–	–	–	–	SCL	SCL	–	–	–	+	CL	–
94	–	–	–	–	++	–	–	–	–	–	CL	–

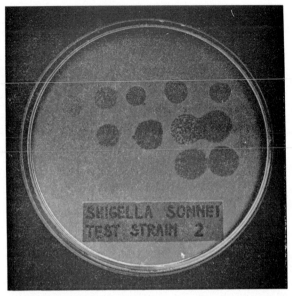

Fig. 6. Lytic patterns of proper phage dilutions, routine test dilutions (= RTD) on selected control test strains of *Shigella sonnei* (see Table XV) as carried out at the reference laboratory for phage typing of *Shigella sonnei*. *S. sonnei* test strain 2.

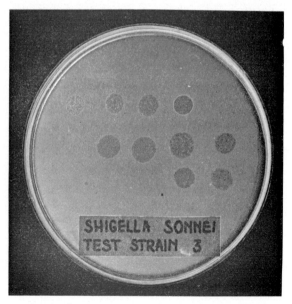

Fig. 6 (*cont.*). *Shigella sonnei*, test strain 3.

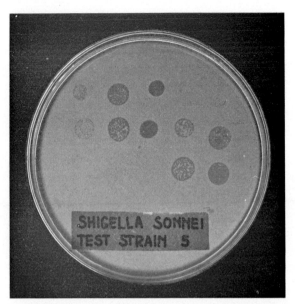

FIG. 6 (*cont.*). *Shigella sonnei*, test strain 5.

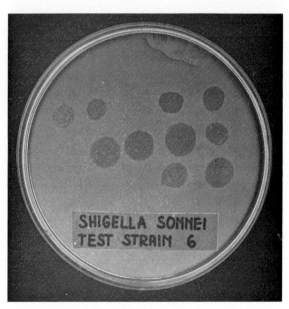

FIG. 6 (*cont.*). *Shigella sonnei*, test strain 6.

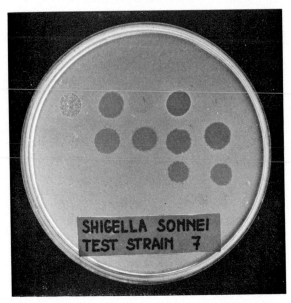

FIG. 6 (*cont.*). *Shigella sonnei*, test strain 7.

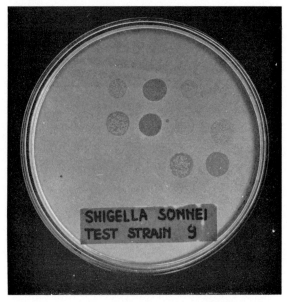

FIG. 6 (*cont.*). *Shigella sonnei*, test strain 9.

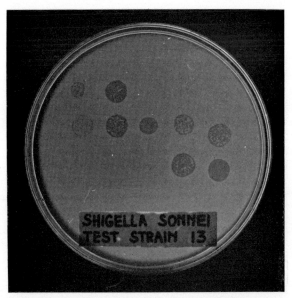

FIG. 6 (*cont.*). *Shigella sonnei*, test strain 13.

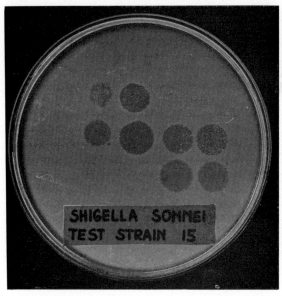

FIG. 6 (*cont.*). *Shigella sonnei*, test strain 15.

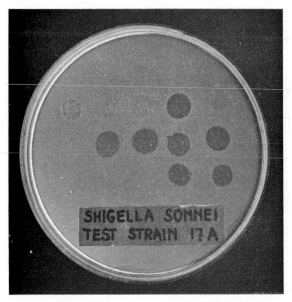

FIG. 6 (*cont.*). *Shigella sonnei*, test strain 17A.

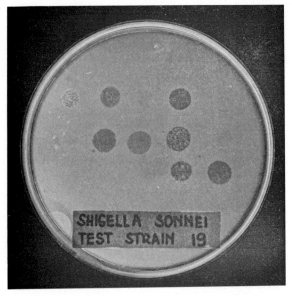

FIG. 6 (*cont.*). *Shigella sonnei*, test strain 19.

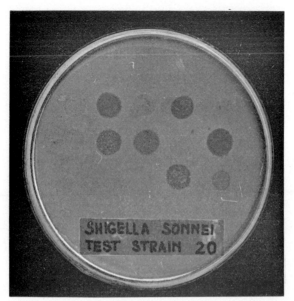

FIG. 6 (*cont.*). *Shigella sonnei*, test strain 20.

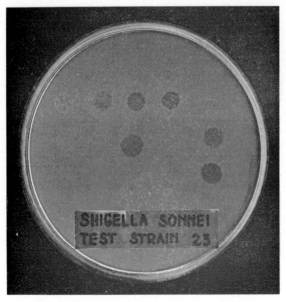

FIG. 6 (*cont.*). *Shigella sonnei*, test strain 23.

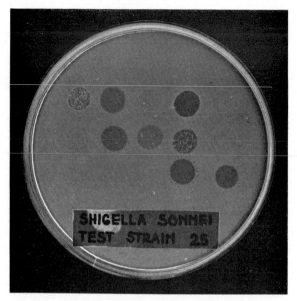

Fig. 6 (*cont.*). *Shigella sonnei*, test strain 25.

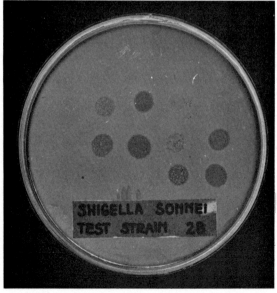

Fig. 6 (*cont.*). *Shigella sonnei*, test strain 28.

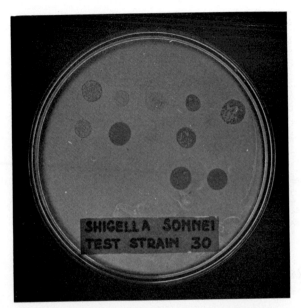

FIG. 6 (*cont.*). *Shigella sonnei*, test strain 30.

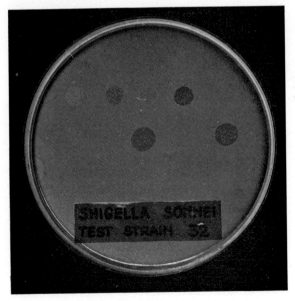

FIG. 6 (*cont.*). *Shigella sonnei*, test strain 32.

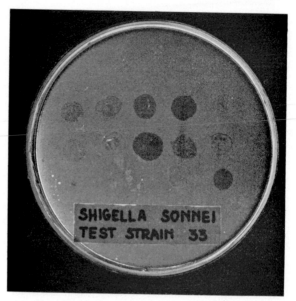

Fig. 6 (*cont.*). *Shigella sonnei*, test strain 33.

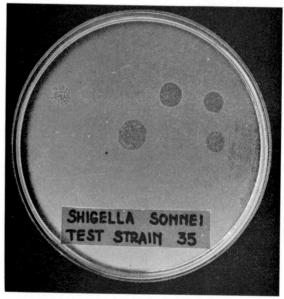

Fig. 6 (*cont.*). *Shigella sonnei*, test strain 35.

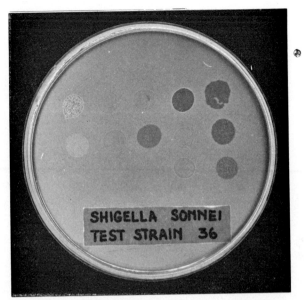

Fig. 6 (*cont.*). *Shigella sonnei*, test strain 36.

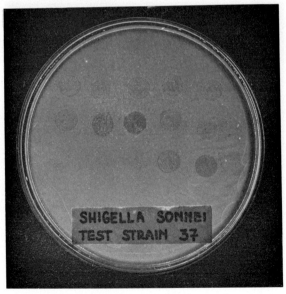

Fig. 6 (*cont.*). *Shigella sonnei*, test strain 37.

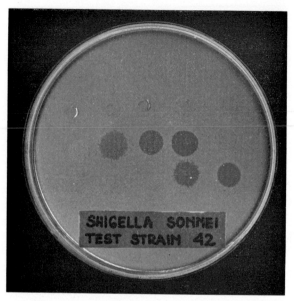

FIG. 6 (*cont.*). *Shigella sonnei*, test strain 42.

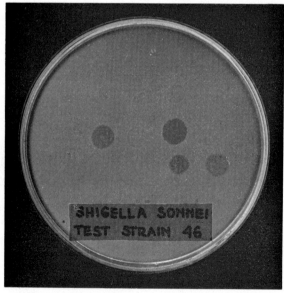

FIG. 6 (*cont.*). *Shigella sonnei*, test strain 46.

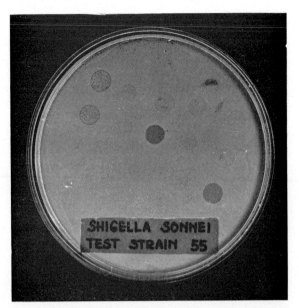

FIG. 6 (*cont.*). *Shigella sonnei*, test strain 55.

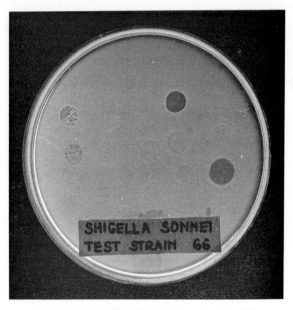

FIG. 6 (*cont.*). *Shigella sonnei*, test strain 66.

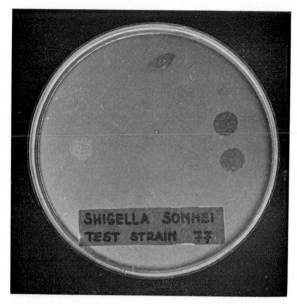

FIG. 6 (*cont.*). *Shigella sonnei*, test strain 77.

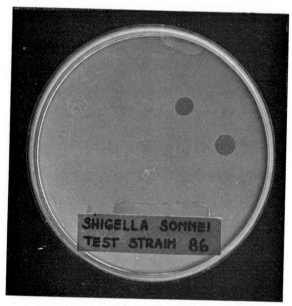

FIG. 6 (*cont.*). *Shigella sonnei*, test strain 86.

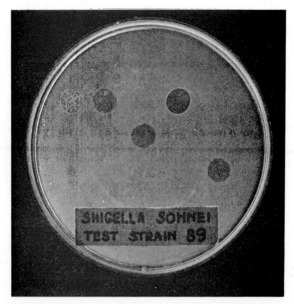

FIG. 6 (*cont.*). *Shigella sonnei*, test strain 89.

FIG. 6 (*cont.*). *Shigella sonnei*, test strain 94.

TABLE XVI

Critical dilutions (CD) and routine test dilutions (RTD) of
***S. sonnei* typing phages**

Phage designations	Ten-fold dilution of freshly prepared phage suspension commonly yielding		Relative dilution of CD suspension to give RTD
	CD	RTD	
I	10^{-3}	10^{-2}	$\times 10^{-1}$
II	10^{-3}	10^{-3}	Undiluted
III	10^{-2}	10^{-1}	$\times 10^{-1}$
IV	10^{-4}	10^{-3}	$\times 10^{-1}$
IVA			
V	10^{-3}	10^{-2}	$\times 10^{-1}$
VI	10^{-3}	10^{-2}	$\times 10^{-1}$
VII	10^{-4}	10^{-3}	$\times 10^{-1}$
VIII	10^{-3}	10^{-3}	Undiluted
VIIIA			
IX	10^{-4}	10^{-4}	Undiluted
X	10^{-2}	10^{-1}	$\times 10^{-1}$
XI	10^{-4}	10^{-3}	$\times 10^{-1}$
XII	10^{-4}	10^{-4}	Undiluted
F12	10^{-6}	10^{-6}	Undiluted
$\gamma 66$	10^{-2}	10^{-1}	$\times 10^{-1}$

culture is made from each colony type. The tubes are incubated overnight at 37°C.

On the second day, the colonies are thoroughly mixed. Then, 9 ml diameter plates are flooded, excess fluid taken off, and the plates dried at 37°C for 10–15 min with the lids off.

The RTD bacteriophage preparations are applied with, for example, the semi-automatic phage applicator of Lidwell (1959) (Fig. 2).

TABLE XVII

Plaque count giving plaques of size compatible with routine test dilution (Hammarström, 1949)

Plaque size (mm)	Plaque number per drop zone	
	Lower limit	Upper limit
0·05	1000	6000
0·1	100	1500
0·2	50	400
$\geqslant 0·3$	20	150

After phage application, plates are dried at room temperature without lids for 10–15 min before incubation at 37°C (to 38·5°C) for 5 h with the agar down and the lids up.

7. *Recording of results*

Phage typing patterns have been recorded as phage types (Table XVIII), although it is now preferred to report patterns instead of types. Variation in one single phage reaction has the unfortunate effect that the bacterium becomes referred to a completely different type. Reporting patterns makes it easier for the recipient of the report to see slight differences in the reaction and thereby better include in the interpretation of results the unavoidable experimental error or details of the method.

The total number of phage patterns observed is very large. Among 5386 cultures from world-wide sources, 3223 different lytic patterns have been noted when any difference is accepted. Among these, a very substantial number have been found only in one or a few strains. Only about $\frac{1}{4}$ occur with any significant frequency.

Such large data masses are difficult to assess unless modern computer technology is utilised. This has been done at the National Bacteriological Laboratory, Stockholm, for all cumulative typing results at the laboratory, including the data of Hammarström (Kallings and Sjöberg, 1975). For this purpose, the phage reactions were coded as follows:

Phage reaction	Computer score
Unknown	0
Negative	1
Less than 40 normal plaques ($+$, $\pm$)	2
Less than 40 large plaques ($+^{\mathrm{L}}$, $\pm^{\mathrm{L}}$)[a]	3
40–80 plaques ($++$)	4
More than 80 plaques ($+++$)	5
Semiconfluent lysis (SCL)	6
Opaque lysis (OL)	7
Confluent lysis (CL)	8

[a] L = large plaques.

Each strain is identified by accession number, source, sampling data, geographical origin (by numerical code), and the reaction strength of each typing phage.

The lytic patterns of 3178 cultures have been compared by computer to see which differences in pattern reactions could be expected within the framework of the errors inherent in the typing technique itself. Table

TABLE XVIII

Phage types of *S. sonnei* according to Hammarström (1949) with extensions

Serotype	Group	I	II	III	IV	V	VI	VII	VIII	IX	X	XI	XII
2	0	CL	CL	CL	CL	CL	CL	CL	CL	CL	CL	CL	CL
3	1	–	CL 0·1 1700	CL	CL	CL	CL	CL	CL	CL	CL	CL	CL
91		CL	CL	CL	CL	CL	CL	CL	CL	CL	CL	CL	CL
6		CL	CL	–	CL	CL	CL	CL	CL	CL	CL	CL	CL
4		CL	CL	CL	CL 0·1 1700	CL	CL	CL	CL	CL	CL	CL	CL
5	2	–	–	CL	CL	CL	CL	CL	CL	CL	CL	CL	CL
7		–	CL	–	CL	CL	CL	CL	CL	CL	CL	CL	CL
9		–	CL	CL	CL	CL	CL	–	CL	CL	CL	CL	CL
10		–	CL	CL	CL	CL	CL	CL	CL	CL 0·05 (1000)	CL	CL	CL
82		(×)	CL	CL	CL	CL	CL	CL	CL	CL	–	CL	CL
69		CL	CL 0·1 2000	CL	CL	CL	CL	CL	CL	CL	CL	–	CL
(14)		CL	CL	+	–	CL	CL	CL	CL	CL	CL	CL	CL
(14a)		CL	+ +	±	–	CL	CL	CL	CL	CL	CL	CL	CL
12		CL	CL	±	–	CL	CL	CL	CL	CL	CL	CL	CL
13		–	–	–	CL	CL	CL	CL	CL	CL	CL	CL	CL
15		–	0·05 300	CL	CL	–	CL	CL	CL	CL	CL	CL	CL

	1	2	3	4	5	6	7	8	9	10	11	12
90	CL	+	+	CL	CL	CL	CL	0.5 6000	CL	CL	–	–
70	CL	CL	CL	CL	CL	0.1 100	CL	CL	CL	CL	–	–
16	CL	–	CL	CL	CL	CL	CL	CL	CL	CL	–	–
17	CL	CL	CL	CL	CL	CL	CL	CL	–	–	CL	0.6 (20)
17a	CL	CL	CL	CL	CL	CL	CL	CL	0.1 500	–	CL	–
19	CL	CL	CL	CL	CL	CL	–	CL	CL	–	CL	–
85	CL	CL	CL	+	CL	CL	0.1 300	CL	CL	–	CL	–
20	CL	CL	CL	+	CL	0.05 (1000)	CL	CL	CL	–	CL	–
(21)	CL	CL	CL	CL	–	CL	CL	CL	CL	–	CL	–
22	CL	CL	CL	+	CL	–	CL	CL	–	+	CL	–
76	CL	CL	CL	CL	CL	0.1 100	CL	CL	–	CL	CL	–
33	0.8 20	CL	CL	–	CL	CL	CL	CL	0.1 1700	CL	CL	–
23	CL	CL	CL	–	CL	–	CL	CL	CL	CL	CL	–
30	CL	CL	CL	CL	CL	CL	–	CL	–	–	CL	CL
95	150	CL	CL	CL	–	0.05 3000	CL	CL	CL	–	CL	CL
(24)	CL	CL	CL	CL	CL	CL	CL	CL	–	–	–	0.4 (40)
25	CL	CL	CL	CL	CL	CL	–	CL	CL	–	–	–
26	CL	CL	CL	CL	CL	–	CL	CL	–	–	–	–
27	CL	CL	CL	CL	CL	CL	CL	–	0.1 1700	CL	0.05 2000	–

TABLE XVIII (*continued*)

Serotype	Group	Phage type											
		I	II	III	IV	V	VI	VII	VIII	IX	X	XI	XII
28		—	0·01 1000	CL	CL	0·3 150	CL	++	CL	CL	—	CL	(++)
81		150	—	CL	CL	0·2 200	CL	CL	CL	CL	—	CL	CL
29	4	—	—	CL	CL	CL	CL	—	CL	—	CL	CL	CL
(61)		—	++	—	—	CL	—	CL	CL	CL	CL	CL	CL
(31)		—	CL	—	—	CL	CL	CL	CL	CL	CL	0·05 6000	CL
87		—	CL	—	CL	CL	—	CL	—	CL	CL	CL	CL
(32)		—	CL	—	CL	CL	CL	—	CL	—	CL	CL	CL
34	CL	CL	0·05 2000	—	—	—	CL	CL	CL	CL	CL	CL	CL
75	CL	CL	CL	—	—	CL	—	CL	—	CL	CL	CL	CL
35	CL	CL	CL	—	—	CL	CL	0·2 400	CL	0·2 500	CL	CL	CL
36	CL	CL	CL	—	—	CL	CL	—	CL	—	CL	CL	CL
37		—	—	—	—	CL	—	CL	CL	CL	CL	CL	CL
38		—	0·1 2000	—	—	CL	CL	CL	—	CL	CL	CL	CL
39		—	0·05 2000	—	CL	—	—	CL	CL	CL	CL	CL	CL
62		—	0·1 1200	—	CL	0·3 125	CL	CL	CL	CL	—	CL	CL
(63)		— 1400	0·1	—	CL	CL	—	CL	—	CL	CL	CL	CL

(41)		—	—	CL	CL	CL	CL	0.3 / 125	CL	CL	0.2 / 400	—
79	5	CL	CL	CL	—	CL	—	CL	—	CL	CL	—
84		—	CL	CL	CL	CL	—	CL	—	CL	CL	—
78		CL	CL	—	CL	—	CL	CL	0.1 / 500	—	CL	—
89		CL	CL	—	CL	—	—	CL	CL	—	CL	—
51		—	—	CL	CL	CL	CL	1.3 / 100	—	—	—	CL
96		CL	CL	CL	—	CL	—	CL	—	—	0.1 / 300	× + +
(42)		CL	CL	CL	CL	CL	—	—	—	—	0.05 / 2000	—
(43)	6	CL	CL	—	CL	CL	—	—	—	—	0.05 / 2000	—
44		CL	—	CL	CL	CL	CL	0.3 / 125	—	—	0.1 / 1000	×
45		CL	—	+	CL	CL	CL	0.3 / 125	—	—	—	—
46		CL	—	CL	CL	CL	CL	—	—	—	—	—
74		CL	—	CL	CL	CL	CL	0.3 / 150	—	0.05 / 1000	0.2 / 150	—
47		CL	+	CL	CL	0.05 (1000) CL	—	CL	—	—	—	×
(92)		CL	CL	CL	—	CL	—	CL	0.3 / 150	—	0.1 / 1700	×
93		CL	CL	CL	—	CL	—	CL	—	—	0.1 / 1700	×
(48)		CL	—	CL	CL	CL	—	CL	—	—	—	—
83		CL	0.05 / 25	CL	CL	—	+	CL	—	—	CL	—

TABLE XVIII (continued)

Serotype	Group	I	II	III	IV	V	VI	VII	VIII	IX	X	XI	XII
								Phage type					
(41)		—	CL	—	—	CL	CL	0·05 / 6000	—	0·05 / 6000	CL	CL	CL
(50)		—	CL	—	—	CL	CL	—	—	—	CL	CL	0·8 / 60
(65)		—	—	—	—	—	±	—	—	+	—	—	×
(52)		—	0·05 / 2000	0·1 / 1400	0·1 / 50	—	0·1 / 1400	CL	0·05 / 6000	CL	CL	CL	CL
58		—	0·05 / (1000)	0·05 / 4000	—	—	—	CL	0·05 / 5000	CL	CL	CL	0·8 / 60
53	7	—	0·05 / 2000	—	—	—	0·1 / 1400	CL	—	CL	CL	CL	CL
(64)		—	—	0·05 / 6000	—	—	0·05 / (1000)	CL	+	CL	CL	CL	0·8 / 30
71		—	0·1 / 1000	—	—	—	0·05 / 3000	CL	—	CL	CL	CL	CL
73		—	0·05 / 1000	—	—	—	0·05 / 2000	CL	0·05 / 2000	CL	CL	CL	CL
(54)		—	0·1 / 1000	—	—	0·3 / 100	CL	CL	—	CL	—	CL	CL
55		—	—	—	—	CL	—	—	CL	—	CL	CL	CL
(56)		—	CL	—	—	CL	0·1 / 1200	—	—	—	CL	CL	CL
66		—	CL	—	—	CL	—	—	—	—	CL	CL	×
88		—	CL	—	—	CL	(1000)	0·05 / 6000	—	0·05 / 6000	CL	+	CL
67		—	CL	—	—	CL	±	—	—	—	CL	—	++

(60)	—	—	—	—	—	—	CL	0·05 / 6000	CL	0·05 / 6000	CL	0·8 / 50
												CL
(57)	—	—	—	—	—	0·1 / 1200	CL	—	CL	—	CL	CL
77	—	—	—	—	0·3 / 100	CL	—	—	—	CL	CL	CL
(59)	—	—	—	—	CL	CL	—	CL	—	0·05 (1000)	CL	CL
72 (8)	CL	CL	—	—	CL	—	—	—	—	CL	0·4 / 100	CL
86	CL	CL	—	—	CL	CL	—	—	—	—	—	CL
(94)	—	—	—	—	0·3 / 100	CL	—	—	—	—	CL	CL
(68)	—	—	—	—	CL	—	—	—	0·05 (1000)	+ +	0·05 (1000)	CL
80 (9)	—	—	—	—	CL	—	—	—	—	—	CL	0·3 / 20

— = Negative reaction or reaction with number of plaques less than 100 for plaque size under 0·05 mm and less than ten for plaque size of 0·1 mm or larger.

± = Negative reaction with weak growth inhibition.

± ± = Reaction with less than semiconfluent lysis (<SCL).

+ = Single plaques.

+ + = Semiconfluent lysis (SCL).

CL = Confluent lysis.

The numerical entries are one upper and one lower for each reaction. The upper ones, with decimal points, indicate the plaque diameter in millimetres. The lower figures indicate the approximate number of plaques (at RTD). For entries in parentheses, negative reactions are the rule, but some production batches show the reactions indicated.

× = Reaction with single plaques having a peripheral lytic zone.

TABLE XIX

Scheme for accepted differences from ideal phage reaction ($\times$) according to phage typing scheme

Phage score of the one culture	Phage score of the other culture							
	CL	OL	SCL	$+++$	$++$	$+^{L}, \pm^{L}$	$+, \pm$	$-$
CL		$\times^{a}$	$\times$	$\times$				
OL	$\times$		$\times$	$\times$				
SCL	$\times$	$\times$		$\times$	$\times$	$\times$		
$+++$	$\times$	$\times$	$\times$		$\times$			
$++$			$\times$	$\times$			$\times$	
$+^{L}, \pm^{L}$			$\times$				$\times$	
$+, \pm$					$\times$	$\times$		$\times$
$-$							$\times$	

Phage reaction scores: CL, confluent lysis; OL, opaque lysis; SCL, semiconfluent lysis; $+++$, >80 isolated plaques; $++$, 40–80 isolated plaques; $+^{L}$, $\pm^{L}$, <40 isolated plaques, large; $+$, $\pm$, <40 isolated plaques.

^{a} The $\times$ symbols signify compared phage reactions which are not differentiated, i.e. not big enough to be considered significant.

XIX demonstrates the principle used in comparing two typing patterns, e.g. of the same strain typed twice or two isolates from the same epidemic. A difference is recorded as significant only if combinations of reactions corresponding to the boxes in Table XIX which are not marked by an $\times$ appear. The reaction of the one isolate is found along the vertical axis and the other strain along the horizontal axis. If, then, one strain has a CL-reaction and the other a SCL-reaction, this combination corresponds to a $\times$ in Table XIX and the difference is not considered significant. However, if the one has CL and the other a $++$ reaction, the difference in reactions is big enough to be recorded as significant.

Table XX shows what the consequences would be of accepting differences between phage typing patterns at different levels of differences for 3178 typing patterns. If any difference, however small, is regarded as a real difference, the 3178 strains could be divided into 2156 different patterns. This corresponds to N8–8 levels of differences—in Table XX (cf. Table XXI). If only two levels of differences are employed, there are alternatives as to what may be regarded as insignificant differences. In example N2A, the reactions $-$, $+$, $+^{L}$, and $++$ are all regarded as sufficiently similar to be referred to the same code, whereas the reactions $+++$, SCL, OL, and CL are given the same rank and belong to the other category. This corresponds to the example in Table XX. Analogously, the

TABLE XX

Example of computer analysis of some phage patterns of cultures from different countries when discrimination alternative N8 is used to differentiate between phage typing patterns of 3178 isolates

I	II	III	IV	IVA	V	VI	VII	VIII	VIIIA	IX	X	XI	XII	F12	γ66	No. of cultures	Isolated in
1	8	1	1	6	6	1	1	1	1	1	4	8	1	3	1	50	Hungary, Japan, Poland, Roumania
1	8	1	1	6	6	1	1	1	1	1	6	8	1	1	1	39	Japan
1	8	8	8	6	8	8	8	8	8	8	6	8	8	6	4	28	USA, France
4	8	8	6	8	8	8	8	6	6	8	6	8	8	8	6	27	Czechoslovakia, USSR
6	8	8	6	8	8	8	8	6	6	8	6	8	8	8	6	27	USSR
1	6	1	6	6	8	8	8	4	6	8	2	2	6	3	4	19	Israel
1	8	1	4	8	8	2	1	1	1	1	6	8	1	6	6	18	Czechoslovakia, USSR
1	8	1	1	6	6	8	1	1	1	1	6	8	1	1	1	16	Japan
2	8	8	6	8	8	8	8	6	6	8	6	8	8	8	6	16	USSR

Phage reactions code per typing phage[a]

[a] Phage pattern coded according to definition of N8 in Table XXI (e.g. 1 corresponds to a negative reaction, 8 to confluent lysis and so on).

example N2B considers the reactions – and + as belonging to one and the same category, and the remaining reaction strength codes to the other category so that no difference is recorded for any reactions ranging from $+^L$ to CL. The number of different patterns in the alternative N2A is 673 and with the alternative N2B there are 481 different patterns.

The most important patterns ensuing with the discrimination alternatives with codes N8 are indicated in Table XXI and with the discrimination alternative N2A in Table XXII. In the body of the tables each reaction level as defined in Table XX is indicated by a numerical code. It is seen that several phage patterns are represented by strains from more than one

TABLE XXI

Consequences of recognising various phage reaction differences as significant when phage patterns of 3178 strains of *S. sonnei* are compared at different levels of discrimination between patterns

Discrimination alternative A: $N8^a$ = all degrees of phage reaction differences are accepted as discriminating between patterns. The eight levels of discrimination obtained with N8 are

N8: 1 –
    2 +
    3 $+^L$
    4 + +
    5 + + +
    6 SCL
    7 OL
    8 CL

With N8, the 3178 cultures are divided into 2156 patterns.

Discrimination alternative B: $N2A$ = only strong reaction differences are accepted as discriminating between patterns. The strong reactions (more than 80 plaques, i.e. + + +, SCL, OL, and CL) are combined into one level of type reaction discrimination, the other patterns are referred to the other level:

N2A: 1 –, +, $+^L$, + +
     2 + + +, SCL, OL, CL

With N2A, the 3178 cultures are divided into 673 patterns.

Discrimination alternative C: $N2B$ = all reactions of more than 40 plaques are combined at the one level of discrimination (positive reaction) and the remainder are combined at the other level of discrimination (negative reaction)

N2B: 1 –, +
     2 $+^L$, + +, + + +, SCL, OL, CL

With N2B, the 3178 cultures are divided into 481 patterns.

^{a} Alternative labelled by number of possible choices.

TABLE XXII

Example of computer analysis of some phage patterns of cultures from different countries when discrimination alternative *N2A* is used to differentiate between phage typing patterns of 3178 isolates

I	II	III	IV	IVA	V	VI	VII	VIII	VIIIA	IX	X	XI	XII	F12	γ66	No. of cultures	Isolated in
1	2	1	1	2	2	1	1	1	1	1	2	2	1	1	1	145	GDR, France, Hungary, Japan, Poland, Sweden
1	2	1	1	2	2	1	1	1	1	1	1	2	1	1	1	142	GDR, Hungary, Japan, Poland, Roumania, USSR
2	2	2	2	2	2	2	2	2	2	2	2	2	2	2	2	141	Chad, GDR, France, Hungary, Iraq, Israel, Scotland, Sweden, Tunisia, USA, USSR
1	2	1	1	2	2	1	1	1	1	1	1	2	1	2	1	118	Czechoslovakia, GDR, Hungary, Israel, Poland, Roumania, Sweden, USA, USSR
1	2	2	2	2	2	2	2	2	2	2	2	2	2	2	2	113	Czechoslovakia, GDR, England, Hungary, Malaya, Sweden, USA, USSR

Phage reaction code per typing phage[a]

[a] Phage pattern coded according to definition of *N2A* in Table XXI (i.e. code 1 corresponds to reactions − or +, and code 2 corresponds to all other reaction strengths).

country. The lesser discrimination afforded by the N2A scheme compared to N8 is reflected in a larger number of strains within each of the patterns listed for the former discrimination alternative.

One word of caution is necessary regarding the size of plaques in relation to RTD. Phages may exhibit large plaques on their host-propagating strains, but small plaques on other strains. In some instances, small plaques with one phage may be a characteristic of a particular infecting strain so that it may serve as an additional marker of the epidemic strains. Occasionally, a phage pattern is not easily interpreted or converted into any given pattern. Phage patterns not listed among the reference types are designated NST (no specific type) or NC (not characterised). Strains without any phage reactions are indicated by NT (non-typable).

8. Phage type frequencies

The most common types at the National Bacteriological Laboratory, Stockholm, are shown in Table XXIII; a few types dominate. The

TABLE XXIII

Frequency of dominant phage types of *S. sonnei* in Sweden during 1946–1966

Phage type	%
5	25
3	23
65	7
7	6
62	5
2	3·1
12a	2·7
14a	2·7
26	2·4
66	1·8
12	1·8
53	1·7

absolute frequency of each type is less important, since the type distribution changes over the years. Although Hammarström (1949) ascribed most of his strains to a designated type, only 40% of the 1055 strains typed at the Bacteriological Laboratory, Stockholm, since 1954 belong to his original patterns.

With the proposed international typing set for *S. sonnei*, non-typable strains are, indeed, very rare (Aldová, *et al.* 1977).

VII. REPRODUCIBILITY AND STABILITY OF TYPE PATTERNS

A. Molecular basis of instability

Due to a high frequency of lysogenisation and frequent phage type changes upon loss of lysogenising phage, or uptake of extrachromosomal DNA (phage, R-factors, colicin factors, etc.), spontaneous lysotype modifications are to be expected. The question is to what extent such changes occur and whether practical implications are recognisable. One problem is associated with the gradual change in phage typing pattern upon storage. Strains should be typed shortly after isolation.

Even if phase II colonies are subcultured a mixture of both phase I and phase II variants may develop. Aldóva *et al.* (1975) noted that type instability and changes on storage particularly affect certain phage types.

B. Type stability *in vitro*

Hammarström (1949) investigated type stability by a variety of methods:

(1) periodic retyping of the same strains,
(2) retyping after serial passage,
(3) testing several R-form colony variants of the same strains.

Reduced sensitivity to the phages was observed upon retyping. Phages that were originally reactive often did not later attack the same bacteria. Such changes could follow mutation or selection after five to eight months *in vitro*. Hammarström (1949) found a change in 3% of the cultures and in 14% after storage for 27 months.

Junghans (1961) only noted changes in 6% among 1732 strains after two to three years. In 2·5% increased resistance against phages had developed. Reportedly, less change is observed if strains are subcultured every six months—after passage on endo-agar on which atypical colonies are quite easily avoided morphologically. Storage as freeze dried cultures may conserve typing patterns. Even lyophilisation, though, is not without its problems, since it may induce loss of phages (Bergan and Midtvedt, 1975).

Often, survival of *Shigella* outside the body does not change the phage type as evidenced by storage in water and soil samples for two to six weeks, and in various media, at different temperatures for up to six months (Grunow, 1965b).

C. Type stability *in vivo*

The question of type stability *in vivo* can only be evaluated in patients who are isolated from epidemic reservoirs and thus do not run the risk of contracting *Shigella* strains during the study period. Mixed infections, though rare, may lead to confusing results.

The question may be elucidated in two ways: (1) serial isolation from the same individuals, and (2) comparison of strains from several patients who are clearly involved in the same epidemic.

Hammarström (1949) pursued both approaches. Among 1451 strains from 987 patients, only 0·9% of the isolates (ten persons—1%) reflected instability. Rare changes have also been observed by Ziesché and Rische (1973) and Aldová and Sucha (1973).

Grunow (1965c) and Ziesché and Rische (1973) among others have observed changes in serial isolates from the same persons. *In vivo*, it is difficult to say why more than one type develops and to determine whether it is due to reinfection, mixed initial infection, or spontaneous shift in phage type. Extrachromosomal DNA may lead to phage restriction, modification, or development of new phage receptors (Grunow, 1965e).

D. Typing method and reproducibility

Phage typing procedures must be carefully standardised. Pattern variation may be due to trivial technical evaluations. It is often preferable to type strains which are to be compared simultaneously. Thereby, the influence of slight differences in phage suspension densities and typing medium are avoided.

With meticulous standardisation of technique, phage typing of *Shigella* is sufficiently reproducible to be of considerable epidemiological value.

VIII. EPIDEMIOLOGICAL USE

Typing must be assessed on achievements in elucidating epidemics.

One relevant parameter of usefulness is the percentage of typable strains. Aldová and Sucha (1973) with the Hammarström phages obtained 17% non-typable *S. sonnei*. High numbers of apparently untypable strains result if the strains are not clearly in phase II, i.e. only R-form colonies (not phase I + S-form) are to be selected, Passage on endo-agar up to five times may be necessary to obtain proper colonies (Grunow, 1965d). With such precautions, *bona fide* untypable strains are extremely rare as evidenced from the number of reports with 100% typability.

A clinically relevant question is whether there is any correlation between phage type and either antibiotic sensitivity or pathogenicity. There have been reports that certain lysotypes have particular antibiotic sensitivity patterns or biotypes but this applies only within restricted geographical areas (Szturm-Rubinsten et al., 1974). Accordingly, such a relationship is probably more apparent than real, due to local dominance of one or a few epidemic strains. In general, there is no relationship between lysotype and antibiotic susceptibility, biotype, colicin type, or pathogenicity (Grunow, 1965a; László and Kerekes, 1969; Mulczyk *et al.*, 1967a, b Rische *et al.*, 1965; Rische, 1966). Accordingly, colicin typing may be useful in subdividing a lysotype which dominates in a particular locality (Meitert *et al.*, 1975; Pfeifer and Krüger, 1968). In contrast, there is an association between serogroup and sensitivity to phages in many instances. Still, the situation on this point requires further elucidation. László *et al.* (1973a, b) reported that there was no correlation between phage type and serotype.

If one or a few lysotypes dominate, a differentiation may be achieved by the addition of local phages (Trifonová and Bratová, 1975; Quynh, 1967, 1969; Ziesché, 1975).

Continuous surveillance in Sweden has indicated that the original lysotypes of Hammarström occur with steadily decreasing frequency and that new types emerge steadily (Fig. 7). In 1971, only seven out of 24 different patterns belonged to a previously recognised type. Corresponding figures for 1972, 1973, and 1974 were 6/17, 4/17, and 6/21 respectively (Kallings *et al.*, 1968). A similar shift has been noted in France and Africa (Szturm-Rubinsten, 1972).

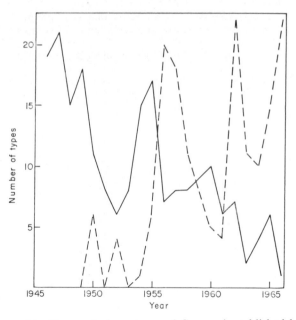

FIG. 7. Annual incidence of phage types of *S. sonnei* established by Hammarström (——) and lytic patterns not belonging to recognised types (– – – –) (Kallings *et al.*, 1968).

Such a situation necessitates periodic, although cautious, modification of the international typing sets.

Another point of significance is the discriminative ability of individual phages. This decides the ability of the phages to differentiate bacterial cultures from separate outbreaks. Several phages in the international working set of *S. sonnei*, e.g. phages IVA, V, and XI, lyse so many bacterial isolates that they have little discriminative value. A number of phages have similar lytic spectra (Table XXIV). The *S. sonnei* phages IVA and V, for instance, lyse the same strains 91% of the time, a frequency which would seem incompatible with their inclusion in the same typing set.

The question of phenetic similarity of the phages in terms of lytic spectra of all phages may be efficiently assessed simultaneously by numerical grouping techniques, as has been done for phages from *Pseudomonas aeruginosa* (Bergan, 1972).

In geographical areas where shigellosis is frequent, phage typing is a powerful tool for identifying sources of infection and tracing routes of transmission. When only few imported cases occur, as is the current situation in Scandinavia, typing is an aid in determining whether the cases are unrelated.

TABLE XXIV

Frequency with which pairs of typing phages both lyse the same of 5386 *S. sonnei* isolates

Phages	Frequency of coinciding lysis (%)
II–IVA	79
II–V	80
II–XI	76
IV–V	75
IVA–V	91
IVA–VIIIA	75
IVA–XI	86
IVA–F12	80
V–VIIIA	77
V–XI	87
V–F12	82
VIIIA–XI	75
IX–XII	78
XI–F12	81

IX. OPINION ON PLACE OF *SHIGELLA* PHAGE TYPING

Lysotyping should be carried out on strains of known serotype.

Lysotyping would have priority over biotyping or colicin typing. The latter may be useful to break down larger groups of strains with the same lytic type.

Phage typing of *Shigella* is recommended only for national, or regional reference laboratories of several collaborating nations.

X. FUTURE TRENDS

The work of the international panels studying the phage typing of *S. flexneri* and *S. sonnei* is approaching definition of suitable typing sets for both species. When agreement is reached, the situation should be followed closely to detect changes in the bacterial population. Such changes may conceivably be observed over the years and may require addition to or exclusion from the typing sets. Computers are useful in such continuous surveillance and in evaluating the discriminatory ability of the typing phages.

Computers are also useful aids in discrimination between patterns. Further development may give additional support to the proposal (*vide supra*) that phage typing patterns should be reported instead of using the rigid system of phage types.

Areas that need scientific elaboration are:

Reproducibility of the reactions of individual phages.

Evaluation of how many phage reactions may be tolerated as acceptable for source identity of epidemiologically related strains.

Validation of the techniques used, such as propagation and temperatures of incubation.

The global impact of *Shigella* infections will continue on a high scale for the unforeseeable future. Shigellosis is under the constant surveillance of the World Health Organisation (WHO). In many countries, *S. dysenteriae* is a more severe health problem due to the high frequency and the gravity of the disease as compared with industrialised areas. Consequently, carefully standardised and internationally studied typing sets for S. *dysenteriae* need to be developed.

ACKNOWLEDGEMENTS

I would like to thank Dr L.-O. Kallings, Stockholm, Sweden, and Dr S. Ślopek Wroclaw, Poland, for generously having made several figures available for this Chapter. Dr Kallings has also contributed valuable pieces of information and has been available for discussions for which I am grateful.

REFERENCES

Aldová, E., Schön, E. and Sucha, J. (1977). *J. Hyg. Epidem. Microbiol. Immun.* **21**, 33–41.

Aldová, E. and Suchá, J. (1973). *J. Hyg. Epidem. Microbiol. Immun.* **17**, 184–201.

Aldová, E., Suchá, J. and Horák, V. (1975). *In* "Proc. VI Int. Colloquium on Phage Typing and Other Laboratory Methods for Epidemiological Surveillance", pp. 80–97. Institut für experimentelle Epidemiologie, Wernigerode.

Arkwright, J. A. (1920). *J. Path. Bact.* **23**, 358–359.

Beer, W. and Seltmann, G. (1973). *Z. allg. Mikrobiol.* **13**, 107–113.

Bercovici, C., Iosub, C., Besleagá, V., Freund, S. and Popa, S. (1967). *J. Hyg. Epidem. Microbiol. Immun.* **11**, 137–146.

Bercovici, C., Iosub, C., Besleagá, V. and Popa, S. (1972). *J. Hyg. Epidem. Microbiol. Immun.* **16**, 282–292.

Bergan, T. (1972). *Acta path. microbiol. scand.* **80**, 189–201.

Bergan, T. and Midtvedt, T. (1975). *Acta path. microbiol. scand.* **83**, 1–9.

Bojlén, K. (1934). "Dysentery in Denmark". Biacho Luno, Copenhagen.

Burnet, F. M. (1933). *J. Path. Bact.* **36**, 307–318.

Burnet, F. M. and McKie, M. (1930). *J. Path. Bact.* **33**, 637–646.

Carpenter, K. P. (1974). *In* "Bergey's Manual of Determinative Bacteriology", 8th edn (R. E. Buchanan and N. E. Gibbons, Eds), pp. 318–321. Williams and Wilkins Company, Baltimore.

Chanishvili, L. G. and Chanishvili, T. G. (1975). *In* "Proc. VI Int. Colloquium on Phage Typing and Other Laboratory Methods for Epidemiological Surveillance", pp. 103–107. Institut für experimentelle Epidemiologie, Wernigerode.

Chantemesse, M. M. and Widal, F. (1888). *Gaz. med. Paris* **5**, 185–187.

Chernova, V. N. (1971). *Z. Mikrobiol. (Moscow)* **48**, 31–35.

Dmitriev, B. A., Backinowsky, L. V. and Kochetkow, N. K. (1973). *Eur. J. Biochem.* **34**, 513–518.

Dmitriev, B. A., Backinowsky, L. V., Lvov, V. L., Knirel, Y. A. and Kochetkov, N. K. (1975a). *Carbohyd. Res.* **41**, 329–333.

Dmitriev, B. A., Backinowsky, L. V., Lvov, V. L., Kochetkov, N. K. and Hofman, I. L. (1975b). *Eur. J. Biochem.* **50**, 539–547.

Edwards, P. R. and Ewing, W. H. (1972). "Identification of *Enterobacteriaceae*", 3rd edn. Burgess Publishing Company, Minneapolis.

Emelyanov, P. I. (1966). *Z. Mikrobiol. (Moscow)* **43**, 91–95.

Eskridge, R. W., Weinfeld, H. and Paigen, K. (1967). *J. Bact.* **93**, 835–844.

Ewing, W. H. (1949). *J. Bact.* **57**, 633–638.

Ewing, W. H. (1963). *Int. Bull. bact. Nomencl. Taxon.* **13**, 95–110.

Ewing, W. H. and Carpenter, P. (1966). *Int. J. syst. Bact.* **16**, 145–149.

Ewing, W. H., Jaugstetter, J. E., Martin, W. J., Sikes, J. V. and Wathen, H. G. (1971). "Biochemical Reactions of *Shigella*". US Department of Health, Education, and Welfare, Center for Disease Control, Atlanta, Georgia.

Eylan (1974)

Financsek, I. and Kétyi, I. (1976). *Acta microbiol. Acad. Sci. hung.* **23**, 317–324.

Financsek, I., Kétyi, I., Sasak, W., Jankowski, W., Janczura, E. and Chojnacki, T. (1976). *Infect. Immun.* **14**, 1290–1292.

Flexner, S. (1900a). *Br. med. J.* **II**, 917–920.

Flexner, S. (1900b). *Zbl. Bakt. I Abt. Orig.* **28**, 625–631.

Gemski, P., Koeltzow, D. E. and Formal, S. B. (1975). *Infect. Immun.* **11**, 685–691.

Giese, H. (1967). Zbl. Bakt. I Abt. Orig. 207, 334–339.
Goldhar, J., Eylan, E. and Goldschmied-Reouven, A. (1973). Med. Microbiol. Immun. 159, 63–72.
Goldhar, J., Eylan, E. and Goldschmied-Reouven, A. (1974). Med. Microbiol. Immun. 159, 233–242.
Goldhar, J. and Eylan, E. (1974). Med. Microbiol. Immun. 160, 59–63.
Goldhar, J., Eylan, E. and Barber, C. (1975). Zbl. Bakt. I Abt. Orig. 230, 336–342.
Goldhar, J., Barber, C. and Eylan, E. (1976). Zbl. Bakt. I Abt. Orig. 236, 89–98.
Gromkova, R. (1966). Arch. Roum. Path. exp. 25, 333–339.
Gromkova, R. (1967). Zbl. Bakt. I Abt. Orig. 203, 74–78.
Gromkova, R. and Trifonova, A. (1967). Zbl. Bakt. I Abt. Orig. 204, 212–216.
Grunow, R. (1965a). Zbl. Bakt. I Abt. Orig. 197, 354–361.
Grunow, R. (1965b). Zbl. Bakt. I Abt. Orig. 198, 432–438.
Grunow, R. (1965c). Zbl. Bakt. I Abt. Orig. 198, 438–448.
Grunow, R. (1965d). Zbl. Bakt. I Abt. Orig. 198, 448–463.
Grunow, R. (1965e). Zbl. Bakt. I Abt. Orig. 198, 463–477.
Grunow, R. and Köhler, F. (1965). Z. ges. Hyg. 11, 699–705.
Hammarström, E. (1947). Lancet 1, 102–103.
Hammarström, E. (1949). "Phage-typing of Shigella sonnei". The State Bacteriological Laboratory, Stockholm.
Hannecart-Pokorni, E., Godard, C. and Beumer, J. (1976). Ann. Microbiol. 127B, 3–14.
Helmholz, M. (1966). Z. ges. Hyg. 12, 727–732.
Horn, G. and Tschäpe, H. (1969). Arch. Roum. Path. exp. Microbiol. 28, 896–900.
Horn, G. and Taubeneck, U. (1970). Z. allg. Mikrobiol. 10, 103–119.
Istrati, G. (1960a). Arch. Roum. Path. exp. Microbiol. 19, 95–101.
Istrati, G. (1960b). Arch. Roum. Path. exp. Microbiol. 19, 85–94.
Istrati, G., Meitert, T. and Ciufeco, C. (1962). Arch. Roum. Path. exp. Microbiol. 21, 288–294.
Istrati, G., Meitert, T. and Ciufeco, C. (1964). Arch. Roum. Path. exp. Microbiol. 23, 593–598.
Jankowski, S., Mulczyk, M. and Mleczko, J. (1974). Acta Microbiol. Polon. Ser. A 6, 275–282.
Junghans, R. (1958). Zbl. Bakt. I Abt. Orig. 173, 50–54.
Junghans, R. (1961). Zbl. Bakt. I Abt. Orig. 182, 191–200.
Kallings, L. O., Lindberg, A. A. and Sjöberg, L. (1968). Arch. Immun. Ther. exp. 16, 280–287.
Kallings, L. O. and Sjöberg, L. (1975). In "Proc. VI Int. Colloquium on Phage Typing and Other Laboratory Methods for Epidemiological Surveillance", pp. 58–72. Institut für experimentelle Epidemiologie, Wernigerode.
Kauffmann, F. (1954). Enterobacteriaceae. Munksgaard, Copenhagen.
Kochetkov, N. K., Dmitriev, B. A. and Backinowsky, L. V. (1976). Carbohyd. Res. 51, 229–237.
Kolbe, C. (1969). Z. ges. Hyg. 15, 883–885.
Korsakova, M. P. and Sabrodina, O. L. (1954). Z. Microbiol. (Moscow) 4, p. 79.
Kruse, W., Ritterhaus, Kemp and Metz [sic] (1907). Z. Hyg. Infekt. Krankh. 57, 417.
Krzywy, T., Kucharewicz-Krukowska, A. and Ślopek, S. (1970). Arch. Immun. Ther. exp. 18, 597–616.
Krzywy, T., Kucharewicz-Krukowska, A. and Ślopek, S. (1971). Arch. Immun. Ther. exp. 19, 15–45.

Krzywy, T. and Ślopek, S. (1974). "Morphology and Ultrastructure of *Shigella* and *Klebsiella* Bacteriophages". Polish Medical Publishers, Warsaw.

Krzywy, T., Ślopek, S. and Kucharewicz-Krukowska, A. (1972). *Zbl. Bakt. I Abt. Orig.* **219**, 313–323.

Kucharewicz, A. (1959). *Arch. Immun. Ther. Dosw.* **7**, 347–365.

László, V. G. and Kerekes, L. (1969). *Acta microbiol. Acad. Sci. hung.* **16**, 309–317.

László, V. G., Mich, H. and Hajnal, A. (1973a). *Acta microbiol. Acad. Sci. hung.* **20**, 135–146.

László, V. G., Milch, H., Rudnai, O. and Kubinyi, L. (1973b). *Acta microbiol. Acad. Sci. hung.* **20**, 147–157.

László, V. G. and Milch, H. (1975). *In* "Proc. VI Int. Colloquium on Phage Typing and Other Laboratory Methods for Epidemiological Surveillance", pp. 31–34. Institut für experimentelle Epidemiologie, Wernigerode.

László, V. G. and Rimanóczy, I. (1976). *Acta microbiol. Acad. Sci. hung.* **23**, 259–270.

Lewin, B. (1977). "Gene Expression", Vol. 3. "Plasmids and Phages", pp. 160–270. John Wiley & Sons, New York.

Lidwell, O. M. (1959). *Mon. Bull. Minist. Hlth* **18**, 49–52.

Lindberg, B., Lönngren, J., Romanowska, E. and Rudén, U. (1972). *Acta chem. scand.* **26**, 3808–3810.

Ludford, C. G. (1953). *Aust. J. exp. Biol.* **31**, 545–552.

Marcuse, K. (1931). *Klin. Wschr.* **10**, 732–734.

Massa, M. (1931). *Z. Immun. Forsch.* **70**, 525–536.

Mayr-Harting, A. (1952). *J. gen. Microbiol.* **7**, 382–396.

Meitert, T., Ciufeco, C. and Istrati, G. (1968). *Arch. Roum. Path. exp. Microbiol.* **27**, 803–808.

Meitert, T., Istratı, G., Ciufeco, C., Sulea, I. and Baron, E. (1969). *Arch. Roum. Path. exp. Microbiol.* **28**, 976–977.

Meitert, T., Ciufecu, C., Sulea, I. T., Pencu, E., Dinculescu, E. and Merlaub, I. (1975). *Arch. Roum. Path. exp. Microbiol.* **34**, 59–66.

Meselson, M. and Yuan, R. (1968). *Nature, Lond.* **217**, 1110–1114.

Metzger, M., Mulczyk, M., Piotrowska, I. and Rudnicka, I. (1958). *Arch. Immun. Ther. Dosw.* **6**, 621–637.

Milch, H., László, G., Ślopek, S. and Mulczyk, M. (1968). *Arch. Immun. Ther. exp.* **16**, 265–279.

Mulczyk, M. and Ślopek, S. (1961). *Arch. Immun. Ther. Dosw.* **9**, 745–750.

Mulczyk, M., Ślopek, S. and Marcinowska, H. (1967a). *Arch. Immun. Ther. exp.* **15**, 609–611.

Mulczyk, M., Ślopek, S. and Marcinkowska, H. (1967b). *Epidemiol. Rev. (Warsaw)* **21**, 97–100.

Mulczyk, M. and Lachowicz, T. M. (1969). *Arch. Immun. Ther. exp.* **17**, 565–572.

Nass, W. (1965). *Arch. Hyg.* **149**, 635–642.

Ørskov, I., Ørskov, F., Jann, B. and Jann, K. (1977). *Bact. Rev.* **41**, 667–710.

Oye, E. van, Pfeifer, I. and Krüger, W. (1968). *Arch. Immun. Ther. exp.* **16**, 452–458.

Pfeifer, I. and Krüger, W. (1968). *Arch. Immun. Ther. exp.* **16**, 471–473.

Pruneda, R. C. and Farmer, J. J. (1977). *J. clin. Microbiol.* **5**, 66–74.

Quynh, D. N. (1967). *Z. ges. Hyg.* **13**, 863–865.

Quynh, D. H. (1968). *Zbl. Bakt. I Abt. Orig.* **207**, 456–463.

Quynh, D. H. (1969). *J. Hyg. Epidem. Microbiol. Immun.* **13**, 358–365.

Rantasalo, I. and Uotila, B. (1961). *Annls Med. exp. Biol. Fenn.* **39**, 415–416.

Rauss, K. and Kontrohr, T. (1964). *Path. Microbiol.* **27**, 310–333.

Report of the *Enterobacteriaceae* Subcommittee of the Nomenclature Committee of the International Association of Microbiological Societies (1958). *Int. Bull. Bact. Nomencl. Taxon.* **8**, 25–70.

Rische, H. (1966). *Ann. Immun. hung.* **9**, 217–227.

Rische, H. (1968a). *Arch. Immun. Ther. exp.* **16**, 392–401.

Rische, H. (1968b). *In* "*Enterobacteriaceae*-Infektionen. Epidemiologie und Laboratoriumsdiagnostik' (J. Sedlák and H. Rische, Eds), pp. 165–250. VEB Georg Thieme, Leipzig.

Rische, H., Mates, M. and Horn, G. (1965). *Zbl. Bakt. I Abt. Orig.* **198**, 167–171.

Rische, H., Horn, G. and Quynh, D. N. (1969). *Arch. Roum. Path. exp. Biol.* **28**, 852–856.

Rohde, C. (1975). *In* "Proc. VI Int. Colloquium on Phage Typing and Other Laboratory Methods for Epidemiological Surveillance", pp. 395–402. Institut für experimentelle Epidemiologie, Wernigerode.

Sartorius, F. (1932). *Z. Immunol. Forsch.* **74**, 313–333.

Sartorius, F. and Reploh, H. (1931). *Zbl. Bakt. I Abt. Orig.* **122**, 135–139.

Schön, E. and Maniková, Z. (1975). *In* "Drug-inactivating Enzymes and Antibiotic Resistance" (S. Mitsuhashi, L. Rosival and V. Krčméry, Eds). Avicenum, Czechoslovak Medical Press, Prague.

Seltmann, G. (1972). *Z. allg. Mikrobiol.* **12**, 497–520.

Seltmann, G. and Beer, W. (1972). *Z. allg. Mikrobiol.* **12**, 423–433.

Shiga, K. (1898). *Zbl. Bakt. I Abt. Orig.* **24**, 817–828.

Ślopek, S. (1968). *In* "*Enterobacteriaceae*-Infektionen. Epidemiologie und Laboratoriumsdiagnostik" (J. Sedlák and H. Rische, Eds), pp. 375–444. VEB Georg Theime, Leipzig.

Ślopek, S. (1973). *In* "Lysotypie und Andere Spezielle Epidemiologische Laboratoriumsmethoden" (H. Rische, Ed.), pp. 215–243. VEB Gustav Fischer Verlag, Jena.

Ślopek S. and Metzger, M. (1958). *Schweiz. Z. Path. Bakt.* **21**, 32–53.

Ślopek, S. and Mulczyk, M. (1961). *Zbl. Bakt. I Abt. Orig.* **181**, 478–481.

Ślopek, S., Mulczyk, M. and Kucharewicz-Krukowska, A. (1961). *Arch. Immun. Ther. Dosw.* **9**, 751–755.

Ślopek, S., Mulczyk, M. and Krukowska, A. (1968a). *Arch. Immun. Ther. exp.* **16**, 512–518.

Ślopek, S., Krukowska, A. and Mulczyk, M. (1968b). *Arch. Immun. Ther. exp.* **16**, 519–532.

Ślopek, S., Prozondo-Hessek, A. and Mulczyk, M. (1969). *Arch. Roum. Path. exp. Microbiol.* **28**, 974–975.

Ślopek, S., Durlakowa, I., Jankowski, S., Kucharewicz-Krukowska, A., Lachowicz, T. M. and Lachowicz, Z. (1972). *Arch. Immun. Ther. exp.* **20**, 61–72.

Ślopek, S., Durlakowa, I., Krucharewicz-Krukowska, A., Krzywy, T., Ślopek, A. and Weber, B. (1973). *Arch. Immun. Ther. exp.* **21**, 1–161.

Sonne, C. (1915). *Zbl. Bakt. I Abt. Orig.* **75**, 408–456.

Sonnenschein, K. (1925a). *Münch. med. Wschr.* **11**, 1443–1444.

Sonnenschein, K. (1925b). *Zbl. Bakt. I Abt. Orig.* **95**, 257–261.

Stelzner, A. and Urbach, H. (1969). *Arch. Hyg.* **153**, 532–537.

Strobel, M. and Nomura, M. (1966). *Virology* **28**, 763–765.

Szturm-Rubinsten, S. (1957). *Ann. Inst. Pasteur* **92**, 652–658.

Szturm-Rubinsten, S. (1964). *Ann. Inst. Pasteur* **106**, 114–122.

286   T. BERGAN

Szturm-Rubinsten, S. (1965). *Ann. Inst. Pasteur* **109**, 921–932.
Szturm-Rubinsten, S. (1972). *Ann. Inst. Pasteur* **122**, 407–414.
Szturm-Rubinsten, S., Piéchaud, D., Gasser, A. and d'Hauteville, H. (1974). *Bull. Soc. Pathol. exotique* **67**, 564–573.
Tee, G. H. (1955). *J. Hyg.* **53**, 54–62.
Thjøtta, T. (1918). *Medicinsk Revue (Bergen)* **35**, 393–398.
Thjøtta, T. (1919). *J. Bact.* **4**, 355–378.
Thjøtta, T. and Waaler, E. (1932). *J. Bact.* **24**, 301–316.
Thomen, L. F. and Frobisher, M. (1945). *Am. J. Hyg.* **42**, 225–253.
Trifonová, A. and Bratoevá, M. (1975). *In* "Proc. VI Int. Colloquium on Phage Typing and Other Laboratory Methods for Epidemiological Surveillance", pp. 73–79. Institut für experimentelle Epidemiologie, Wernigerode.
Tschäpe, H. and Rische, H. (1970a). *Zbl. Bakt. I Abt. Orig.* **214**, 91–100.
Tschäpe, H. and Rische, H. (1970b). *Z. ges. Hyg.* **16**, 662–664.
Vaillard, L. and Dopter, C. (1903). *Ann. Inst. Pasteur* **131**, 463–491.
Vaneeva, N. P., Kushnarev, V. M., Tsvetkova, N. V. and Khomenko, N. A. (1976). *Z. Mikrobiol. Epidemiol. Immunol.* **47**, 26–28.
Waaler, E. (1935). "Studies on the Dissociation of the Dysentery Bacilli". *Det norske Vitenskabs Akademis Skrifter*, I. Matem. Naturv. Klasse, No. 2. Oslo.
Wheeler, K. M. and Burgdorf, A. L. (1941). *Am. J. publ. Hlth* **31**, 325–331.
Ziesché, K. (1975). *In* "Proc. VI Int. Colloquium on Phage Typing and Other Laboratory Methods for Epidemiological Surveillance", pp. 98–102. Institut für experimentelle Epidemiologie, Wernigerode.
Ziesché, K. and Rische, H. (1973). *In* "Lysotypie und Andere Spezielle Epidemiologische Laboratoriumsmethoden", pp. 245–341. VEB Gustav Fischer Verlag, Jena.

CHAPTER VI

The Application of Fluorescent Antibody Techniques to the Identification of *Actinomyces* and *Arachnia*

MARY ANN GERENCSER

*Department of Microbiology, Medical Center, West Virginia University,
Morgantown, W.V. 26506, U.S.A.*

I. INTRODUCTION

Actinomyces has been associated with disease in animals and in man since the 1870s. The early history of these bacteria and the confusion concerning their nomenclature have been reviewed in several publications (Smith,

1975; Slack and Gerencser, 1970a, 1975; Bowden and Hardie, 1973). At present, five species are included in the genus *Actinomyces*. Of these, *A. israelii* is the principal agent of human actinomycosis, but *A. naeslundii*, *A. viscosus* and *A. odontolyticus* are also found in human infections (Georg, 1974; Slack and Gerencser, 1975; Mitchell *et al.*, 1977). *A. bovis*, the type species of the genus and the agent of bovine actinomycosis, has never been demonstrated in or isolated from human materials. The numerous reports of *A. bovis* in human actinomycosis found in the literature are due to the practice of lumping all *Actinomyces* under the single species name *bovis*. In addition to *A. bovis*, *A. israelii* and *A. viscosus* have been found in animal infections (Georg, 1974; Georg *et al.*, 1972). In addition to causing classical actinomycosis, species of the genus *Actinomyces* are thought to be involved in periodontal disease and in some types of dental caries (Slack and Gerencser, 1975).

  Arachnia propionica will be included in this discussion because it causes clinically typical actinomycosis and must be differentiated from *A. israelii*. This organism, originally described as *Actinomyces propionicus*, was reclassified because it differs significantly in carbohydrate metabolism and cell wall composition from *Actinomyces*.

  In the past ten years considerable progress has been made in the classification of the family *Actinomycetaceae* and the genus *Actinomyces*. The classification of the family as it appears in the eighth edition of "Bergey's Manual of Determinative Bacteriology" (1974) is shown in Table I. This classification differs from that of the seventh edition of the Manual in that the genus *Nocardia* has been removed from the family and the genera *Arachnia*, *Bifidobacterium*, *Bacterionema* and *Rothia* have been added. In the genus *Actinomyces*, one species, *A. baudetii*, was deleted and placed on the list of *nomina dubia* while *A. naeslundii*, *A. odontolyticus* and *A. viscosus* were added. Modification of the description of the genus to include catalase positive organisms permitted the reclassification of *Odontomyces viscosus* as *Actinomyces viscosus*. Three species are listed in the Manual as of uncertain affiliation. Of these, *A. eriksonii* should be placed in the genus *Bifidobacterium* while the status of *A. humiferus* and *A. suis* is still uncertain.

  It is already evident that further modifications of the classification shown in Table I will be needed. All recent evidence seems to support the earlier suggestions that *A. naeslundii* and *A. viscosus* are varieties of a single species. For the purposes of this discussion, the classification will be used in its present form and *A. naeslundii* (catalase negative) and *A. viscosus* (catalase positive) will be considered separate species. The status of *A. suis* needs to be clarified as does that of *A. humiferus*, a soil organism. An additional species, *Actinomyces meyerii*, has been recognized since the publication of Bergey's Manual. Strains of the organism classified as

Actinobacterium meyerii by Prevot have been studied (Holmberg and Nord, 1975; Smith, 1975) and found to be different from other *Actinomyces* species.

Identification of *Actinomyces* and *Arachnia* isolates is often difficult because of problems in obtaining and maintaining pure cultures, a requirement for special media and procedures such as end product analysis, and

TABLE I

Family Actinomycetaceae†

Family I. *Actinomycetaceae*

Genus I.   *Actinomyces*
    Species 1. *A. bovis*
    Species 2. *A. odontolyticus*
    Species 3. *A. israelii*
    Species 4. *A. naeslundii*
    Species 5. *A. viscosus*

    Species *incertae sedis*
        1. *A. eriksonii*
        2. *A. humiferus*
        3. *A. suis*

Genus II.   *Arachnia*
    Species 1. *A. propionica*

Genus III.   *Bifidobacterium*
    Species 1–11 are described

Genus IV.   *Bacterionema*
    Species 1. *B. matruchotii*

Genus V.   *Rothia*
    Species 1. *R. dentocariosa*

† Adapted from "Bergey's Manual of Determinative Bacteriology", 8th edn, 1974.

the need for an extensive set of biochemical tests. Identification is further complicated by the wide range of strain variation within each species. Since biochemical identification presents problems, serological identification of *Actinomyces* has been attempted for many years. Numerous reports since 1960 indicate that the fluorescent antibody (FA) technique is a convenient tool for species identification of *Actinomyces* and that it also reveals serotypes within each species.

II. ISOLATION AND IDENTIFICATION OF *ACTINOMYCES* AND *ARACHNIA*

A. Isolation

Procedures for isolation of *Actinomyces* and *Arachnia* from clinical material have been described by Smith (1975) and by Slack and Gerencser (1975). Briefly, pus or sputum is inspected for the presence of "sulphur granules". If granules are present, they are removed, washed in two or three changes of sterile saline, and used for making smears and cultures. If granules are not found, smears and cultures are made from well-mixed pus or exudate or from bits of purulent material in sputum. Smears should be stained with Gram and acid-fast stains. Cultures should be inoculated with material from all cases of suspected actinomycosis and from other specimens in which Gram-positive diphtheroidal or filamentous organisms are seen in smears.

Inoculate two plates each of blood agar and brain–heart infusion agar, and a tube of fluid thioglycollate broth containing $0\cdot1-0\cdot2\%$ sterile rabbit serum. Incubate one plate of each medium aerobically with carbon dioxide $(5-10\%)$ and the second plate of each medium anaerobically with carbon dioxide $(5-10\%)$. Plates should be examined after 24–48 h and again after five to seven days incubation for the presence of colonies suggestive of *Actinomyces*. If plate cultures are negative, the thioglycollate broth culture is Gram stained. When Gram-positive diphtheroidal rods are seen in the broth, they are subcultured to blood agar plates which are treated like the primary isolation plates.

B. Identification

After pure cultures are obtained, identification is based on morphology and biochemical tests. *Actinomyces* and *Arachnia* are Gram-positive rods which are usually slender and irregularly staining with V, Y or T forms present in most smears. Long filaments with or without branching may be present. Separation of *Actinomyces* and *Arachnia* from morphologically similar genera is usually based on oxygen requirements, end products of glucose fermentation, and a few selected biochemical tests. Both *Actinomyces* and *Arachnia* are anaerobes or facultative anaerobes. They do not produce indole, nor liquefy gelatin, and with the exception of *A. viscosus* they are catalase negative. *Actinomyces* produce succinic, lactic, acetic and formic (trace) acids from the fermentation of glucose. *Arachnia propionica* produces propionic, acetic and formic acid from glucose. It is most easily distinguished from *Propionibacterium* species by the catalase test.

Biochemical reactions which are useful in identifying *Actinomyces* and *Arachnia* are shown in Table II. Additional information on biochemical

TABLE 11

Biochemical reactions of *Actinomyces* and *Arachnia*†

Test§	*Actinomyces*						*Arachnia propionica*
	bovis	*odontolyticus*	*meyerii*	*viscosus*	*naeslundii*	*israelii*	
Catalase	−(0)	−(0)	−(0)	+(100)	−(0)	−(0)	−(0)
Indole	−(0)	−(0)	−(0)	−(0)	−(0)	−(0)	−(0)
Nitrate reduction	−(6)	+(98)	−(0)‡	+(87)	+(92)	d(54)	+(100)
Nitrite reduction	n.d.	−(0)	n.d.	d(57)	−(18)	−(0)	n.d.
Esculin hydrolysis	+(93)	d(73)	−(0)	+(95)	+(93)	+(96)	−(0)
Starch hydrolysis††	+(100)	−(0)	−(0)	−(0)	−(0)	−(0)	−(0)
Gelatin liquefaction	−(0)	−(0)	−(0)	−(0)	−(0)	−(0)	−(0)
Urea agar	−	−	n.d.	+	+	−	−
Adonitol	−(0)	−(0)	−(0)	−(0)	−(0)	−(0)	+(100)
Arabinose	−(0)	d(25)	d(60)	−(0)	−(2)	d(64)§§	−(0)
Dulcitol	−(0)	−(0)	n.d.	−(0)	−(0)	−(0)‡‡	−(0)
Glucose	+(100)	+(100)	+(100)	+(100)	+(100)	+(100)	+(100)
Glycerol	d(57)††	d(36)††	−	+(96)	+(90)	−(0)	d(29)
Lactose	+(93)	d(56)	+(100)	d(86)	d(80)	+(90)	+(100)
Mannitol	−(10)	−(0)	−(0)	−(2)	−(0)	d(69)	+(100)
Raffinose	−(0)	d(19)	−(0)	+(98)	+(94)	d(55)	+(100)
Salicin	(36)††	d(49)	−(0)	d(77)	d(79)	d(66)	−(14)
Starch	+(96)	+(93)	−	d(64)	d(37)	d(40)	+(79)
Sucrose	+(81)	+(98)	+	+(99)	+(100)	+(99)	+(100)
Trehalose	−	d(52)	−(0)	d(86)	d(85)	+(100)	+(100)
Xylose	d(57)	d(70)	+(100)	−(0)	−(2)	+(94)	−(0)

† Data from Slack and Gerencser (1975); data for *A. meyerii* from Smith (1975) and Holmberg and Nord (1975).

‡ Holmberg and Nord record + nitrate reduction for *A. meyerii*.

§ +, positive reaction; −, negative reaction; d, different types. Figures in parentheses indicate the percentage of positive strains when information is available. n.d., no data.

†† +, wide zone of clear hydrolysis, other species marked − may give a narrow (less than 10 mm) zone of hydrolysis.

‡‡ According to Georg (1974) most strains of *A. israelii* ferment dulcitol.

§§ Serotype 1 strains of *A. israelii* are usually +; all serotype 2 strains are −.

characteristics, strain variability, and methods for doing biochemical tests can be found in Holmberg and Hallender (1973), Georg (1974), "Bergey's Manual of Determinative Bacteriology" (1974), Slack and Gerencser (1975), Smith (1975), and Holmberg and Nord (1975). It should be pointed out that biochemical reactions of *Actinomyces* are highly dependent upon the medium and method. This should always be considered when comparing results obtained with an unknown strain with published results.

III. PRODUCTION OF FLUORESCENT ANTIBODY REAGENTS

A. Production of antiserum

Over the years a variety of methods for preparing immunising antigens and several different injection schedules have been used in making *Actinomyces* antiserum. The methods currently being used in our laboratory will be given in some detail along with information about other methods. The cultures which have been used at West Virginia University for antiserum production are listed in Table III.

1. *Preparation of immunising antigens*

Cultures to be used as immunisation antigens are grown in Trypticase Soy Broth (BBL) or Brain–Heart Infusion Broth (BBL or Difco). The medium is prepared in 500 ml amounts in 1 litre screw-capped flasks and autoclaved. The caps are tightened as soon as the flasks are removed from the autoclave. Each flask is inoculated with 5·0–10·0 ml of a 2–3 day old culture growing in the same medium. A stream of oxygen-free gas containing 95% N_2–5% CO_2 is run into the flask for approximately 1 min after which the cap is again tightened. The inoculated flasks are incubated at 35°C for 3–5 days with periodic shaking by hand. When good growth is obtained, the cells are harvested by centrifugation, washed three times in phosphate buffered saline, pH 7·2 (FTA Buffer, BBL) and resuspended in the same buffer to a concentration equal to that of a No. 8 MacFarland standard. When necessary, the suspensions are homogenised on a Vortex type mixer or by sonication for 15–30 s. This may be needed for any organism but is especially useful with *A. israelii* strains. The suspension is distributed in 4·0–5·0 ml amounts in screw-capped vials and frozen at −70°C. An aliquot is removed, allowed to thaw and warmed to room temperature for each injection.

Cultures killed by the addition of formalin to the medium to a final concentration of 0·5% and then washed and resuspended in formalinised saline (0·5% formalin in 0·85% NaCl) are equally useful as immunisation antigens.

TABLE III

Strains of *Actinomyces* and *Arachnia* used for antiserum production

Species	WVU number	Other	Serotype
		Strain designation†	
A. bovis	116§	ATCC 13683; CDC X521; Pine 1	1
	292		2
A. odontololyticus	867§	ATCC 17929; CDC X363	1
	482	ATCC 29323	2
A. israelii	46§	ATCC 12102; CDC X523, CDC W855	1
	307	CDC W1011, ATCC 29322	2
A. naeslundii	45	ATCC 12104; CDC X454	1
	1523	CDC W 1544	2
	820	N 1600‡	3
A. viscosus	745§	ATCC 15987; CDC X603	1
	371	ATCC 19246; CDC W859	2
Actinomyces sp.	963		Type 963
Arachnia propionica	471§	ATCC 14157; CDC X364	1
	346	CDC W904, ATCC 29326	

† WVU—West Virginia University culture collection; ATCC—American Type Culture Collection; CDC—Center for Disease Control.
§ Type, cotype or neotype strain.
‡ N1600—received from S. Bellack, Lincoln State School, Ill.

A variety of media have been used for growing *Actinomyces* to be used as immunising antigens. Slack *et al.* (1951) found that antibodies to culture medium ingredients interfered with precipitin tests and described a non-antigenic peptone dialysate medium (PDT) for growing cultures to be used for immunising antigens. When antigens are prepared with this medium, longer growth times (seven days) and larger volumes of medium (3–4 litres) are needed but there is no danger of producing antobodies to medium constituents. Our recent studies have shown that cultures grown in Trypticase Soy Broth (BBL), Brain–Heart Infusion Broth (BBL or Difco) or Actinomyces Broth (BBL) can be used for immunisation if the cells are thoroughly washed to remove medium constituents. In one experiment, rabbits were immunised with concentrated Actinomyces Broth (AB). The resulting antiserum produced precipitation lines in immunodiffusion tests with the homologous antigen showing that at least some of the medium constituents were antigenic. When the antiserum was conjugated with fluorescein isothiocyanate and used to stain *Actinomyces* grown in AB or PDT, the results were uniformly negative. Since antigenic components of the medium do not seem to interfere with FA results, media giving faster growth and higher cell yields than PDT are now used routinely for growing immunising antigens. These media, Trypticase Soy Broth, Brain–Heart Infusion Broth, and Actinomyces Broth, have also been used for this purpose by other workers (Lambert *et al.*, 1967; Bellack and Jordan, 1972; Holmberg and Forsum, 1973).

2. Immunisation schedules

Rabbits are bled prior to immunisation, the serum collected and stored frozen. Two immunisation schedules used in our laboratory are shown in Table IV. Schedule 1 is the one now used regularly and generally gives good results although an occasional rabbit does not respond. Similar schedules have been used by other workers (Lambert *et al.*, 1967).

The second schedule which has also been used to produce good antiserum is shown for comparison since it employs an adjuvant and subcutaneous injections as well as intravenous injections.

While both of these schedules and others reported in the literature (Bellack and Jordan, 1972; Bowden *et al.*, 1976) yield usable antiserum, there is no evidence to indicate that any one injection schedule is optimal. This is illustrated by the variability in antiserum produced in response to the same lot of antigen given by the same injection schedule. Some of this is undoubtedly due to variations in individual animal response, but this is not the sole explanation. Antiserum differences are most noticeable when used in immunodiffusion or similar tests, but FA conjugates are also affected. This problem has been discussed by Bowden and Hardie (1973), Landfried (1966, 1972) and by Holmberg *et al.* (1975a).

3. Evaluation of the antiserum

Antibody titres, either during the course of the immunisation or at the completion of the schedule, can be determined by the agglutination, precipitin or by indirect FA tests. Unfortunately none of these tests correlates well with subsequent FA titres of fluorescein isothiocyanate conjugates made from the serum. They do, however, indicate that an antibody response has occurred and serve to weed out animals with very poor titres. In general, a serum which has a high titre in agglutination tests or by indirect FA will give a satisfactory conjugate. As a rough guide, we have found that antiserum with an indirect FA titre of 1:500 or greater, usually yields a usable conjugate. It should be pointed out that there is no direct correlation between the indirect titre of the whole serum and the titre of the conjugate.

TABLE IV
Immunisation schedules for antiserum production

Schedule	Week	Intravenous injection (ml) on day					Other injections
		1	2	3	4	5	
1	1	1·0		1·0		1·0	None
	2	1·0		1·0		1·0	
	3	1·0		1·0		1·0	
	4	1·0		1·0		1·0	
	5	←———— Rest ————→					
	6	←————Bleed†————→					
	7	1·0		1·0		1·0	
	8	←———— Rest ————→					
	9	←———— Bleed ————→					
2	1	2·0	2·0	3·0	3·0	4·0	1·0 ml antigen–
	2	←———— Rest ————→					adjuvant‡ mixture
	3	2·0	2·0	3·0	3·0	4·0	given subcutane-
	4	←———— Rest ————→					ously 2–3 days
	5	2·0	2·0	3·0	3·0	4·0	before and 1 day
	6	←———— Rest ————→					after intravenous
	7	←———— Bleed ————→					series in first week
		For booster injections repeat week 1, rest 1 week and bleed					

† Bled by cardiac puncture; serum tested for antibody production.
‡ Adjuvant mixture contains equal quantities of antigen and Freund's adjuvant.

B. Conjugation with Fluorescein Isothiocyanate (FITC)

The methods which we use at present for fractionating antiserum and for conjugating the resulting globulin are essentially those which have been described in great detail by Hebert *et al.* (1972). Information on the preparation and evaluation of fluorescein conjugates can also be found in many other publications including those of Nairn (1964), Goldman (1968), White (1970), Brighton (1970) and Cherry (1974).

The factors influencing conjugate quality can be evaluated by the physicochemical methods described by Hebert *et al.* (1972). The actual performance of a conjugate is tested by staining both homologous and heterologous antigen systems and by testing it for non-specific staining of mammalian cells. High staining titres are desirable, both for economy in the use of reagents and because dilution alone effects a reduction or sometimes elimination of non-specific staining.

1. *Fractionation of rabbit antiserum* (adapted from Hebert *et al.*, 1972, and Cherry, 1974)

The following procedure is for the fractionation of 10·0 ml of antiserum:

(a) Add 10·0 ml 70% saturated $(NH_4)_2SO_4$ to the serum while stirring gently.

(b) Mix well and allow to stand at 4°C overnight or at 25°C for 4 h.

(c) Centrifuge and discard supernatant.

(d) Resuspend the globulin precipitate in distilled water to a total volume of 10·0 ml.

(e) Add 10·0 ml 70% saturated $(NH_4)_2SO_4$; mix, centrifuge and discard supernatant.

(f) Resuspend globulin precipitate to a total volume of 10·0 ml in distilled water and repeat step (e).

(g) Resuspend the globulin to a total volume of 10·0 ml or less if concentration is desired in distilled water.

(h) Dialyse against frequent changes of 0·9% NaCl, pH 8·0 to remove the $(NH_4)_2SO_4$. The presence of sulphate can be detected by adding a small amount of saturated $BaCl_2$ to an equal volume of dialysate. There should be no cloudiness in an aliquot of an overnight dialysate. Excess $(NH_4)_2SO_4$ may also be removed by passing the globulin through a Sephadex G-25 column.

2. *Conjugation with fluorescein isothiocyanate (FITC)* (adapted from Hebert *et al.*, 1972, and Cherry, 1974)

(a) Determine the protein concentration of the globulin. The biuret method is satisfactory for this purpose.

(b) Determine the total protein content of the globulin and calculate the amount of FITC needed for the desired F/P ratio. To achieve F/P ratios in the range of 10–12 requires 15 μg/mg FITC while ratios of 15–20 require 25 μg/mg FITC. These amounts apply when the globulin has a protein concentration between 10 and 30 mg/ml.

(c) For direct labelling of globulin, bring all reagents to 25°C.

(d) Dissolve the FITC in buffer (pH 9·0, 0·1 M Na_2HPO_4), using one-half the volume of the globulin to be labelled; for 10 ml globulin, dissolve FITC in 5·0 ml.

(e) Place the globulin in a flask and add 0·2 M Na_2HPO_4 in a volume, equal to 1/4 that of the original; stir gently; for 10 ml globulin, use 2·5 ml Na_2HPO_4.

(f) Add the FITC solution while stirring gently.

(g) Immediately adjust to pH 9·5 by dropwise addition of 0·1 M Na_3PO_4.

(h) Add enough 0·85% NaCl to bring the total volume to twice the original volume of globulin; mix gently.

(i) Allow to react at 25 °C for 2·5 h without shaking.

(j) Remove any precipitate which forms by centrifugation and dialyse against pH 7·6 phosphate-buffered saline to remove unreacted fluorescein. Unreacted fluorescein can also be removed by passage through a Sephadex column.

(k) Centrifuge again if necessary. Add merthiolate (1:10 000) and store at 4 °C or freeze in small aliquots. Repeated freezing and thawing should be avoided.

3. *Evaluation of the conjugate*

Performance assays of conjugates are done by staining homologous and heterologous antigens (in this case cell suspensions of *Actinomyces*) with two-fold dilutions of conjugate. Background staining in such smears should be noted as a guide to the presence of unreacted fluorescein and to possible non-specific staining. Evaluation of non-specific staining can be done by staining mammalian cell culture smears (Pittman *et al.*, 1967) or by staining leukocytes from venous blood (Holmberg and Forsum, 1973). FA staining titres which allow dilution of the conjugate are highly desirable since such dilution is probably the most efficient method of reducing or removing true non-specific staining. Cross-reactions due to common antigens in the bacteria can be removed by appropriate absorption of the conjugate.

Other tests should also be done on conjugates, if possible. The F/P ratio should be determined using the method of Wells *et al.* (1966) or that of McKinney *et al.* (see Hebert *et al.*, 1972). Cellulose acetate strip electrophoresis (Hebert *et al.*, 1971; Hebert *et al.*, 1972) yields valuable information as to the purity of the globulin labelled and the presence of unreacted FITC.

Optimal F/P ratios for *Actinomyces* conjugates are not well defined. The optimal ratio for conjugates made by direct labelling of $(NH_4)_2SO_4$ fractionated antiserum is in the range of 15–20 (Bragg, pers. comm.). Our experience also suggests that while conjugates with F/P ratios of approximately 10 have good titres, those with ratios between 15 and 20 usually have higher staining titres. It should be remembered that F/P ratios determined on conjugates prepared by this method reflect average labelling of the globulin since the globulins are not homogeneously labelled.

Holmberg and Forsum (1973) determined the optimal F/P ratio for *Actinomyces* conjugates. In this case, purified IgG was labelled and then fractionated on a DEAE cellulose column to yield homogeneous fractions with varying F/P ratios. In this study, conjugates with F/P ratios of 10 or less produced higher titres with homologous antigens and less non-specific staining than did conjugates with higher F/P ratios.

IV. REACTIONS OF *ACTINOMYCES* WITH FITC CONJUGATED ANTISERA

A. Staining procedures

The direct staining method has been used almost exclusively in our studies of *Actinomyces*. The antigens consist of cells grown in BHI broth or on BHI agar slants. The broth cultures are centrifuged, the supernatant decanted and the cells resuspended in FTA buffer (BBL) to make a barely turbid suspension. Agar grown cells are washed from the slant and suspended in the same buffer to the same concentration. The addition of a few glass beads to the tube, followed by vortexing, is helpful in obtaining smooth suspensions especially of *A. israelii*. Antigen suspensions may be refrigerated and used for several weeks but the cells do not continue to stain at maximum intensity indefinitely. For this reason, antigen suspensions used for control purposes are renewed at least monthly.

Smears are prepared by spreading a small loopful of the suspension on a circumscribed area of a microscope slide. Smears are allowed to air dry and are gently heat fixed. One drop of conjugate is placed on each smear and the slides are incubated for 30 min at room temperature under an inverted Petri dish containing moist filter paper. Excess conjugate is drained off and the slides washed in phosphate buffered saline pH 7·2 (FTA Buffer, BBL), for 10 min, counterstained with Evans Blue (1·0% aqueous) for 5 min and finally rinsed for 1 min each, in two changes of carbonate buffer, pH 9·0.

Alternatively, smears are counterstained with Bovine Albumin Rhodamine Labelled Counterstain (BBL). In this case, the rhodamine counterstain is added to each conjugate before it is applied to the smears and washing is carried out by placing the slides in two changes of pH 7·2 phosphate buffered saline for 5 min. each, followed by two changes of carbonate buffer, pH 9·0, 1 min each. The rhodamine albumin counterstain has been used most often on direct smears from dental plaque, faeces and other clinical material.

Slides are mounted in buffered glycerol saline, pH 9·0, and examined under a suitable microscope. Equipment for examination of FA slides will not be discussed here as such information is available in a number of publications (Nairn, 1964; Goldman, 1968; Johnson, 1970; Cherry, 1974). Originally, we used a Leitz Ortholux microscope equipped with an HBO 200 mercury lamp, cardioid darkfield condenser, BG-12 primary filter and an OG-1 eyepiece filter. Now we are using the same microscope with an incident light (Ploem) illuminator and the filter combination $2 \times$ KP 490, K 510.

Slides are examined and graded visually, 1 to 4+ fluorescence, at each dilution of the serum tested. The titre of a conjugate is the highest dilution showing 4+ fluorescence. The working titre is one, two-fold dilution less than this titre.

B. Reactions of *Actinomyces* species and *Arachnia propionica*

The number of antigens involved in FA reactions of *Actinomyces* and *Arachnia* and their chemical composition are not known, but they are generally thought to be cell wall antigens. Cisar *et al.* (1978) have demonstrated antigens on the fine fibrils found on the surface of *A. viscosus* using an indirect peroxidase-labelled antibody technique. Such surface antigens could be involved in FA reactions. There is no evidence that the extracellular polysaccharides produced by some *Actinomyces* play a role in FA serotyping.

Although Reed (1972) reported that the major cell wall antigen of one strain of *A. viscosus* was a polypeptide, most studies have found the major cell wall antigens to be neutral carbohydrates (Bowden and Hardie, 1973; Bowden *et al.*, 1976). Bowden *et al.* (1976) demonstrated that *Actinomyces* have more than one cell wall carbohydrate determinant and that charged antigens may also be associated with the cell wall. The results of FA studies and serological studies using various soluble antigens, suggests that at least some of the same antigens are detected in the different systems (Lambert *et al.*, 1967; Slack *et al.*, 1969; Bowden and Hardie, 1973). Levine *et al.* (1978) demonstrated identical antigenic determinants on three antigens of *A. viscosus* (ATCC 19246) which were adherent to the cells and secreted into the medium.

Actinomyces species have been placed into six serological groups on the basis of FA reactions (Slack and Gerencser, 1975). These serological groups designated by capital letters were found to correspond to species as determined by other criteria. Therefore species names are generally used rather than serogroup designations. The serogroups are as follows; group A—*A. bovis*; group B—*A. naeslundii*; group C—*A. eriksonii*; group D—*A. israelii*; group E—*A. odontolyticus* and group F—*A. viscosus*. Group C may be eliminated from this serological grouping since *A. eriksonii* has been reclassified as *Bifidobacterium*.

1. *Unabsorbed conjugates*

The results which we have obtained with *Actinomyces* and *Arachnia* are shown in Table V. The interspecies cross-reactions are shown in this table without regard to serotypes within the species which will be discussed later.

TABLE V

Serological cross-reactions between species of *Actinomyces* and *Arachnia*

Antigens	FITC conjugated antiserum					
	A. *bovis*	*A.* *odontoly-* *ticus*	*A.* *israelii*	*A.* *naeslundii*	*A.* *viscosus*	*Ar.* *propionica*
A. bovis	■	−	−	−	−	−
A. odontolyticus	−	■	−	−	▨	−
A. israelii	−	−	■	▨	▨	−
A. naeslundii	−	−	▨	■	■ or ▨	−
A. viscosus	−	▨	▨	■ or ▨	■	−
Ar. propionica	−	−	−	−	−	■

■, Positive at titre of serum, cross-reaction between *A. naeslundii* and *A. viscosus* is serotype and strain variable.
▨, Positive at less than working titre of serum, strain variable.
−, Negative.

A. bovis does not cross-react with other *Actinomyces* species or with *Arachnia*. *A. odontolyticus* is also quite distinct serologically, showing only a minor cross-reaction with one serotype of *A. viscosus*. We have not studied *A. myerii*, but Holmberg and Nord (1975) found no cross-reactions between *A. meyerii* and other species of *Actinomyces*. In contrast, *A. israelii*, *A. naeslundii* and *A. viscosus* have some common antigens and show varying degrees of cross-reactivity. With a few exceptions involving *A. naeslundii* and *A. viscosus* these cross-reactions are of low titre and strain variable. Similar interspecies cross-reactions have been reported by Lambert *et al.* (1967), Georg *et al.* (1972), Bellack and Jordan (1972) and Holmberg and Forsum (1973). Dilution of the antiserum may be sufficient to prevent these interspecies cross-reactions but for diagnostic purposes we prefer to use absorbed sera.

Using indirect FA staining, Schaal and Pulverer (1973) found a wider range of interspecies cross-reactions than is shown in Table V. For example, *A. bovis* showed cross-reactions with both *A. israelii* and *A. naeslundii*. Species-specific sera suitable for indirect FA staining were prepared by extensive absorption.

Despite its close morphological resemblance to *Actinomyces*, *Arachnia propionica* is serologically distinct and does not cross-react with any of the *Actinomyces* species by either direct FA staining (Table V, Holmberg and Forsum, 1973) or indirect FA staining (Schaal and Pulverer, 1973).

Each of the species, except *A. meyerii*, has been divided into at least two serotypes based on FA reactions. The serotypes of *A. bovis* show little or no cross-reaction with each other although cross-reactions have been shown by immunodiffusion (Georg *et al.*, 1964). Among the strains studied, all serotype 2 strains produce filamentous microcolonies (rough *A. bovis*) while serotype 1 strains produce smooth microcolonies.

Most isolates of *A. odontolyticus* do not cross-react with other *Actinomyces* (Slack and Gerencser, 1975; Holmberg and Forsum, 1973). In 1971, Slack, Landfried, and Gerencser reported the isolation of a second serotype of *A. odontolyticus* which had a strain-variable, low-titred cross-reaction with serotype 1 strains and a strong cross-reaction with *A. viscosus*, serotype 1 antiserum. The existence of serotype 2 has been confirmed by

TABLE VI

Cross-reactions of *Actinomyces israelii* serotypes

	FITC conjugated antiserum				
Antigens	*A. israelii* 1	*A. israelii* 2	*A. naeslundii* 1	*A. viscosus* 2	*Actinomyces* sp. type 963
A. israelii, 1	■	□	□	□	□
A. israelii, 2	□	■	–	–	–
Actinomyces sp. type 963	□	–	□	–	■

■, Homologous reaction.

□, Reciprocal cross-reaction; all cross-reactions are strain variable and generally low-titred.

Holmberg and Nord (1975) and Bragg and Kaplan (1976). Bragg and Kaplan (1976) found a one-way cross-reaction between the *A. odontolyticus* serotypes and confirmed the cross-reaction between *A. odontolyticus*, serotype 2 and *A. viscosus*, serotype 1 antiserum.

Two serotypes of *A. israelii* were reported by Lambert *et al.* (1967) and later confirmed by Brock and Georg (1969) and by Slack *et al.* (1969). The cross-reactions of *A. israelii* serotypes are shown in Table VI. Our studies and those of Holmberg and Forsum (1973) indicate a reciprocal cross-reaction between the two serotypes while Brock and Georg (1969) found only a one-way cross-reaction in which serotype 1 conjugates stained serotype 2 antigens. Serotype 1 strains show more cross-reactions with other species than do serotype 2 strains. Antigenic variation among strains of serotype 1 has been shown by FA (Brock and Georg, 1969; Slack *et al.*, 1969) and by immunodiffusion tests (Bowden and Hardie, 1973), but at present these differences are not thought sufficient to warrant further serotype differentiation. Brock and Georg (1969) reported that 95% of the clinical isolates submitted to the Center for Disease Control for identification belonged to serotype 1. In contrast, Holmberg (1976) isolated serotype 2 more frequently than serotype 1 in a study of dental plaque flora. Further study is needed to determine the actual proportions of the two serotypes in the normal oral flora and in clinical disease.

TABLE VII
Cross-reactions of *Actinomyces naeslundii* serotypes

| | FITC conjugated antiserum | | | | | | |
| | A. naeslundii | | | Actinomyces sp. | A. israelii | A. viscosus | A. viscosus |
Antigens	1	2	3	type 963	1	1	2
A. naeslundii, 1	■	□	□	□	□	⊟	□
A. naeslundii, 2	□†	■	□	−	−	−	□†
A. naeslundii, 3	□†	□†	■	−	−	−	□†
Actinomyces sp. type 963	□	⊟	−	■	□	−	−

■, Homologous reaction.
□, Reciprocal cross-reaction.
⊟, One-way cross-reaction.
−, Negative.
†, Cross-reaction usually high titred; other cross-reactions are low-titred; all cross-reactions are strain variable.

Three serotypes of *A. naeslundii* (Table VII) have been reported, but only serotype 1, as represented by strain ATCC 12104, has been studied to any extent. Our studies of *A. naeslundii* serotype 1 show antigenic differences within the serotype analogous to those seen in *A. israelii* serotype 1. Some of the variant strains which can be separated from type 1 by absorption may represent new serotypes, but as with *A. israelii*, at present these strains are included in serotype 1.

Serotype 2 of *A. naeslundii* was described by Bragg *et al.* in 1972 on the basis of four strains which cross-reacted with both *A. naeslundii* serotype 1 and *A. viscosus* serotype 2. Three of the strains were similar to serotype 1, while the fourth strain (CDC W 1544) seemed to be more closely related to *A. viscosus* serotype 2 (L. Georg, pers. comm.). In a more recent report, Bragg *et al.* (1975) determined that only the last strain (W 1544) should be considered serotype 2 and that the other three strains should be considered untyped. In our studies of the four original serotype 2 strains, we also found that three of them were closely related to serotype 1 and by our criteria would be considered serotype 1 variants while the fourth strain (W 1544, WVU 1523) was definitely different. Thus, our results confirm those of Bragg *et al.* (1975). Only two or three additional strains of serotype 2 have been identified.

Jordan *et al.* (1974) described strain N16 (WVU 820) as a new serotype of *A. naeslundii* serotype 3. This strain is more closely related to serotype 2 than to serotype 1. Like serotype 2, one of the problems in studying this serotype is the very small number of strains available. We now have five additional strains in our collection which belong to serotype 3. These strains cross-react strongly with serotype 2 (W 1544) but like N16 can be separated from it by absorption.

Cross-reactions between *A. naeslundii* serotypes and other species are shown in Table VII. *A. naeslundii* serotype 1 shows a wide range of cross-reactions which are highly strain dependent and vary with the particular lot of antiserum being used. In general, *A. naeslundii,* serotype 1 strains cross-react with *A. israelii* serotype 1 and with both serotypes of *A. viscosus* (Lambert *et al.,* 1967; Slack *et al.,* 1969; Slack and Gerencser, 1975). Serotypes 2 and 3 show high titred cross-reactions with *A. viscosus* serotype 2.

A group of organisms, listed in Tables VI and VII as *Actinomyces* sp. serotype 963 must be considered at this point. In a study of dental plaque isolates, we found a group of eight strains of *Actinomyces* which fell into a clearly delineated serological group. These organisms had some biochemical reactions resembling *A. israelii,* but their overall pattern of reactions was more nearly that of *A. naeslundii.* When these strains were first described (Gerencser and Slack, 1976), they were tentatively designated *A. naeslundii* serotype 4. Since we now feel that it was premature to give this group a species name, we have included them in tables showing reactions of both *A. naeslundii* and *A. israelii* under the accession number of the strain used for antiserum production as *Actinomyces* sp. serotype 963. The serotype 963 strains cross-react with *A. israelii* serotype 1 and *A. naeslundii* serotype 1 and have a weak one-way cross-reaction with *A. naeslundii* serotype 2. Unlike the *A. naeslundii* serotypes, *Actinomyces* sp. serotype 963 has very limited cross-reactions with *A. viscosus* serotype 2.

TABLE VIII

Cross-reactions of *Actinomyces viscosus* serotypes

Antigens	FITC conjugated antiserum							
	A. viscosus 1	*A. viscosus* 2	*A. naeslundii* 1	*A. naeslundii* 2	*A. naeslundii* 3	*A. odontolyticus* 2	*A. israelii* 1	*Actinomyces* sp. 963
A. viscosus, 1	■	–	–	–	–	–	–	–
A. viscosus, 2	□	■	□	□	□	–	□	–

■, Homologous reaction.
□□, Reciprocal cross-reaction.
□□, One-way cross-reaction.
–, Negative.
All cross-reactions are strain variable and usually low-titred.

A. viscosus strains are divided into two serotypes as shown in Table VIII. Gerencser and Slack (1969) showed that rodent strains (serotype 1) were serologically distinct from human isolates (serotype 2). Both serotypes cross-reacted with *A. naeslundii* and to a lesser extent with *A. israelii* but had only a one-way cross-reaction with each other in which *A. viscosus* serotype 1 conjugate stained *A. viscosus* serotype 2 cells. The two serological types of *A. viscosus* and their cross-reactions have been confirmed by Bellack and Jordan (1972) and by Holmberg and Forsum (1973). Like the other species, antigenic variation can be demonstrated in strains of *A. viscosus* serotype 2. Georg *et al.* (1972) studied *A. viscosus* isolates from dogs, pigs and a goat. The FA reactions of these strains were not identical to those of either rodent or human strains although high-titred cross-reactions occurred. This suggests that further serotypes may occur among animal strains. In our collection of isolates from human dental plaque, there are strains of *A. viscosus* which stain poorly or not at all with available conjugates, also suggesting additional serotypes among human strains.

Arachnia propionica can be divided into two serotypes. These serotypes were described by Gerencser and Slack (1967) with only a single strain (WVU 346) representing serotype 2. The serotypes cross-react at low titres in FA tests but the sera are easily made type specific by dilution or preferably by absorption. Brock *et al.* (1973) confirmed the existence of two serotypes in a study of 11 cases of actinomycosis due to *Arachnia propionica*. Three of the 11 strains isolated in this study belonged to serotype 2. Genetic data from an extensive study of *Propionibacterium* and *Arachnia* by Johnson and Cummins (1972) has raised a question concerning the serotype 2 strains of *Arachnia propionica*. They found no DNA homology between *Arachnia propionica* serotype 1 strains and strain WVU 346 (VPI 5067) but a high degree of homology was found between serotype 1 strains. On the other hand, in the recent numerical taxonomy study of Holmberg and Nord (1975), strain WVU 346 clustered with other strains of *Arachnia propionica* on a phenetic basis.

2. Absorbed conjugates

The cross-reactions previously discussed are those which occur when unabsorbed conjugates are tested at varying dilutions. Many of the cross-reactions are eliminated by simple dilution of the conjugate to its working titre. However, to ensure species specificity and to avoid missing the occasional strain which reacts with a conjugate at low dilutions only, absorbed conjugates are generally used for direct staining of clinical specimens and for confirming the identification of pure cultures. Absorbed conjugates are prepared so that there are no interspecies cross-reactions. Additional absorbed conjugates are needed for serotyping.

C. Preparation of absorbed conjugates

1. Batch absorption method

Cells for absorption are grown in TSB broth for 3–5 days, harvested by centrifugation, washed twice in FTA buffer and resuspended in a very small amount of buffer. This heavy suspension is stored at 4°C until used, at which time the cells are packed by centrifugation and measured. Conjugated antiserum (1·0 ml to 0·1 ml packed cells) is added, the cells are suspended evenly in the serum and the mixture incubated at 50°C for 1 h. After overnight refrigeration, the conjugate is centrifuged to remove the cells and the process repeated with fresh cells for a total of three absorptions. The conjugate is then tested with the absorbing strain and if absorption is complete the serum is titred with the homologous strain. Absorbed serum is clarified by high speed centrifugation and preserved with merthiolate (1:10 000). If desired, the serum may be filter sterilised and stored frozen without a preservative.

2. Column absorption method

McKinney and Thacker (1976a, b) described a bacterial cell column absorption method for preparing serotype specific *Streptococcus mutans* conjugates. Bragg *et al* (1977) have adapted the column absorption technique for the preparation of species and serotype specific reagents for *Actinomyces*, *Arachnia*, and *Rothia*. The following procedure for the preparation of immunoabsorbent columns is based on the report of Bragg *et al.* (1977) and on additional details kindly supplied by Sandra Bragg.

Preparation of bacterial-cell columns

(1) Add formalin (1% final concentration) to a broth culture of the organism to be used and allow the culture to stand at least 2 h.

(2) Harvest cells by centrifugation and wash once in 0·85% saline containing 0·5% formalin.

(3) Resuspend cells in acetone and centrifuge at 1000 rev/min for 15 min. Measure packed cells and decant acetone.

(4) Resuspend cells in 0·01 M phosphate buffered saline (PBS), pH 7·4, containing 1:10 000 merthiolate, using 10·0 ml PBS for each 1·0 ml packed cells.

(5) Label cells with FITC (5 mg FITC/mg packed cells) by mixing a 0·1% solution of FITC in 0·1 M Na_2HPO_4 with the cell slurry. Raise pH to 9·5 and incubate at room temperature overnight. Centrifuge, decant excess FITC and wash labelled cells once in PBS, pH 7·4.

Labelling of cells with FITC may be omitted but labelled cells adhere to columns better than do unlabelled cells.

(6) Resuspend cells in 0.1 M PBS, pH 7.6 containing 0.1% NaN$_3$ using 10.0 ml PBS for 1.0 ml cells. Make a slurry of trimethylaminoethyl (TEAE) cellulose in PBS-NaN$_3$ using 1.0 g TEAE for each 1.0 ml packed cells (measured in (3)). Mix cell suspension with TEAE slurry.

(7) With all reagents at room temperature, pour the TEAE-cellulose bacterial cell slurry into a column with a $\frac{1}{2}$–1 in plug of nonionic cellulose powder (Whatman CF-1) at the bottom. When all the slurry is in the column, allow PBS-NaN$_3$ to flow until effluent and buffer head are free of cells. Then add another plug of cellulose powder to the top of the column, followed by a disc of Whatman No. 1 filter paper. Column may be stored at 22°C

Use of bacterial cell column

(1) For use, wash prepared column with desorbing buffer (0.05 M NaH$_2$PO$_4$ in 1.0% NaCl adjusted to pH 2.3 with HCl), and then equilibrate with 0.1 M PBS, pH 7.6, followed by 0.05 M PBS, pH 7.4 until a stable baseline (280 nm on ultraviolet monitor) is obtained.

(2) Add antiserum to the column and elute with pH 7.4 PBS. The effluent should be monitored by absorbance at 280 nm. After the first protein fraction (F1) is eluted, the antibody bound to the column is eluted with pH 2.3 desorbing buffer (F2). The F1 and F2 fractions are obtained by pooling the appropriate tubes. Either or both of the fractions may be conjugated with FITC by standard techniques.

(3) To prepare column for re-use, wash with desorbing buffer and equilibrate with pH 7.4 PBS. If it is not to be used for several days, wash with PBS-NaN$_3$ and store at 22°C. For re-use repeat step (1).

Whole unfractionated antiserum, (NH$_4$)$_2$SO$_4$ fractions of antiserum and IgG fractions of immune serum may be used on bacterial cell columns. Depending on the absorbing organism and the antibody used, the columns may be used to prepare pure antibody fractions for conjugation or to remove cross-reacting antibody from immune serum.

3. *Species specific conjugates*

Species specific conjugates for identification consist of pooled sera, each of which has been absorbed to remove cross-reactions. The individual components of each pool are absorbed separately and then pooled. The sera which we have used in the pools are shown in Table IX. To conserve absorbed sera, pure culture isolates are usually screened with unabsorbed, diluted conjugate pools and then checked with absorbed sera as necessary. The unabsorbed conjugate pools contain the sera listed in Table IX, diluted so that each serum is present in the pool at its working titre.

Sera prepared as described in Table IX have been used successfully in our laboratory for identifying *Actinomyces* and very similar sera have been used by others (Holmberg and Forsum, 1973; Holmberg, 1976). Recent studies by S. Bragg (pers. comm.) and Fillery *et al.* (1978) suggest that an ideal antiserum for identification should contain antibodies to more than one strain of each serotype. These studies also suggest that absorption with more than one isolate may sometimes be necessary to make antisera species specific, particularly for *A. naeslundii* and *A. viscosus*.

TABLE IX

Pools of absorbed antiserum used for speciating *Actinomyces* and *Arachnia*

Species	Antiserum pool	Serotype	Absorbed with
A. bovis	WVU 116†	1	Not absorbed
	WVU 292	2	Not absorbed
A. odontolyticus	WVU 867	1	Not absorbed
	WVU 482	2	*A. viscosus*, 745
A. israelii	WVU 46	1	*A. naeslundii*, WVU 45; *A. viscosus*, WVU 371; *Actinomyces* sp., WVU 963; *P. avidum*
	WVU 307	2	Not absorbed
A. naeslundii	WVU 45	1	*A. israelii*, WVU 46; *A. viscosus*, WVU 371
	WVU 820	3	*A. viscosus*, WVU 371
	WVU 1523	2	*A. viscosus*, WVU 371
Actinomyces sp. serotype 963	WVU 963		*A. israelii*, WVU 46; *A. viscosus*, WVU 371; *A. naeslundii*, WVU 1523
A. viscosus	WVU 745	1	*A. naeslundii*, WVU 45
	WVU 371	2	*A. naeslundii*, WVU 45; WVU 1523 and WVU 820
Ar. propionica	WVU 471	1	Not absorbed§
	WVU 346	2	Not absorbed

† West Virginia University Culture Collection numbers.
§ May be absorbed with *P. acnes*.

4. *Serotype specific conjugates*

To prepare conjugates for serotyping within a species, additional or different absorptions are needed. Serotype specific conjugates are shown in Table X. Since *Actinomyces* sp. strain WVU 963 is not now assigned to a species, the conjugates used for this serotype are the same for speciation and serotyping. This is not true in other cases. The sera are meant to be

TABLE X

Serotype specific *FITC* conjugates for *Actinomyces* and *Arachnia*

Serotype	Antiserum	Absorbed with
A. bovis, 1	WVU 116†	*A. bovis*, WVU 292
A. bovis, 2	WVU 292	*A. bovis*, WVU 116
A. odontolyticus, 1	WVU 867	*A. odontolyticus*, WVU 482
A. odontolyticus, 2	WVU 482	*A. odontolyticus*, WVU 867
A. israelii, 1	WVU 46	*A. israelii*, WVU 307; *Actinomyces* sp., WVU 963 (*P. avidum*)
A. israelii, 2	WVU 307	*A. israelii*, WVU 46
A. naeslundii, 1	WVU 45	*A. naeslundii*, WVU 1523 and WVU 820
A. naeslundii, 2	WVU 1523	*A. naeslundii*, WVU 45 and WVU 820
A. naeslundii, 3	WVU 820	*A. naeslundii*, WVU 45 and WVU 1523
Actinomyces sp. serotype 963	WVU 963	*A. israelii*, WVU 46; *A. naeslundii*, WVU 1523; *A. viscosus*, WVU 371
A. viscosus, 1	WVU 745	*A. viscosus*, WVU 371
A. viscosus, 2	WVU 371	*A. viscosus*, WVU 745
Ar. propionica, 1	WVU 471	*Ar. propionica*, WVU 346
Ar. propionica, 2	WVU 346	*Ar. propionica*, WVU 471

† WVU—West Virginia University Culture Collection numbers.

used in sequence, so that cultures are first speciated and then serotyped. Serotype specific sera are not pooled. If it is desired to prepare a conjugate which is both species and serotype specific, the absorptions indicated in both tables should be used. This requires multiple absorptions of a single serum and may result in loss of titre. We have made such sera using batch absorption methods, but they are less satisfactory for routine culture identification than sequential use of two sera and much more difficult to prepare. Column methods should make preparation of these reagents more practical. Conjugates absorbed to be both species and serotype specific may be of value in direct staining of clinical material and would be most useful in ecological surveys.

D. Reactions with morphologically similar bacteria

Actinomyces show very few cross-reactions with other genera and those which do occur are generally strain variable and low titred (Lambert *et al.*, 1967; Holmberg and Forsum, 1973; Slack and Gerencser, 1975). Only minor cross-reactions have been reported between *Actinomyces* and *Arachnia*. Holmberg and Forsum (1973) observed 1+ fluorescence of *Arachnia propionica* antigen with low dilutions of *A. israelii* serotype 1 antiserum, but the reciprocal reaction was negative. Gerencser and Slack (1969) reported that 15 of 22 strains of *A. viscosus* stained with undiluted *Arachnia propionica* serotype 2 conjugate but not with the same conjugate at its working titre. Again the reciprocal reaction was negative.

Cross-reactions with other genera in the family *Actinomycetaceae* are also infrequent. Holmberg and Forsum (1973) found no cross-reactions between *Rothia dentocariosa* and *Actinomyces* species while Slack and Gerencser (1969) found such cross-reactions to be limited to *A. viscosus*. Undiluted *R. dentocariosa* ATCC 17931 antiserum stained 7 of 22 strains of *A. viscosus*. In contrast, Bragg (pers. comm.) found strong cross-reactions between some strains of *R. dentocariosa* and both *A. israelii* serotype 1 and *A. israelii* serotype 2. Absorption of *A. israelii* serotype 1 antiserum with *A. israelii* serotype 2 to make it serotype specific also removed the cross-reaction with *Rothia*, but *A. israelii* serotype 2 antiserum required absorption with *Rothia*. No cross-reactions have been reported between *Actinomyces* and either *Bifidobacterium* or *Bacterionema* (Lambert *et al.*, 1969; Slack *et al.*, 1969; Holmberg and Forsum, 1973).

Cross-reactions with other genera include reactions with *Propionibacterium* and *Corynebacterium pyogenes*. Both *A. israelii* and *Arachnia propionica* may cross-react with *P. acnes*. We find that some freshly isolated strains of *P. acnes* stain with high dilutions of *Arachnia propionica* antiserum. Bragg (pers. comm.) found high-titred cross-reactions, requiring absorption, between *A. israelii* serotype 1 and a strain of *P. avidum*. Holmberg and Forsum (1973) found very weak cross-reactions between *A. israelii* antiserum and *P. acnes* and no cross-reaction with *P. avidum*. We have encountered cross-reactions between *C. pyogenes* and *A. odontolyticus* antiserum as did Bragg and Kaplan (1976). Holmberg and Forsum (1973) did not find any cross-reactions between *Actinomyces* or *Arachnia* and various species of *Corynebacterium, Lactobacillus, Nocardia,* and *Mycobacterium* or with *Leptotrichia buccalis* and *Streptomyces azureus*.

The variation in results obtained by different investigators has been pointed out to emphasise that each batch of FA reagents must be evaluated for sensitivity and specificity and that information such as that in Tables IX and X must be considered as only a guide to production of specific reagents. Different or additional absorptions may be necessary to make any particular lot of antiserum specific.

V. USE OF FA IN IDENTIFICATION

A. Identification of pure cultures

Slack, *et al.* (1961) first reported the use of FA for identifying and grouping *Actinomyces* serologically. Later Slack and Gerencser (1966,

1970b) used the technique to revise the initial serological groups and eliminate organisms recognized as anaerobic diphtheroids. These results showed that the serological groups paralleled species as identified by morphological and biochemical criteria which meant that FA could be used as a tool for the rapid identification of *Actinomyces* species. In 1967, Gerencser and Slack showed that *Arachnia propionica* could also be distinguished serologically using FA. This species had been largely unrecognised before this time, partly because of the difficulty of separating it from *A. israelii*. FA has been used widely for identifying species of *Actinomyces*, some examples of which will be given below.

Lambert *et al.* (1967) identified *A. israelli* and *A. naeslundii* and separated these species from other *Actinomyces* and from corynebacteria using FA. FA was also used in identifying *A. viscosus* isolated from animals (Georg *et al.*, 1972) and in identifying *Arachnia propionica* from cases of human actinomycosis (Brock *et al.*, 1973). Bellack and Jordan (1972) were able to identify *A. viscosus* from rodent sources and to confirm the serological differences between human and rodent strains with a combination of agglutination tests and FA. Holmberg and Forsum (1973) identified all five species of *Actinomyces* and *Arachnia propionica* in a group of Gram-positive filamentous organisms isolated from dental plaque.

Schaal and Pulverer (1973) showed that indirect FA staining may also be used to identify *Actinomyces*. By using absorbed sera, the authors identified 499 of 526 actinomycete cultures isolated from clinical specimens. This included 461 strains of *A. israelii*, 23 of *A. naeslundii* and 14 of *Arachnia propionica*.

B. Identification in clinical material

Both *Actinomyces* and *Arachnia* have been identified in direct smears of various tissues and exudates and in tissue sections. Slack *et al.* (1966) reported the diagnosis of two cases of actinomycosis by observation of the causative organism in exudate from the lesions. In both cases, an *Actinomyces* was isolated from the lesion. We have also identified *Arachnia propionica*, *A. odontolyticus* and *A. israelii* in two cases of lacrimal caniliculitis by direct FA staining of homogenised clinical material and by culture (Slack and Gerencser, 1975). In each case an *Actinomyces* and *Arachnia propionica* were present. The mixed infections were first recognised by the examination of the direct FA stains. In the case involving *A. israelii* and *Arachnia propionica*, the presence of two organisms would not have been suspected without the previous FA staining because all the colonies on the isolation plates were identical.

Diagnosis of a case of actinomycosis of the kidney was made possible by the use of FA, when cultures were not available. Formalin-fixed tissue fragments of the kidney were sent to us by Dr D. J. Guidry from a case where stained sections of kidney and brain showed filamentous bacteria. Smears made from the homogenised tissue showed filaments which stained brightly with *A. israelii* conjugates but not with conjugates for other species.

Blank and Georg (1968) used FA for rapid and specific identification of *A. israelii* and *A. naeslundii* in smears of homogenised granules, pustular material and tissue suspensions of human tonsils. They found that 30 of 116 specimens contained *A. israelii* when observed directly but only 12 of these yielded a positive culture. They also found that FA was useful in monitoring the presence of *Actinomyces* in mixed primary cultures from clinical material. Hotchi and Schwarz (1972) demonstrated *A. israelii* and *A. naeslundii* in formalin-fixed tonsillar tissue using FA. *A. naeslundii* was found in nine patients, *A. israelii* in 26 patients and both species in eight patients when one or more sections from tonsils containing granules were examined.

Georg *et al.* (1972) demonstrated the presence of *A. viscosus* in hepatic tissue removed from a dog at necropsy using FA staining. Staining of the tissue with other *Actinomyces* conjugates was negative and *A. viscosus* was isolated from the animal. Altman and Small (1973) diagnosed a case of actinomycosis in a non-human primate by direct FA staining of formalin-fixed tissue. Filaments in the tissue stained with FITC conjugates for *A. israelii* but not with those for *A. bovis* and *A. viscosus*. In this case, cultures were not available so only the FA staining made a definitive diagnosis possible.

The indirect staining technique has also been used successfully on clinical material. Schaal and Pulverer (1973) used FA on 528 cases of suspected actinomycosis. They found *A. israelii* in 452 cases, *Arachnia propionica* in 14 cases, *A. naeslundii* in 11 cases and both *A. israelii* and *A. naeslundii* in four cases.

C. Identification in dental plaque

Using direct FA staining of smears of homogenised plaque or calculus Slack *et al.* (1971) and Collins *et al.* (1973) showed that *Actinomyces* were a consistent part of the microbial flora and that more than one species was present in all the specimens studied. In studies of 75 plaque or calculus specimens *A. israelii* was found in 72, *A. naeslundii* in 65, *A. odontolyticus* in 46, *A. viscosus* in 64, and *Arachnia propionica* in 50 (Slack *et al.*, 1971; Collins *et al.*, 1973; Gerencser and Slack, unpublished data). All specimens had at least two species and a number of them contained all five. *A. israelii* was the most frequent species and *A. odontolyticus* the least frequent. In all these studies we obtained about 75% agreement between culture and FA results with only a single culture from each specimen.

Holmberg and Forsum (1973) also observed four *Actinomyces* species and *Arachnia* in direct smears of dental plaque from 20 individuals with periodontal disease. They found *A. israelii*, *A. naeslundii* and *Arachnia propionica* primarily in subgingival specimens while *A. viscosus*, *A. naeslundii* and *A. odontolyticus* were found in supragingival sites. In this study, 88% agreement between direct FA and culture was obtained after repeat culture of some FA positive, culturally negative specimens.

In a later study of Gram-positive rods in dental plaque, Holmberg (1976) studied 446 isolates. Of these, 75 were identified as *A. viscosus*, 92 as *A. naeslundii*, 14 as *A. odontolyticus*, 51 as *A. israelii* and 33 as *Arachnia propionica* on the basis of biochemical tests and serology. Fifteen of the 446 isolates were lost before biochemical tests could be done, but eight of them were identified by FA staining as *Actinomyces*. Both serotypes of *A. israelii* were isolated, but only one serotype of each of the other species, namely *A. viscosus*, two; *A. naeslundii*, one; *A. odontolyticus*, one and *Arachnia propionica*, one, were identified. All the organisms identified as *Actinomyces* or *Arachnia* by biochemical tests reacted in the appropriate antiserum. Direct smears of the plaque were also examined by FA in this study. The overall agreement between cultural and FA identification was 79%.

VI. OTHER SEROLOGICAL TECHNIQUES

A. Cell wall agglutination

Other serological techniques have been used for identifying *Actinomyces*. The agglutination test was widely used by early workers (reviewed by Slack *et al.*, 1951) and is still used to some extent (Bellack and Jordan, 1972). Agglutination tests are convenient to use but are not completely satisfactory because it is difficult to produce homogeneous antigens which will not autoagglutinate. Agglutination tests using cell walls instead of whole cells have been used for studying *Actinomyces* (Cummins, 1962; in Slack and Gerencser, 1975) and *Arachnia* (Johnson and Cummins, 1972). The species of *Actinomyces* could be separated by cell wall agglutination although high-titred cross-reactions occurred with *A. naeslundii* serotype 1 cell walls and both *A. israelii* serotype 1 and *A. viscosus* serotype 2 antisera. Serotype 1 strains of *Arachnia propionica* did not react in cell wall agglutination tests with the serotype 2 strain.

Bowden and Hardie (1973) and Bowden *et al.* (1976) demonstrated cell wall associated antigens which were generally species specific using trypsin or pronase treated cell walls in agglutination tests. Specific titres were higher with trypsin treated walls and more cross-reactions were demonstrated with these antigens. Pronase treatment lowered homologous titres and removed some of the cross-reactions.

B. Detection of soluble antigens

Immunodiffusion tests (ID) using acetone precipitated culture medium supernatant antigens were first described by King and Meyer (1963). Like FA, serological groupings obtained by ID tests paralleled species identification by other methods so that this test can be used for identification. In general, ID is about equal to FA in sensitivity but is less species specific (Lambert *et al.*, 1967). The number of precipitation lines produced in homologous systems and the degree of cross-reaction between different organisms is highly variable. This is due to lack of standardisation of the antigen content of the crude culture medium antigens and to variability between different lots of antiserum (Slack *et al.*, 1969; Bowden and Hardie, 1973; Slack and Gerencser, 1975; Bowden *et al.*, 1976).

Bowden and Hardie (1973) and Bowden *et al.* (1976) found that pronase-treated autoclaved extracts of whole cells were useful for serological identification of *A. israelii* serotypes 1 and 2, *A. naeslundii*, and *A. viscosus* serotype 2. These antigens were used in capillary tube or gel precipitin tests with antisera known to react with cell wall carbohydrate antigens. The antisera were prepared by injecting rabbits with heat-treated whole cells in combination with Freund's complete adjuvant. The precipitin tests distinguished between the three species and confirmed the strong cross-reaction between *A. naeslundii* and *A. viscosus* serotype 2.

Holmberg *et al.* (1975a, b) performed a detailed analysis of cytoplasmic antigens of *A. israelii* and developed a standard antigen–antibody system for use with crossed immunoelectrophoresis (CIE). They used two standard antisera, one containing pooled antiserum to whole cells of *A. israelii* and one with pooled antiserum to cell lysates. The standard antigen was a pool of sonic lysates prepared from six strains of *A. israelii*. With this system ten cytoplasmic antigens were found in *A. israelii*. When single strains were tested by CIE, cytoplasmic antigens from type 1 strains had at least five individual antigens and some strains had all ten. Serotype 2 strains contained up to six of the individual antigens. Using CIE and an intermediate gel, antiserum from other species of *Actinomyces*, *Arachnia*, *Propionibacterium* and several fungi were tested against the standard *A. israelii* antigen. The cytoplasmic antigens prepared by sonication were species specific and did not react with any of the heterologous antisera tested. Common antigens between *A. israelii* and both *A. naeslundii* and *P. acnes* were found in cytoplasmic antigens prepared by other extraction procedures.

VII. SEROLOGICAL DIAGNOSIS OF ACTINOMYCOSIS

The various serological tests which have been used successfully to identify *Actinomyces* in culture or in tissue have been tested in attempts to develop a serological test for actinomycosis. Georg *et al.* (1968) used acetone precipitated culture media antigens and the immunodiffusion test to detect antibodies in sera from known cases of localised and disseminated actinomycosis, from clinically normal persons, and from patients with other diseases. Precipitating antibodies were found in cases of systemic disease but were not always demonstrable in cases of localised actinomycosis. Cross-precipitating antibodies were formed in other diseases, particularly in pulmonary tuberculosis, so that the test could not be relied on for a presumptive diagnosis of actinomycosis.

Indirect FA tests have not been used for the diagnosis of clinical actinomycosis but Gilmour and Nisengard (1974) showed titres to *A. israelii* and *A. naeslundii* as well as to other dental plaque bacteria in people with localised gingivitis and periodontitis which suggests that this technique would not be useful for the diagnosis of clinical actinomycosis. It was not possible to distinguish between two classes of periodontal disease on the basis of specific serum titres to either of the *Actinomyces* species or the other bacteria tested.

The CIE studies of Holmberg *et al.* (1975b) describe the first serological test which is apparently specific for actinomycosis. Using a standardised preparation of antigen from *A. israelii*, nine known cases of actinomycosis had antibodies to one or more antigens in the preparation. Antibodies were not found in patients with other systemic disease or in patients with periodontal disease. Further study of this test seems warranted.

VIII. ANTIGENS OF *ACTINOMYCES*

Perhaps the most extensive studies of *Actinomyces* antigens are those of Bowden and Hardie (1973), Bowden *et al.* (1976), Bowden and Fillery (1978) and Fillery *et al* (1978). These workers have demonstrated two groups of antigens in *A. israelii*, *A. naeslundii*, *A. viscosus* and *A. odontolyticus*. Group 1 antigens are carbohydrate polymers associated with the cell wall. Group 2 antigens are charged, polypeptide-containing antigens found in acid and autoclave extracts of whole cells and in culture medium supernatants. Both groups of antigens, while similar in chemical composition from different species, contain some antigenic components which are species or serotype specific.

Group 1 antigens are complex carbohydrates containing glycerol, but not teichoic acids. These antigens are pronase resistant and are not mobile at pH 8·6. They can be detected in acid and autoclave extracts of whole cells as well as in purified wall preparations.

The charged antigens (group 2) are more complex than the group 1 antigens. The charged antigens were trypsin resistant, pronase sensitive, and acid and heat stable. Trypsin-treated cell walls seemed to contain both groups of antigens, but pronase-treated walls contained only the group 1 antigens. Charged antigens from acid extracts were very similar to culture medium supernate antigens and showed a more complex pattern in precipitin tests than autoclave extracts.

Fractionation of the wall carbohydrate antigens of *A. israelii* yielded two components, one which was serotype 2 specific and one which cross-reacted with serotype 1. The charged antigens showed differences among strains of serotype 1 similar to those previously reported on the basis of FA studies (Brock and Georg, 1969; Slack *et al.*, 1969).

A more detailed study of the wall carbohydrate antigens of *A. israelii* confirmed the presence of a specific carbohydrate component in *A. israelii* serotype 2. Common carbohydrate antigens found in all strains of *A. israelii* were not found in other oral Gram-positive rods. A comparison of stock strains and recent isolates showed the presence of carbohydrate antigens in fresh isolates of *A. israelii* serotype 1 which were not present in the stock strains. The strains could be identified as *A. israelii* on the basis of the common antigen. Differences apparently similar to the serotype differences in *A. israelii* exist in the carbohydrate antigens of *A. naeslundii* and *A. viscosus*. Cell wall carbohydrate from human *A. viscosus* has two components; one was specific for *A. viscosus* serotype 2 and one was common to *A. viscosus* serotype 2 and *A. naeslundii*.

Other studies of the antigenic composition of *Actinomyces* have been devoted primarily to *A. viscosus*. Hammond *et al.* (1976) demonstrated several antigens in Rantz and Randall extracts of cell walls of *A. viscosus* T14 by immunoelectrophoresis. One of these antigens, which was not present in extracts from an avirulent mutant of strain T14 was isolated and characterised. The high molecular weight antigen was composed of two parts; a polysaccharide with 6-deoxytalose as the major antigenic component and a small peptide. The virulence associated 6-deoxytalose containing antigen was found in other strains of *A. viscosus* and in *A. naeslundii*, but not in other *Actinomyces* species.

More recently, Cisar *et al.* (1978) showed a quantitative rather than a qualitative difference in strains T14 V and T14 AV and presented evidence that the virulence antigen is associated with fibrils present on the cell surface of *A. viscosus*. The chemical nature of the virulence associated antigen on the fibrillar layer is not known, but it is not the 6-deoxytalose containing antigen described by Hammond *et al.* (1976).

Brown *et al.* (1978) found quantitative and qualitative differences in the chemical composition of antigens from the cell walls of *A. viscosus* T14 and T14 AV. These differences were similar to those reported earlier for extracts of whole cells (Callihan and Birdsell, 1977). The methods used for extraction of antigens from whole cells or to prepare cell walls influenced the number of antigens detected although common and specific antigens were found in all preparations. In a companion study, Powell and Birdsell (1978) studied Lancefield extracts of *A. viscosus* T14V and T14 AV by two-dimensional immunoelectrophoresis and Laurell Rocket immuno-electrophoresis. Extracts from T14V and T14 AV cells grown in a chemically defined medium appeared to be identical while extracts from cells grown in supplemented tryptic soy broth showed distinct serological differences.

Reed *et al.* (1978) have studied an electronegative antigen (ENA) from sonicates of *A. viscosus* and *A. naeslundii*. The antigens from the two species were identical in gel diffusion tests. The ENA contained protein, both neutral- and amino-sugars but little methyl pentose. This antigen is thought to be a surface antigen.

Baker (1978) extracted three antigens from *A. viscosus* (M100) cell walls. Antigen 1, extracted at pH 3·3, migrated as a single band in polyacrylamide gel electrophoresis (PAGE) and stained with both carbohydrate and protein stains. Antigen 2 was extracted at pH 2·3 and antigen 3 at pH 1·8. Antigen 3 also migrated as a single band in PAGE and stained as a protein. Antigens 1 and 3 elicited an antibody response in rabbits, but only antigen 2 stimulated lymphocyte transformation.

These studies on the antigens of *A. viscosus* are only a few examples of the kinds of studies in progress on the antigenic composition of *Actinomyces*. They illustrate the difficulties encountered in attempting to correlate studies in various laboratories. The studies from one laboratory (Callihan and Birdsell, 1977; Powell and Birdsell, 1978; Brown *et al.*, 1978) emphasise the fact that the antigens detected are dependent on the age of the culture, the composition of the culture medium and the method of antigen extraction as well as the serological test used for their detection.

IX. DISCUSSION AND CONCLUSIONS

The introduction of fluorescent antibody techniques has helped to solve several problems connected with *Actinomyces* and actinomycosis. First, FA staining provides the most efficient method yet devised for the identification of pure cultures of *Actinomyces* and *Arachnia*. Results of FA studies are available long before complete biochemical testing can be done and identification by FA may be more accurate. FA can also be used to identify strains which do not survive in culture long enough for complete biochemical studies or which are non-viable when received by a reference laboratory.

Secondly, the application of FA to direct smears of clinical material frequently permits a diagnosis of actinomycosis to be made before cultures are available. Perhaps more important, it permits a diagnosis to be made when cultures are negative or when appropriate cultures were not carried out. Even today, a diagnosis of actinomycosis is often based on the demonstration of granules containing Gram-positive filaments in a tissue specimen, although a variety of bacteria may resemble *Actinomyces* morphologically and produce similar granules in tissue (Slack and Gerencser, 1975). In fact such, often dubious, cytological evidence coupled with the clinical symptoms form the basis for the diagnosis in the majority of cases. The regular use of available FA techniques in the diagnosis of actinomycosis would provide a more accurate diagnosis for the individual patient and would also provide epidemiological data which is now lacking.

With the increasing interest in oral microbiology, caries and periodontal disease, it should be pointed out that FA is a valuable tool for studying oral ecology. It can be used for the identification of *Actinomyces* in plaque, calculus and other oral samples and for following experimental infections.

The advantages of FA are so obvious that it is sometimes easy to overlook the limitations of the method but these should be kept in mind. Identification of any bacterium by FA is based on only two parameters, morphology and the specificity of a serological reaction. As in any other serological test, cross-reactions due to related antigens may occur and the reagents used may not contain antibodies to all strains of a given organism. In the case of *Actinomyces*, all available evidence indicates that species specificity of the antiserum can be achieved. Some isolates which do not stain with available FA reagents may represent new serotypes or species. This means that while positive FA reactions are reliable, negative reactions of organisms resembling *Actinomyces* morphologically cannot be used to rule out the possibility of *Actinomyces*. This does not seriously detract from the value of the method since the great majority of clinical isolates do give positive FA reactions. In fact, the greatest disadvantage of FA at the moment is the lack of commercially available reagents for *Actinomyces*.

FA which has been so successful in providing accurate identification of the pathogenic agent, is apparently not the method of choice in developing a serological test for actinomycosis. A serological test for the disease would further improve the ability of the laboratory to aid in the diagnosis of actinomycosis. Other serological tests, such as the CIE test described by Holmberg *et al.* (1975b), seem more promising for the purpose.

For the future, the value of FA tests will be improved by the continued identification of new serotypes and by continued improvement of FA reagents. Studies on the antigenic make-up of *Actinomyces* (Holmberg *et al.*, 1975; Bowden *et al.*, 1976) will help to identify the species and serotype specific antigens and show which of these are involved in the FA reaction. With this information, it should be possible to produce improved reagents using antiserum made against purified antigens.

ACKNOWLEDGEMENTS

The author would like to express her appreciation to Dr John M. Slack for the many years of active collaboration on studies with *Actinomyces* and for his help in the preparation of the manuscript. Studies on *Actinomyces* at West Virginia University were supported in part by Public Health Service grants A1-01801 from the National Institute of Allergy and Infectious Disease and DE-02675 from the National Institute of Dental Research.

REFERENCES

Altman, N. H. and Small, J. D. (1973). *Lab. Anim. Sci.* **23**, 696–700.

Baker, J. J. (1978). *J. dent Res.* **51**, Special Issue A, 357.

Bellack, S. and Jordan, H. V. (1972).   *Archs Oral Biol.* **17**, 175–182.

Blank, C. H. and Georg, L. K. (1968). *J. Lab. clin. Med.* **71**, 283–293.

Bowden, G. H. and Hardie, J. M. (1973). In "Actinomycetales: Characteristics and Practical Importance" (G. H. Sykes and F. A. Skinner, Eds.), pp. 277–299. Academic Press, London.

Bowden, G. H., Hardie, J. M. and Fillery, E. D. (1976). *J. dent. Res.* **55**, Special Issue A, A192–A204.

Bowden, G. H. and Fillery, E. D. (1978). Proc. Int. Symp. Secretory Immune System and Caries Immunity, Birmingham, Alabama.

Bowden, G. H. and Fillery, E. D. (1978). *In* "Secretory Immunity and Infection" (J. H. McGhee, J. Mestecky and J. L. Babb, Eds), pp. 685–693. Plenum Press, New York and London.

Bragg, S., Georg, L. and Ibrahim, A. (1972). *Abst. Ann. Meeting ASM* **72**, 38.

Bragg, S., Kaplan, W. and Hageage, G. (1975). *Abst. Ann. Meeting ASM* **75**, 86.

Bragg, S. and Kaplan, W. (1976). *Abst. Ann. Meeting ASM* **76**, 94.

Bragg, S., Elliot, L. and Kaplan, W. (1977). *Abst. Ann. Meeting ASM* **77**, 128.

Brighton, W. D. (1970). In "Standardization in Immunofluorescence" (E. J. Holborow, Ed.), pp. 55–62. Blackwell, Oxford.

Brock, D. W. and Georg, L. K. (1969). *J. Bact.* **97**, 581–588.

Brock, D. W., Georg, L. K., Brown, J. M. and Hicklin, M. D. (1973). *Am. J. clin. Pathol.* **59**, 66–77.

Brown, D. A., Wheeler, T. T., Grow, T. E. and Birdsell, D. C. (1978). *Abst. Ann. Meeting ASM* **78**, 158.

Buchanan, R. E. and Gibbons, N. E. (Ed.), (1974). "Bergey's Manual of Determinative Bacteriology", 8th edn, pp. 659–681. Williams and Wilkins Co., Baltimore.

Callihan, D. R. and Birdsell, D. C. (1977). *Abst. Ann. Meeting ASM* **77**, 18.

Cherry, W. B. (1974). In "Manual of Clinical Microbiology", 2nd edn (E. H. Lennette, E. H. Spaulding and J. P. Truant, Eds), pp. 29–44. ASM, Washington.

Cisar, J. O., Vatter, A. E. and McIntire, F. C. (1978). *Infect. Immun.* **19**, 312–319.
Collins, P. A., Gerencser, M. A. and Slack, J. M. (1973). *Arch. Oral Biol.* **18**, 145–153.
Cummins, C. S. (1962). *J. gen. Microbiol.* **28**, 35–50.
Fillery, E. D., Bowden, G. H. and Hardie, J. M. (1978). *Caries Research,* **12**, 299–312.
Georg, L. (1974). In "Anaerobic Bacteria, Role in Disease", (A. Balows, R. M. DeHaan, V. R. Dowell, Jr. and L. B. Guze, Eds), pp. 237–256. Charles C. Thomas, Springfield.
Georg, L. K., Coleman, R. M., and Brown J. M. (1968). *J. Immunol.* **100**, 1288–1292.
Georg, L. K., Robertstad, G. W. and Brinkman, S. A. (1964). *J. Bact.* **88**, 477–490.
Georg, L. K., Brown, J. M., Baker, H. J. and Cassell, G. H. (1972). *Am. J. Vet. Res.* **33**, 1457–1470.
Gerencser, M. A. and Slack, J. M. (1967). *J. Bact.* **94**, 109–115.
Gerencser, M. A. and Slack, J. M. (1969). *Appl. Microbiol.* **18**, 80–87.
Gerencser, M. A. and Slack, J. M. (1976). *J. dent. Res.* **55**, Special Issue A, A184–A191.
Gilmour, M. N. and Nisengard, R. J. (1974). *Arch Oral Biol.* **19**, 959–968.
Goldman, M. (1968). "Fluorescent Antibody Methods". Academic Press, New York.
Hammond, B. F., Steel, C. F. and Peindl, K. S. (1976). *J. dent. Res.* **55**, Special Issue A, A19–A25.
Hebert, G. A., Pittman, B. and Cherry, W. B. (1971). *Ann. N.Y. Acad. Sci.* **177**, 54–68.
Hebert, G. A., Pittman, B., McKinney, R. M. and Cherry, W. B. (1972). "The Preparation and Physicochemical Characterisation of Fluorescent Antibody Reagents". U. S. Department of Health, Education and Welfare, CDC, Atlanta.
Holmberg, K. (1976). *Arch Oral. Biol* **21**, 153–160.
Holmberg, K. and Hallender, H. O. (1973). *J. gen. Microbiol.* **76**, 43–63.
Holmberg, K. and Forsum, U. (1973). *Appl. Microbiol.* **25**, 834–843.
Holmberg, K. and Nord, C. E. (1975). *J. gen. Microbiol.* **91**, 17–44.
Holmberg, K., Nord, C. E. and Wadström, T. (1975a). *Infect. Immun.* **12**, 387–397.
Holmberg, K., Nord, C. E. and Wadström, T. (1975b). *Infect. Immun.* **12**, 398–403.
Hotchi, M. and Schwarz, J. (1972). *Archs Pathol.* **93**, 392–400.
Johnson, G. D. (1970). In "Standardization in Immunofluorescence" (E. J. Holborow, Ed.), pp. 11–14. Blackwell, Oxford.
Johnson, J. L. and Cummins, C. S. (1972). *J. Bact.* **109**, 1047–1066.
Jordan, H. V., Bellack, S., Keyes, P. H. and Gerencser, M. A. (1974). *J. dent. Res.* **53,** 73.
King, S. and Meyer, E. (1963). *J. Bact.* **85**, 186–190.
Lambert, F. W., Brown, J. M. and Georg, L. K. (1967). *J. Bact.* **94**, 1287–1295.
Landfried, S. (1966). M.S. Thesis, West Virginia University.
Landfried, S. (1972). Ph.D. Dissertation, West Virginia University.
Levine, M., Rodriquez, P., King, R. and McCallum, R. E. (1978). *J. dent. Res.* **57**, Special Issue A, 202.
McKinney, R. M. and Thacker, L. (1976a). *J. dent. Res.* **55**, Special Issue A, A50–A57.
McKinney, R. M. and Thacker, L. (1976b). *Infect. Immun.* **13**, 1161–1169.

Mitchell, P. D., Hintz, C. S. and Haselby, R. C. (1977). *J. clin. Microbiol.* **5,** 658–660.

Nairn, R. C. (Ed.) (1964). "Fluorescent Protein Tracing", 2nd edn. Williams and Wilkins Co., Baltimore.

Pittman, B., Hebert, G. A., Cherry, W. B. and Taylor, G. C. (1967). *J. Immunol.* **98,** 1196–1203.

Powell, J. T. and Birdsell, D. C. (1978). *Abst. Ann. Meeting ASM* **78,** 158.

Reed, M. J. (1972). *J. dent. Res.* **51,** 1193–1202.

Reed, M. J., Levine, M. J., Black, P. and Sadowski, G. (1978). *J. dent. Res.* **57,** Special Issue A, A356.

Schaal, K. P. and Pulverer, G. (1973). *Zbl. Bakt. Abt. I Orig. A* **225,** 424–430.

Slack, J. M. and Gerencser, M. A. (1966). *J. Bact.* **91,** 2107.

Slack, J. M. and Gerencser, M. A. (1970a). In "The Actinomycetales" (H. Prausser, Ed.), pp. 19–27. Gustav Fischer, Jena.

Slack, J. M. and Gerencser, M. A. (1970b). *J. Bact.* **103,** 266–267.

Slack, J. M. and Gerencser, M. A. (1975). "Actinomyces, Filamentous Bacteria, Biology and Pathogenicity". Burgess Publishing Co., Minneapolis.

Slack, J. M., Landfried, S. and Gerencser, M. A. (1969). *J. Bact.* **97,** 873–884.

Slack, J. M., Landfried, S. and Gerencser, M. A. (1971). *J. dent. Res.* **50,** 78–82.

Slack, J. M., Ludwig, E. H., Bird, H. H. and Canby, C. M. (1951). *J. Bact.* **61,** 721–735.

Slack, J. M., Moore, D. W. and Gerencser, M. A. (1966). *W. Va. Med. J.* **62,** 228–231.

Slack, J. M., Winger, A. and Moore, D. W. (1961). *J. Bact.* **82,** 54–65.

Smith, L. Ds. (1975). "The Pathogenic Anaerobic Bacteria", 2nd edn. Charles C. Thomas, Springfield.

Wells, A. F., Miller, C. E. and Nadel, M. K. (1966). *Appl. Microbiol.* **14,** 271–275.

White, R. G. (1970). In "Standardization in Immunofluorescence" (E. J. Holborow, Ed.), pp. 49–53. Blackwell, Oxford.

CHAPTER VII

Serological Identification of Atypical Mycobacteria

WERNER B. SCHAEFER†

*National Jewish Hospital and Research Center,
Denver, Colorado 80206, U.S.A.*

I. INTRODUCTION

During the past 25 years it has become evident that tuberculosis-like diseases due to mycobacteria other than tubercle bacilli are more frequent than was assumed earlier. The mycobacteria concerned are heterogeneous. They have been called "atypical mycobacteria" (McDermott, 1965) and

† Deceased.

tentatively classified into four groups: I, photochromogens; II, scoto-chromogens; III, non-photochromogens; and IV, rapid growers (Runyon, 1959).

A more precise classification into species with common cultural and bio-chemical properties confirmed by numerical analysis of many strains has been developed (Meissner, 1974).

Serological techniques have also been used as tools for identification. The most discriminating and useful test is the seroagglutination test (Schaefer, 1965). This test is applicable only to smooth colony-forming mycobacteria. The majority of the atypical mycobacteria are of this colony form. The seroagglutination test is based on the presence of species- or type-specific antigens on the surface of the smooth colony-forming bac-teria. Rough colony variants do not have these antigens. Suspensions of rough strains are usually unstable and unsuitable for the agglutination test. *Mycobacterium tuberculosis* and *Mycobacterium bovis* form only rough colonies and cannot be identified by the seroagglutination test.

The methodology of the seroagglutination test will be described in detail and findings shedding some light on the ecology and epidemiology of the atypical mycobacteria will be presented. Results obtained with other immunological tests will also be reviewed. It should be pointed out that the seroagglutination test can be profitably applied only after the strains have been classified by their cultural and biochemical properties. The characteristics of the various species will be described and it will be indi-cated whether the seroagglutination test is applicable to them or not.

II. CLASSIFICATION OF THE ATYPICAL MYCOBACTERIA

A. *Mycobacterium kansasii*

M. kansasii, was described by Buhler and Pollock (1953) as the cause of pulmonary disease indistinguishable from tuberculosis, and called "the yellow bacillus". The name was later changed to *M. kansasii*. The colonies of this species become visible within two weeks of inoculation and incuba-tion at 37°C. They are non-pigmented when grown in the dark, but become yellow when left in daylight for one or two days. To observe the colour change it is preferable to leave the cultures in daylight for a few hours and reincubate them at 37°C overnight. After the reincubation the yellow colour is fully developed. The phenomenon of colour change after exposure to light is called photochromogenicity.

Colonies of *M. kansasii* grown on oleic acid–albumin medium are smooth, slightly wrinkled and with undulate edges. They can be identified by this characteristic appearance. Sometimes only rough colonies resembling those of *M. tuberculosis* are seen. They are distinguished from the latter by their photochromogenic property, which is not displayed by *M. tuberculosis*.

M. kansasii hydrolyses Tween-80 and reduces nitrate within five days of inoculation and incubation at 37°C.

Freshly prepared bacterial suspensions of *M. kansasii* can be identified serologically by agglutination. Older bacterial suspensions may become unstable, showing non-specific clumping which is reduced in normal or heterologous serum, whereas the specific agglutination remains in *M. kansasii* sera. Truly rough strains of *M. kansasii*, however, cannot be characterised by agglutination.

The serotyping of *M. kansasii* is particularly helpful in the identification of very sluggishly photochromogenic strains, non-pigmented (var. *album*) or scotochromogenic variants (var. *aurum*) of *M. kansasii*. The agglutination test is also useful for the differentiation of *M. kansasii* from *M. marinum*.

B. *Mycobacterium marinum*

M. marinum is the causative agent of nodular skin lesions which may appear some weeks after scratches suffered in swimming pools (Schaefer and Davis, 1961; Morgan and Blowers, 1964; Waddington, 1967), beaches and fields (Walker *et al.*, 1962) and after handling tropical fish aquaria (Swift and Cohen, 1962; Adams *et al.*, 1970; Mansson *et al.*, 1970). It was described first by Aronson (1926) as the cause of lesions in salt-water fish. A similar organism was described 25 years later as the cause of skin lesions in men acquired in a swimming pool in Sweden and named *Mycobacterium balnei* (Norden and Linell, 1951; Linell and Norden, 1954). When both species were later recognised as identical, *M. marinum* was accepted as the official name.

M. marinum is photochromogenic. It has an optimal growth at 30–34°C, compared to *M. kansasii* which grows equally well between 30 and 37°C. The lower temperature requirement of *M. marinum* explains why the organism cannot penetrate the lymph nodes or inner organs. *M. marinum* hydrolyses Tween-80, but does not reduce nitrate. The two species *M. marinum* and *M. kansasii* are rapidly differentiated by seroagglutination.

C. *Mycobacterium szulgai*

M. szulgai was discovered by Marks *et al.* (1972), by thin-layer chromatography of lipids according to Marks and Szulga (1965).

Schaefer *et al.* (1973a) subsequently showed strains of this species to be antigenically identical, being agglutinated specifically by *M. szulgai* antisera.

M. szulgai is scotochromogenic when cultured at 37°C, but photochromogenic at room temperature of 25°C. It slowly hydrolyses Tween-80, gives a positive arylsulphatase test after three to five days, and reduces nitrate. Since the original publication on seven strains, 13 more isolates have been found in the U.S.A. (Schaefer, unpublished data).

D. *Mycobacterium avium* and *Mycobacterium intracellulare*

M. avium was described in 1890 by Maffucci as the causative agent of avian tuberculosis. *M. intracellulare* is indistinguishable from *M. avium* in its cultural and biochemical properties, but does not cause tuberculosis in birds. Only in the early 1950s was it noted by Smith and Stergus (1964) at the Battey State Hospital as the apparent cause of pulmonary tuberculosis in man. It was called the "Battey" bacillus. The name was changed to *M. intracellulare* (Runyon, 1967).

M. avium and *M. intracellulare* form three main types of colonies (Figs 1 and 2): (1) The smooth transparent form (T) which is most often found on primary isolation from the infected animal or man. The colonies are

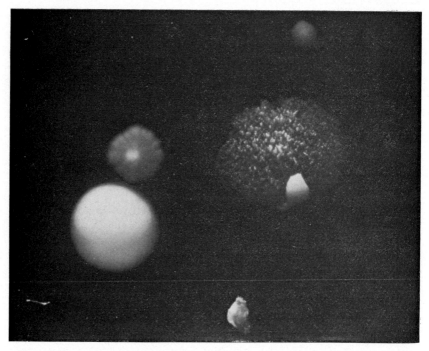

Fig. 1. Opaque, transparent and rough colonies (left to right) of *M. avium* serotype 1. The rough colony encloses a small opaque colony (original magnification × 8). Reproduced from *Path. Microbiol.* (1969) **34**, 317, with permission of the publisher.

either droplike or starlike, spreading thinly on the surface of oleic acid–albumin medium. (2) The smooth dome-shaped opaque form (D) which has a much faster growth rate than the T form and may become the predominant form on repeated transfer on artificial culture media. The T and the D forms share type-specific antigens, give stable suspensions in physiological saline, and are suitable for the agglutination test. (3) The rough form (R), which produces colonies resembling those of *M. tuberculosis*, but differs from *M. tuberculosis* in giving a negative niacin test and being completely resistant to the antitubercular drugs (Schaefer *et al.*, 1970). *M. avium* and *M. intracellulare* are usually non-pigmented or ivory coloured, although they may rarely be yellow. The pigmentation develops slowly when cultures are kept in daylight.

Differentiation between *M. avium* and *M. intracellulare* was originally based on pathogenicity in chicken after intravenous injection of 0·1–1·0 mg culture. Cultures that kill the animals in several weeks are considered to be *M. avium*, whereas non-pathogenic cultures are designated *M. intracellulare*.

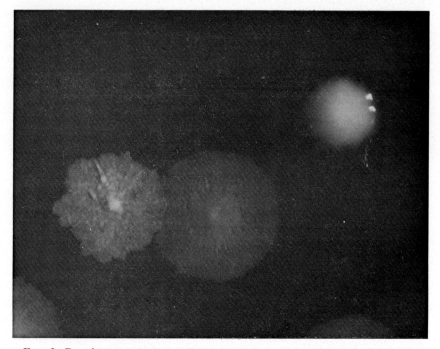

FIG. 2. Rough, transparent, and opaque (left to right) colonies of *M. avium* serotype 2, on oleic acid–albumin agar (original magnification × 8). Reproduced from *Am. Rev. resp. Dis.* (1970) **102**, 501, with permission of the publisher.

Serological study of *M. avium* and *M. intracellulare* revealed that the
avian strains belonged almost exclusively to two types: 1 and 2. Type 2
predominates (Schaefer, 1965). Later a type 3 was discovered in England
(Marks *et al.*, 1969); this has now also been found frequently on the European continent.

M. intracellulare isolated from human patients and domestic animals has
belonged to a variety of serotypes other than *M. avium*. These types were
at first designated by Roman numerals, or personal names. At present they
are designated by Arabic numbers beginning with 4 (Table I) (Wolinsky
and Schaefer, 1973).

While *M. intracellulare* in general is avirulent, strains of the serotypes 4,
5, 6, 8, and 9 exhibit a marked virulence upon intravenous injection (Anz
and Meissner, 1972). This group was therefore designated as the "intermediate" group. On the other hand, there are also attenuated and avirulent
strains of *M. avium*. The loss of virulence in *M. avium* may accompany a
change from the T to the D form. It has been shown that the T forms are
virulent whereas the D colonies isolated from the same strains are avirulent
(Schaefer *et al.*, 1970; Anz and Meissner, 1972).

TABLE I

Numbering system for the serotypes of the *M. avium* complex

	Old designation	New designation
M. avium	1, 2, 3	1, 2, 3
M. intracellulare	IV	4
	V	5
	VI	6
	VII	7
	Davis	8
	Watson	9
	IIIa	10
	IIIb	11
	Howell	12
	Chance	13
	Boone	14
	Dent	15
	Yandle	16
	Wilson	17
	Altmann	18
	Darden	19
	Arnold	20
M. scrofulaceum	Scrofulaceum	41
	Lunning	42
	Gause	43

M. avium and *M. intracellulare* have identical biochemical properties. They give a negative result in the Tween-hydrolysis and the nitrate reductase tests.

E. Mycobacterium scrofulaceum

M. scrofulaceum was described by Prissick and Masson (1956) as the causative agent of cervical lymph node infections in children. It is scotochromogenic, fails to hydrolyse Tween-80 or reduce nitrate, but is, in contrast to *M. intracellulare*, urease positive. The negative result of the Tween-hydrolysis test distinguishes *M. scrofulaceum* from the saprophytic scotochromogens.

M. scrofulaceum has three serotypes with the numbers 41, 42 and 43. The original names of these types were Scrofulaceum, Lunning and Gause (Schaefer, 1968).

F. Mycobacterium xenopi

M. xenopi was isolated by Schwabacher (1959) from a toad. Its occurrence in human infections was reported by Marks and Schwabacher (1965). It is very slow growing and forms microcolonies at 37°C. Its optimal growth temperature range is 42–45°C. The colonies are yellow and give positive results on the three-day arylsulphatase test. In contrast to *M. intracellulare*, *M. xenopi* is sensitive to Streptomycin, Isoniazid and Ethionamide (Bretey and Boisvert, 1969).

M. xenopi has been differentiated serologically from other types of mycobacteria in the agar double diffusion (Beck and Stanford, 1968), and complement fixation tests (Kantor *et al.*, 1970). Differentiation is also possible by chromatographic analysis of lipids (Jenkins *et al.*, 1972). Whether *M. xenopi* can be differentiated from other microcolony forming and non-pigmented mycobacteria is not yet known.

G. Mycobacterium simiae

The name was originally given to a group of atypical poorly growing photochromogenic mycobacteria isolated from monkeys (Karassova *et al.*, 1965). Some of these strains were niacin-positive (Weiszfeiler *et al.*, 1971). Valdivia *et al.* (1971) reported the isolated of niacin-positive Runyon group III strains resembling mycobacteria from sputa in Cuba and named them *M. habana*. Meissner and Schroeder (1975) found the niacin-positive strains of Weiszfeiler and Valdivia to be culturally and antigenically identical and designated them as *M. simiae* type 1. This type differed from *M. intracellulare* by giving positive results in the niacin and urease tests by exhibiting delayed photochromogenicity, and by being specifically agglutinated by *M. simiae* antisera. *M. simiae* 1 has been found repeatedly in human and animal infections (Krasnow and Gross, 1975; Schaefer, unpublished data). Boisvert (1974) found niacin- and urease-positive strains of *M. intracellulare* type 18 which show delayed photochromogenicity.

H. The rapidly growing mycobacteria

Among the mycobacteria that appear in clinical specimens either as contaminants or as opportunist pathogens, there is a group distinguished by rapid growth and lack of pigmentation. This category has been designated as Group IV by Runyon (1959). This group of strains is heterogeneous, but three predominating species have been differentiated by their biochemical and serological characteristics. They are *Mycobacterium fortuitum, Mycobacterium chelonei* (synonyms: *M. abscessus* and *M. borstelense*), and *Mycobacterium peregrinum*. All three give a positive reaction to the three-day arylsulphatase test. *M. fortuitum* and *M. peregrinum* reduce nitrate, whereas *M. chelonei* does not.

The three species can be identified by seroagglutination. *M. fortuitum* sera are usually specific without absorption, whereas *M. chelonei* sera and *M. peregrinum* sera become specific only after absorption by *M. peregrinum* and *M. chelonei* strains respectively. *M. fortuitum* and *M. chelonei* strains often produce unstable suspensions. Such strains can only be identified by agglutinin absorption tests using *M. fortuitum* and *M. chelonei* sera respectively. Truly rough strains are devoid of the species-specific antigens and cannot be identified serologically (Jenkins *et al.*, 1971).

I. *Mycobacterium gastri, Mycobacterium terrae, Mycobacterium triviale* and *Mycobacterium gordonae*

M. gastri was described by Wayne (1966) as a saprophytic mycobacterium. Its colonies resemble those of *M. kansasii*, but are usually nonphotochromogenic. They rapidly hydrolyse Tween-80, lack nitrate reductase, and have a moderately positive catalase test. They are specifically agglutinated by *M. gastri* antisera (Schröder and Magnusson, 1970), and their methanol or phenol extracts give a specific precipitin band with the same sera (Wayne, 1971a, b). Immunodiffusion analyses of culture filtrates show a high degree of antigenic similarity between cytoplasmic antigens of *M. gastri* and *M. kansasii* (Norlin *et al.*, 1969).

M. terrae, also described by Wayne (1966), is a saprophyte producing pale buff to white colonies. It rapidly hydrolyses Tween-80, reduces nitrate moderately, and has a high catalase activity. Antigenically, *M. terrae* is heterogeneous.

M. gordonae is the name for a widely spread group of saprophytic, scotochromogenic mycobacteria. It hydrolyses Tween-80 which readily distinguishes it from *M. scrofulaceum*. *M. gordonae* includes many serotypes, three of which have been identified (Rynearson and Wolinsky, 1967; Wolinsky and Rynearson, 1968; Jenkins et al., 1972).

M. triviale (Kubica et al., 1970) is saprophytic with non-pigmented rough, slowly growing colonies resembling those of M. tuberculosis. In contrast to M. tuberculosis it is niacin negative, has a very high catalase activity, rapidly hydrolyses Tween-80, and is weakly positive in the three-day arylsulphatase test.

III. METHODOLOGY OF SEROTYPING ATYPICAL MYCOBACTERIA

A. Culture of mycobacteria and preparation of suspensions

The mycobacterium is cultured in a tube of Tween 80–albumin medium to obtain dispersed growth. 0·1 ml of ten-fold dilutions (10^{-2}–10^{-6}) of this culture in 0·2% aqueous bovine albumin is inoculated on to oleic acid–albumin 7H11 agar medium in Petri dishes (9 cm in diameter). The inoculum is spread over the agar surface by means of a bent capillary pipette while the plate is revolving on a turntable. The pipette is sterilised by dipping in alcohol and brief flaming. The inoculated plates are enclosed in polyethylene bags and incubated in inverted position at 37°C (M. marinum is incubated at 30–33°C). The plates showing separate colonies (usually 10^{-5} and 10^{-6} dilutions) are examined for colony morphology and purity. For serological study, the bacteria are collected from the surface of the medium by emulsification in phosphate buffered saline at pH 7·0 containing 0·5% phenol (PPBS), the diluent being added slowly. The bacterial suspensions, before being used in the agglutination test, are washed with PPBS in order to eliminate soluble antigens. To be suitable for agglutination the suspensions should have an optical density of 0·3 in tubes of 20 mm outer diameter, using the Coleman Junior Spectrophotometer at 525 nm. This density corresponds roughly to tube No. 2 of the McFarland scale. Spontaneously agglutinating suspensions can sometimes be stabilised by the addition of a heterologous rabbit serum at a dilution of 1:50.

Some rapid growers give unstable suspensions when grown at 37°C, but give stable suspensions when grown at 30°C.

B. Preparation of type-specific antisera

Rabbits are injected intravenously twice weekly with 2 ml of the bacterial suspension of a density similar to that used in the agglutination test. After three days, following the fourth injection, the serum is titrated. If the titre is insufficient, the rabbit is re-injected with 1 ml or less of the bacterial suspension and the serum is retitrated after each of the subsequent injections because of the danger that the animal may die of anaphylactic shock.

When the titre has reached 1:160, the animal is exsanguinated, the serum is collected, preserved with merthiolate 1:10,000, and stored at 4°C.

A method for obtaining type-specific antisera in mice by injecting intraperitoneally the bacteria suspended in Freund's adjuvant and collecting the intraperitoneal exudate has been described by Thoen and Karlson (1970).

C. Serotyping procedure

The antisera may be stored in the refrigerator as 1:10 stock dilutions, with single working dilutions available for each antiserum. Each serotype is represented by two type-specific antisera in order to determine the cross-reactivity of two strains (see Table II). The working dilutions are pipetted into 12 × 75 mm tubes in volumes of 0·5 ml. A saline control tube is also included. To all tubes an equal volume of the bacterial suspension is added. The tubes are briefly shaken, then deposited in an incubator at 37°C. After 3–5 h, the agglutination results are read without shaking the tubes, and read again the following day.

D. Interpretation of results

A strain is classified as a particular serotype if the bacterial suspension is agglutinated by both antisera of this type and is not agglutinated by the antisera of other types. The type-specific agglutination usually becomes visible within the first 3 h of incubation. With some strains it becomes visible only after 6 h, or over-night. Sometimes agglutination remains incomplete. It is then necessary to carry out the agglutinin absorption test. If agglutinated by a single serum only, the specificity of this agglutination should be verified by an absorption test. Certain strains possess such small amounts of the type-specific antigens that they are agglutinated only by the stronger of the two antisera of the homologous type.

E. Agglutinin absorption test

A 0·5 ml portion of the 1:10 diluted antiserum is mixed in a centrifuge tube with an equal volume of a very dense suspension of the bacteria to be tested. This suspension is made by suspending the bacterial sediment, obtained by centrifugation, in a double volume of PPBS. The mixture is kept at 4°C for an hour, centrifuged at high speed, and the supernatant which now corresponds to a dilution of 1:20 is transferred to an agglutination tube, and two-fold dilutions made with equal volumes of PPBS to 1:640. A parallel series of dilutions is made beyond the titre of the unabsorbed serum. A 0·5 ml aliquot of a stable homologous bacterial suspension of the homologous type is added to all tubes. The tubes are incubated at 37°C and the agglutination results read the following day. Strains which

TABLE II

Absorption required to make antisera of *M. avium*, *M. intracellulare* and *M. scrofulaceum* type-specific

Antiserum Type designation		Absorbed by Type designation		Antiserum Type designation		Absorbed by Type designation	
New	Old	New	Old	New	Old	New	Old
1	*M. avium* 1	2	*M. avium* 1	14	Boone	15	Dent
2	*M. avium* 2	1	*M. avium* 1	15	Dent	14	Boone
3	*M. avium* 3	1 and 2	*M. avium* 1 and 2	16	Yandle	17	Wilson
4	*M. intracellulare* IV	0	Howell and Chance	17	Wilson	16	Yandle
6	*M. intracellulare* VI	0		18	Altmann	2	*M. avium* 2
7	*M. intracellulare* VII	12 and 13	Howell and Chance	19	Darden	0	0
8	*M. intracellulare* Davis	21		20	Arnold	0	0
9	*M. intracellulare* Watson	0		21	—	8	Davis
10	*M. intracellulare* IIIa	0		41	Scrofulaceum	20	0
11	*M. intracellulare* IIIb	10		42	Lunning	0	0
12	*M. intracellulare* Howell	13 and 7	Howell	43	Gause	0	0
13	*M. intracellulare* Chance	12 and 7	Howell VII				

1. Types 1 and 2, types 3 and 2, types 7, 12, and 13, types 8 and 21, types 14 and 15 and types 16 and 17 cross–agglutinate. Their sera are therefore absorbed by the bacteria of the cross–reacting type.

2. The bacteria of type 18 are co–agglutinated by type 2 sera, and the bacteria of type 11 by the sera of type 10. The sera of these types are therefore absorbed by the bacteria of the cross–reacting type.

3. Strains of type 9 are co–agglutinated by type 3 sera. Since absorption of type 9 sera by type 3 cells removes most of the agglutinins of these sera, we omit this absorption and consider strains agglutinated by type 9 and type 3 sera as belonging to type 9, whereas bacteria agglutinated only by type 3 sera are identified as type 3.

4. Strains of type 43 often cross–agglutinate with type 20 sera. Since absorption of type 43 sera by type 20 strains removes most of the agglutinins of these sera, we omit this absorption and consider strains agglutinated by type 43 and type 20 as belonging to type 43, whereas strains of type 20 are agglutinated by only type 20 sera and are identified as type 20.

5. Type 5 has been excluded from the typing procedure because of its rare occurrence. (Type 5 has not been observed in human infections but only in animals.)

Antisera and reference strains for all serotypes can be obtained from National Jewish Hospital and Research Center, 3800 East Colfax Avenue, Denver, Colorado 80206, U.S.A. The requested materials are provided free of charge under a contract authorised by the U.S.A. National Institutes of Health, Bethesda, Maryland.

reduce an agglutinin titre 16-fold or more are considered to possess the homologous antigen. An eight-fold reduction of titre is interpreted to indicate that the strain is closely related to, but not quite identical with, the homologous antigen. A two- to four-fold reduction of titre and no reduction of titre is considered to indicate non-identity.

After termination of the agglutination test the tubes undergo routine cleaning procedure. They are kept in a phenolic disinfectant solution overnight, autoclaved, washed and dried for future use.

Several modifications of the agglutination test have been proposed. (1) a slide agglutination test (Engel and Beerwald, 1970), (2) a "simplified" agglutination test using volumes of antisera and antigens of 0·1 instead of 0·5 ml and only a single antiserum for each type (Reznikov and Leggo, 1972), (3) an agglutination inhibition test for the differentiation of *M. avium* from *M. intracellulare* (Richards and Eacret, 1972), (4) a micro-method for serotyping strains of the *M. avium* complex (Thoen *et al.*, 1975).

F. Preparation of absorbed antisera

Because of cross-reactions between mycobacterial species (Schaefer, 1967), absorptions have to be carried out in the following way: 5–8 ml non-absorbed undiluted antiserum is mixed with 1 ml of the sediment of a strain of each coagglutinating type. The mixture is incubated at 4°C overnight and centrifuged at 5 rev/min for 10 min. The serum is decanted and its homologous and heterologous titres are determined by using suspensions of two homologous and two heterologous strains as the antigens. If the serum is not yet type-specific the absorption is repeated. The antiserum is used at the lowest dilution at which it is type-specific. Table II indicates the coagglutinating types for 23 serotypes of *M. avium*, *M. intracellulare* and *M. scrofulaceum*.

IV. CHEMISTRY OF THE TYPE-SPECIFIC ANTIGENS

The specific antigens of the atypical mycobacteria can be extracted from the bacteria by boiling methanol. The bulk of the antigen is contained in the acetone-soluble fraction of the methanol extract (Schaefer, 1964). The antigens are also soluble in phenol (Wayne, 1971a, b).

Work in progress on the purification of the specific antigens (Brokl, Schaefer and Goren, unpublished) indicates that the antigens of *M. intracellulare* serotype 8 are very likely peptidoglycolipids, whereas those from *M. szulgai* and *M. kansasii* may be lipo-oligosaccharides containing only a small portion of a nitrogenous component. The specific substances are obtained from methanol extracts by a combination of solvent partitioning, column chromatography and preparative thin-layer chromatography. Two serologically active fractions from serotype 8 differ in their content of carbohydrate components, and have a similar distribution of simple fatty acids (principally palmitic, stearic, oleic (minor), tuberculostearic, and tetracosanoic) and of amino-acids, principally alanine and threonine. The serologically active component(s) of *M. szulgai* and *M. kansasii* appear to be simpler in structure, with little or no peptide content. Both contain only a small, but similar amount of a nitrogenous, as yet unidentified, substance. This may be a basic amino-acid, or an amino-alcohol.

V. THE PROTOPLASMIC ANTIGENS

In addition to their type-specific surface antigens mycobacteria have numerous protoplasmic antigens. They are present in culture filtrates and in bacterial extracts obtained by sonic disintegration. At least 17 different antigens have been detected in the filtrates of *M. tuberculosis*. The specificities of the protoplasmic antigens from various mycobacterial species have been investigated using the Ouchterlony immunodiffusion technique. These studies have shown that some of the antigens are common to many species, others are common to only some of them and still others are species specific (Lind, 1965).

Stanford and Muse (1969) used the Ouchterlony technique to determine whether the strains of the *M. avium*-Battey group belonged to a single or different species. The *M. avium* antiserum tested against the homologous extract gave 12 precipitin lines. Tested with extracts of other strains the sera gave either 12 precipitin lines (serotype A) or ten precipitin lines (serotype B). Virulence tests in chickens and rabbits indicated that the strains giving 12 precipitin lines (serotype A) were virulent for chickens and rabbits and therefore were *M. avium*. The strains giving ten precipitin lines (serotype B) were either avirulent for chicken and rabbits or virulent for the chicken only. These latter strains therefore were either Battey strains or attenuated *M. avium* strains. Thus, this test enables us to differentiate fully virulent strains from attenuated *M. avium* or *M. intracellulare* strains.

The most remarkable property of the protoplasmic antigens is their ability to elicit delayed hypersensitivity reactions of the tuberculin type. Stottmeyer *et al.* (1969) reported that the tuberculin activity and specificity of purified protein derivatives and protoplasmic extracts of various species of mycobacteria paralleled their antigenic activity and specificity in immunodiffusion tests.

Considerable work has been done, using immunoelectrophoresis (Castelnuovo and Morellini, 1965), ion-exchange chromatography (Kniker,

1965), discontinuous pore gradient gel electrophoresis (Affronti *et al.*, 1972), or two-dimensional polyacrylamide electrophoresis (Wright *et al.*, 1972) to separate the protoplasmic antigens of *M. tuberculosis* and other mycobacteria.

The antigens of the mycobacterial cell walls of various species of atypical mycobacteria and of a strain of *M. tuberculosis* have been studied by Wong *et al.* (1970). A fraction solubilised with lysozyme was relatively species specific in double immunodiffusion and guinea-pig skin tests. This fraction appeared to be a lipopolysaccharide.

VI. ECOLOGY

The atypical mycobacteria are widespread in their distribution, varying from place to place. Bailey *et al.* (1970) found *M. kansasii* to be present in water taps. Bullin *et al.* (1970) found *M. xenopi* at the same source. Beerwerth and Schürmann (1969) found group III strains frequently in arable soil. Reznikov *et al.* (1971a) found them in house dust.

Reznikov *et al.* (1971b) found that the serotypes of *M. intracellulare* predominant in house-dust samples were also predominant in sputum specimens. This finding suggested that the environment was the source of the strains found in the sputa. Kleeberg and Nel (1969, 1973) investigated animal feeds mixed with wood shavings and estimated that at certain farms, pigs daily eat 3 kg of certain feed mixtures containing 5000–7500 cells of *M. intracellulare*. Tsukamura *et al.* (1974) reported that *M. intracellulare* was rare in room dust, but frequent in sputa. In order to compare the antigenic components obtained in different laboratories, a reference system consisting of an *M. tuberculosis* H37Rv culture filtrate and a goat antiserum was established by a committee (Janicki, 1971, 1972). Thirteen bands were identified immunoelectrophoretically and each assigned a specific number. The same preparations in crossed immunoelectrophoresis showed 36 antigen-antibody bands (Wright and Roberts, 1974). In only few instances, however, have the bands been related to components of known chemical composition.

Chase and Kawata (1974) found that guinea-pigs immunised with killed mycobacteria in paraffin oil or living BCG were induced to produce antibodies specific for different mycobacterial antigens contained in H37Rv filtrates. They then used antisera specific for each of the ten antigenic components of these culture filtrates in passive cutaneous anaphylaxis experiments, challenging the animals by intravenous injection of various preparations of tuberculin and PPDs. The capacity of the tuberculin to elicit a reaction at one of the sites where the sera had been injected indicated the presence of the homologous antigens in the tuberculin samples. Using this method they were able to determine which of the antigens were present in a tuberculin preparation and their relative amounts.

Some insight into the distribution of the serotypes of *M. avium*, *M. intracellulare*, and *M. scrofulaceum* has been gained from an investigation of strains isolated from human sources in Wales, New South Wales and the U.S.A. (Table III) during the last five years (Schaefer, unpublished results).

Table II shows that a total of 757 strains was studied and that 73·8% of the strains were identified serologically. The unclassified strains were either rough or belonged to serotypes not represented in the serotyping procedure. It is of interest that the strains of clinical significance from Wales (repeated sputum isolations and strains from cervical lymph nodes) and those from New South Wales (cervical lymph nodes) gave a higher percentage of typeable strains (86·7, 84·5 and 78·0%) than the strains from the U.S.A., the origin of which was often unknown (66·2 and 73·5%).

It appears from this study that the 23 serotypes used in the typing procedure (type 5 was excluded because of its rarity) identified nearly 75% of the strains from human sources. Some regional differences in type distribution were observed: *M. avium* type 2 in Wales accounted for 12·5% of all isolates from sputa or lymph nodes but in the U.S.A. only for 2%. *M. avium* 2 in Wales was 2·9 times more frequent than *M. avium* 1, whereas in the U.S.A. *M. avium* 1 was 3·7 times more frequent than *M. avium* 2.

The *M. scrofulaceum* types (41–43) from Wales accounted for 9·8% of the isolates from cervical lymph nodes, whereas in New South Wales they accounted for 47·6%.

A study of 77 strains isolated in Japan from patients with pulmonary disease (Nemoto *et al.*, 1975) indicated that 26 (33·8%) were untypable because of spontaneous agglutination. Five belonged to the *M. scrofulaceum* types and 46 to the *M. avium-M. intracellulare* complex. Only one strain of *M. avium* 1 and no strains of *M. avium* 2 were found.

VII. EPIDEMIOLOGY

M. avium, especially *M. avium* 2, is undoubtedly the type which is most pathogenic and most easily transmitted to other animals (Schaefer, 1968). The source of infection is usually an old tuberculous chicken with open intestinal lesions the content of which is shed with the faeces on to the soil where these bacteria can persist for many years. The bacteria are picked up from the soil by other domestic animals such as chickens, cattle, and swine. The infection thus, because of the persistence of the bacteria in the soil, easily becomes endemic. *M. avium* 1 plays a similar role, but type 1

TABLE III

Incidence of serotypes of *M. avium*, *M. intracellulare* and *M. scrofulaceum* in strains isolated from sputa or cervical lymph nodes

Serotype new	Designation old	Britain (Wales) Sputum	Britain (Wales) Lymph nodes	Australia (NSW) Lymph nodes	U.S.A. CDC state lab	NJHRC	Totals
1	*M. avium* 1	7	1	0	26	15	49
2	*M. avium* 2	17	6	0	9	2	34
3	*M. avium* 3	1	4	0	0	0	5
4	IV	4	7	1	6	5	23
6	VI	17	4	1	3	2	27
7	VII	2	1	0	12	7	22
8	Davis	10	11	0	10	12	43
9	Watson	6	7	0	15	13	41
10/11	IIIa and b	2	1	0	0	1	4
12	Howell	1	0	2	15	9	27
13	Chance	0	0	0	12	14	26
14	Boone	13	4	6	20	12	55
15	Dent	0	1	0	1	6	8
16	Yandle	4	2	2	14	13	35
17	Wilson	2	0	0	4	4	10

		1	2	3	4	5
18	Altmann	0	0	2	5	8
19	Darden	7	0	33	28	70
20	Arnold	1	0	0	1	2
21	21	1	0	0	0	2
41	Scrofulaceum	2	2	0	3	11
42	Lunning	0	4	14	12	31
43	Gause	1	14	4	5	26
	Identified strains	98	32	200	169	559
	Unclassified strains	15	9	102	61	198
	Total	113	41	302	230	757
	Percentage identified	86·7	78·0	66·2	73·5	73·8

The cultures to be serotyped came from the following sources:

I. The Tuberculosis Reference Laboratory, Cardiff, Wales. Repeated isolations from sputum (column 1). Cervical lymph nodes (column 2).

II. The Clinical and Pathological Research Laboratory in Newcombe (N.S.W.), Australia. Cervical lymph nodes (column 3).

III. The Tuberculosis Laboratory of the Center for Disease Control, Atlanta, Georgia (CDC). The cultures were sent to CDC from various state laboratories (column 4).

IV. The Tuberculosis Reference Laboratory of the National Jewish Hospital and Research Center, 3800 East Colfax Avenue, Denver, Colorado 80206 (column 5).

Information on the source of the cultures involved in studies III and IV was often missing.

seems to be relatively more prevalent in free-living birds, such as ducks
(Schaefer *et al.*, 1973b). *M. avium* 3 appeared to have been limited to
European countries until recently, when a strain of this type was isolated
in the U.S.A. from a tree duck by Thoen (unpublished data, verified by
Schaefer). The bacteriological and epidemiological features of Danish *M.
avium* infections in animals and in man were described by Engbaek *et al.*
(1968), and enzootics of avian tuberculosis in pigs by Kleeberg and Nel
(1969), and Jørgensen *et al.* (1972).

Infections with *M. intracellulare*, *M. scrofulaceum*, and other atypical
mycobacteria are nearly always sporadic and caused in man as well as in
animals predominantly by the serotypes represented in the typing procedure
(Schaefer, 1968; Schaefer *et al.*, 1969). Epidemics have been encountered
only in circumstances where the bacteria could multiply freely in a closed
reservoir such as a swimming pool or water storage tanks. Four such
occurrence have been observed: (1) Swimming pool infections by *Mycobacterium marinum*. They occurred in Sweden (Linell and Norden, 1954),
England (Morgan and Blowers, 1964), and the U.S.A. In the latter instance, it was at a very old, poorly built outdoor swimming pool receiving
its water from natural hot springs and from a river. The water was not
chlorinated and contained 1000 bacteria (*M. marinum*)/ml. Some 300 cases
of skin infection by this organism were found. Most of the lesions were on
the elbows because these were injured when holding on to a rope at the
edge of the pool. These injuries evidently served as the bacterial port of
entry (Schaefer and Davis, 1961). (2) An outbreak of tuberculin reactivity
among chickens was observed on three neighbouring farms in Czechoslovakia and a mycobacterium called *M. brunense* (Kazda, 1967b) later
identified as *M. intracellulare* serotype 8, was the causative agent (Kubin
et al., 1969). The infection was traced to open tanks from which the chickens
received their drinking water. The tanks were heavily contaminated by the
mycobacteria. Elimination of the source of infection terminated the tuberculin sensitivity within a few months (Kazda, 1967a). (3) An outbreak of
lymphadenitis affecting 67% of the pigs of a deep litter piggery in S.E.
Queensland was due to *M. intracellulare* serotype 6. The source of infection
was the deep litter which was never changed and the water from a trough
in the deep litter pen (Tammenagi and Simmons, 1968; Reznikov, 1970).
(4) Finally, contamination of numerous specimens of sputa by *M. intracellulare* serotype 8 has been observed in a hospital and traced to the
presence of the bacteria in the water storage tank (Schaefer, unpublished
observations).

ACKNOWLEDGEMENTS

Our thanks are due to Dr Mayer B. Goren for writing the section on the chemistry
of the type-specific antigens and to Mr Ned Eig for revising the manuscript.

The personal studies reported herein were supported by funds of the National Institute of Allergy and Infectious Diseases, National Institutes of Health, Contract No. AI-02079.

REFERENCES

Adams, R. M., Remington, J. S., Steinberg, G. J. and Seibert, J. S. (1970). *J. Am. med. Ass.* **211**, 457–461.
Affronti, L. F., Wright, G. L. and Reich, M. (1972). *Infect. Immun.* **5**, 474–481.
Anz, W. and Meissner, G. (1972). *Zbl. Bakt. Abt. I Orig.* **22**, 334–342.
Aronson, J. D. (1926). *J. infect. Dis.* **39**, 315–320.
Bailey, R. K., Wyler, S., Dingley, M., Hesse, F. and Kent, G. W. (1970). *Am. Rev. resp. Dis.* **101**, 430–431.
Beck, A. and Stanford, J. L. (1968). *Tubercle* **49**, 226–234.
Beerwerth, W. and Schürmann, J. (1969). *Zbl. Bakt. Abt. 1 Orig.* **211**, 58–69.
Boisvert, H. (1974). *Bull. Soc. path. Exotique* **67**, 458–465.
Bretey, J. and Boisvert, H. (1969). *Rev. Tuberc. Pneumol.* **33**, 337–345.
Buhler, V. B. and Pollak, A. (1953). *Am. J. clin. Path.* **23**, 363–374.
Bullin, C. H., Tanner, E. J. and Collins, C. H. (1970). *J. Hyg. (Camb.)* **68**, 97–100.
Castelnuovo, G. and Morellini, M. (1965). *Am. Rev. resp. Dis.* **92**, Suppl. 29–33.
Chaparas, S. D. and Hedrick, S. R. (1973). *Infect. Immun.* **7**, 770–780.
Chase, M. W. and Kawata, H. (1974). International WHO–Lab Symposium on Standardisation and Control of Allergens Administered to Man, Geneva, 1974. *Devel. Biol. Stand.* **29**, 308–330.
Engbaek, H. C., Vergman, B., Baess, J. and Bentzon, M. W. (1968). *Acta path. microbiol. scand.* **72**, 277–294, 295–312.
Engel, H. W. B. and Beerwald, L. G. (1970). *Am. Rev. resp. Dis.* **101**, 112–115.
Janicki, B. W. (1971). *Am. Rev. resp. Dis.* **104**, 602–604.
Janicki B. W. (1972). *Am. Rev. resp. Dis.* **106**, 142–147.
Jenkins, P. A., Marks, J. and Schaefer, W. B. (1971). *Am. Rev. resp. Dis.* **103**, 179–187.
Jenkins, P. A., Marks, J. and Schaefer, W. B. (1972). *Tubercle* **53**, 118–127.
Jørgensen, J. B., Engbaek, H. C. and Dam, A. (1972). *Acta vet. scand.* **13**, 68–86.
Kantor, J., Tacquet, A. et Debruyne, J. (1970). *Ann. Institut Pasteur, Lille* **31**, 115–130.
Karassova, V., Weiszfeiler, G. and Karzgy, E. (1965). *Acta microbiol. hung.* **12**, 275–282.
Kazda, J. (1967a). *Zbl. Bakt. Abt. I Orig.* **203**, 92–101.
Kazda, J. (1967b). *Zbl. Bact. Abt. I Orig.* **203**, 199–211.
Kleeberg, H. H. and Nel, E. E. (1969). *J. S. Afr. med. Ass.* **40**, 233–250.
Kleeberg, H. H. and Nel, E. E. (1973). *Am. Soc. Belge med. trop.* **83**, 405–416.
Kniker, W. T. (1965). *Am. Rev. resp. Dis.* **92**, Suppl., 19–33.
Krasnow, J. and Gross, W. (1975). *Am. Rev. resp. Dis.* **111**, 357–360.
Kubica, G. D., Silcox, V. A., Kilburn, J. O., Smithwith, R. W., Beam, R. E., Jones, W. D., Jr and Stottmeier, K. D. (1970). *Int. J. synt. Bact.* **20**, 161–174.
Krasnow, J. and Gross, W. (1975). *Am. rev. Resp. Dis.* **111**, 357–360.
Kubin, M., Matuskova, E. and Kazda, J. (1969). *Zbl. Bakt. Abt. I Orig.* **210**, 207–211.
Lind, A. (1965). *Am. Rev. resp. Dis.* **92**, Suppl., 54–62.
Linell, F. and Norden, A. (1954). *Acta. tuberc. scand.*, Suppl. XXXIII, 5–83.

Maffucci, A. (1890). *Zbl. allg. Path. Anat.* **7**, 409.

Mansson, T., Brehmer-Anderson, E., Wittbeck, B. and Grubb, R. (1970). *Acta Derm-vener., Stockh.* **50**, 119–124.

Marks, J. and Schwabacher, H. (1965). *Br. med. J.* **1**, 32–33.

Marks, J. and Szulga, T. (1965). *Tubercle* **46**, 400–411.

Marks, J., Jenkins, P. A. and Schaefer, W. B. (1969). *Tubercle* **50**, 394–395.

Marks, J., Jenkins, P. A. and Tsukamura, M. (1972). *Tubercle* **53**, 210–214.

Meissner, G. (1974). *J. gen. Microbiol.* **83**, 207–235.

Meissner, G. and Schroeder, K. H. (1975). *Am. Rev. resp. Dis.* **111**, 196–200.

Morgan, J. K. and Blowers, R. (1964). *Lancet* **i**, 1034–1036.

McDermott, W. (1965). *Am. Rev. resp. Dis.* **91**, 289.

Nemoto, H., Yugi, H. and Tsukamura, M. (1975). *Jap. J. Microbiol.* **19**, 69–71.

Norden, A. and Linell, F. (1951). *Nature, Lond.* **168**, 826.

Norlin, M., Lind, A. and Ouchterlony, Ö. (1969). *Z. Immunforsch. Allerg., Klin. Immunol.* **137**, 241–248.

Prissick, F. H. and Masson, A. M. (1956). *Can. med. Ass. J.* **75**, 798–803.

Reznikov, M. (1970). *Aust. vet. J.* **46**, 239.

Reznikov, M., Leggo, J. H. and Dawson, D. J. (1971). *Am. Rev. resp. Dis.* **104**, 951–953.

Reznikov, M., Leggo, J. H. and Tuffley, R. E. (1971). *Aust. vet. J.* **47**, 622–623.

Reznikov, M. and Leggo, J. H. (1972). *Appl. Microbiol.* **23**, 189–823.

Richardes, W. B. and Eacret, W. G. (1972). *Appl. Microbiol.* **24**, 318–322.

Runyon, E. H. (1959). *Med. Clin. N. Am.* **43**, 273–290.

Runyon, E. H. (1967). *Am. Rev. resp. Dis.* **91**, 289.

Rynearson, T. K. and Wolinsky, E. (1967). *Am. Rev. resp. Dis.* **96**, 155 (Abstract).

Schaefer, W. B. and Davis, C. L. (1961). *Am. Rev. resp. Dis.* **84**, 837–844.

Schaefer, W. B. (1964). *Bact. Proc. M* **80**.

Schaefer, W. B. (1965). *Am. Rev. resp. Dis.* **92**, 85–93.

Schaefer, W. B. (1967). *Am. Rev. resp. Dis.* **96**, 1165–1168.

Schaefer, W. B. (1968). *Am. Rev. resp. Dis.* **97**, 18–23.

Schaefer, W. B., Brim, K. J., Jenkins, P. A. and Marks, J. (1969). *Br. med. J.* **2**, 412–415.

Schaefer, W. B., Davis, C. L. and Cohn, M. L. (1970). *Am. Rev. resp. Dis.* **102**, 499–506.

Schaefer, W. B., Wolinsky, E., Jenkins, P. A. and Marks, J. (1973a). *Am. Rev. resp. Dis.* **108**, 1320–1326.

Schaefer, W. B., Beer, J. V., Wood, N. A., Boughton, E. Jenkins, P. A. and Marks, J. (1973b). *J. Hyg. (Camb.)* **71**, 549–557.

Schröder, K. H. and Magnusson, M. (1970). *Rev. Tuberc. Pneumol.* **34**, 73–78.

Schwabacher, H. (1959). *J. Hyg. (Camb.)* **71**, 57–67.

Smith, C. E. and Stergus, J. (1964). *Am. Rev. resp. Dis.* **89**, 497–502.

Stanford, J. L. and Muse, R. (1969). *Tubercle* **50**, Suppl., 80–83.

Stottmeyer, K. D., Beam, R. E. and Kubica, G. P. (1969). *J. Bact.* **100**, 201–208.

Swift, S. and Cohen, H. (1962). *New Engl. J. Med.* **267**, 1244–1246.

Tammenagi, L. and Simmons, G. C. (1968). *Aust. vet. J.* **44**, 121.

Thoen, C. O. and Karlson, A. G. (1970). *Appl. Microbiol.* **20**, 847–848.

Thoen, C. O., Jarnagin, J. L. and Champion, N. L. (1975). *J. clin. Microbiol.* **1**, 469–477.

Tsukamura, M., Mizuno, S., Nemoto, M. and Yugi, H. (1974). *Jap. J. Microbiol.* **18**, 271–277.

Valdivia, J. A., Suares, R. M. and Echemendia, M. F. (1971). *Boln Hig. Epid.* **9**, 65–73.

Waddington, E. (1967). *Trans. a. Rep. St. John's Hosp. derm. Soc., Lond.* **53**, 122–124.

Walker, H. H., Shin, F. M., Higaki, M. and Ogafa, J. (1962). *Hawaii med. J.* **21**, 403–409.

Wayne, L. G. (1966). *Am. Rev. resp. Dis.* **93**, 919–928.

Wayne, L. G. (1971a). *J. gen. Microbiol.* **66**, 255–271.

Wayne, L. G. (1971b). *Infect. Immun.* **3**, 36–46.

Weiszfeiler, G., Karrasseva, V. and Karzag, E. (1971). *Acta microbiol. hung.* **18**, 247–252.

Wolinsky, E. and Rynearson, T. K. (1968). *Am. Rev. resp. Dis.* **97**, 1032–1037.

Wolinsky, E. and Schaefer, W. B. (1973). *Int. J. Synt. Bact.* **23**, 182–183.

Wong, K. H., Pickett, M. J. and Froman, S. (1970). *Infect. Immun.* **1**, 400–407.

Wright, G. L., Jr, Affronti, L. F. and Reich, M. (1972). *Infect. Immun.* **5**, 482–490.

Wright, G. L. and Roberts, D. B. (1974). *Am. Rev. resp. Dis.* **109**, 306–310.

CHAPTER VIII

Methods for Bacteriophage Typing of Mycobacteria

WILLIAM B. REDMOND

Veterans Administration Hospital and Emory University School of Medicine, Atlanta, Georgia 30322, U.S.A.

JOSEPH H. BATES

Veterans Administration Hospital and University of Arkansas School of Medicine, Little Rock, Arkansas 72201, U.S.A.

H. W. B. ENGEL

Rijks Instituut Voor de Volksgezondheid, Bilthoven, Netherlands

I. INTRODUCTION

A. Early investigations

Apparently there was little interest in mycobacterial phages before about 1950. The acceptance of the "atypical" mycobacteria into the realm of human infection (tuberculosis) appears to have stimulated several investigators to seek means of identifying and of showing relationships among the "organisms which will not be hustled" (Edson, 1951, p. 148). This time-consuming factor may have discouraged many workers who might otherwise have turned some of their thoughts and activities to the phages. A more likely discouraging factor was the consensus that the tubercle bacillus was a *closed entity* having little relative importance to other acid-fast bacilli, and for which sufficient methods of identification were available (Dubos, 1954).

The first isolations of bacteriophages showing activity on the mycobacteria were followed by attempts to identify various strains or species of mycobacteria by phage typing. Only a few phages exhibited lytic activity on strains other than the saprophytic bacteria which had been used in most cases for the isolations (Hauduroy and Rosset, 1948; Hnatko, 1953; Takeya and Yoshimura, 1957). Phages that were active on more than one or two types appeared to have activity on saprophytic, "atypical" and pathogenic bacteria rendering them of little value as typing phages.

Due to these facts, many workers concluded that there was little or no possibility of utilising the phage typing technique for identifying the species of the genus *Mycobacterium*. When it was later shown that phages specific for pathogenic tubercle bacilli could be isolated, that these phages could be used for identifying these important organisms, and that there was a possibility that human and bovine types might be differentiated, interest in mycobacterial phage typing increased considerably (Redmond, 1963; Šula and Mohelska, 1965).

B. Stimulus by WHO

The Tuberculosis Division of the World Health Organisation and particularly the WHO International Reference Centre for the Diagnosis of Tuberculosis in Prague was instrumental in organising a group of laboratory workers interested in developing methods and techniques for phage typing the mycobacteria. A small group of workers was assembled in Prague at the International Reference Centre in 1965 in order to discuss and outline the problems and possible methods for phage typing mycobacteria (Šula and Mohelska, 1965).

At this first meeting some preliminary results were presented and problems and methods of procedure were discussed. Among the problems recognised as requiring concerted efforts of all the laboratories were:

1. The need for more phages with specific activities on strains of all groups of mycobacteria, especially the pathogenic human, bovine and avian groups.

2. Media for propagating the phages and for use in the typing tests.

3. More appropriate methods of storing, and especially of shipping phages to workers in other laboratories.

4. Better criteria for determining and standardizing methods for reporting the extent of lysis, including the routine test dilution (RTD).

In addition to the problems recognised there were a few positive aspects presented also.

A new medium (liquid and solid) had been formulated (Redmond and Ward, 1966) and tested in several laboratories using comparable methods. It was considered to be adequate for use in further studies. A small controlled experiment carried out in five countries on the same bacterial strains and using the same phages had demonstrated that comparable results could be obtained in laboratories in different countries. Numerous phages that had been isolated, demonstrated the widespread nature of mycobacterial phages. Several lysogenic strains of mycobacteria had been found and temperate phages isolated and studied (Hnatko, 1953; Segawa et al., 1960; Bowman and Redmond, 1959).

Subsequent meetings of the phage group, with several additions to the group, have been held in Utrecht (1967), Atlanta (1969), Göteborg (1971), Montreal (1972), Tokyo (1973), Pisa (1974), and Leicester (1975), Little Rock (1976), and Paris (1977).

The methods that have been worked out and established as having significant possibilities for phage typing of the mycobacteria have been the results of co-operative studies and much testing by the several laboratories involved. In addition, individuals have performed specific research in the field thus adding to the possibilities. This is especially true of isolation of new phages and of morphological and biochemical studies on several of the phages (Buraczewska et al., 1971, 1972; Jones and White, 1968).

Although various species of mycobacteria were shown as early as 1954 to differ in susceptibility to mycobacteriophages (Froman et al., 1954), the principal application of this phenomenon did not develop until more than a decade later. Following their initial discovery by Gardner and Weiser (1947), increasing numbers of phages were described and studied primarily for biochemical characteristics, morphology, and methodology of quantitative techniques. Takeya et al. (1959) first reported that when careful attention was given to the number of phage particles required to produce a single plaque on a lawn of a mycobacterium, marked differences among members of the same species could be shown. They further showed that eight strains of M. *tuberculosis*, attenuated by repeated culture on Calmette's bile medium, were less susceptible to phage than the original virulent strains even though the human strains H37Rv and H37Ra showed identical susceptibility patterns. By 1963, sufficient information had been published on the subject to permit a review article on mycobacteriophages (Redmond, 1963), but no mention of epidemiologic studies was made. In this review the phage GS4E was described regarding its isolation as a mutant from an original phage obtained from garden soil. Phage GS4E was significant in that it was recognised to lyse the M. *tuberculosis* strain H37Rv, but would not lyse any of a group of strains of M. *bovis*. In the same year Murohashi *et al.* (1963) reported the capacity of GS4E to subdivide the human strain of mycobacteria. Little attention was given to these early reports, and most investigators considered that mycobacteriophages which lysed one member of a species would lyse all others. Thus, the continuing emphasis was on isolating and characterising phages that would be useful in taxonomy and in genetic studies of mycobacteria. However, as improved biochemical techniques were developed that allowed a more precise characterisation of mycobacteria, the expectation of utilising phage typing as a taxonomic tool waned.

II. MEDIA

Probably the most important factor leading to the variations and diversity of results obtained in early attempts to phage type mycobacteria was the medium used in the tests. It soon became apparent from experiments employing the same phages and bacteria in various laboratories that the composition of the medium caused wide variations.

Several components commonly found in culture media have an inhibitory action on phage. These may have little or no inhibitory effect on bacterial growth. Probably the most important of these factors are those that interfere with adsorption of the phage and its subsequent introduction into the bacterial cell. Many, or most, of the media that have been found to be quite adequate for primary isolation of mycobacteria, e.g. Lowenstein–Jensen (L–J) Medium, those used primarily in assaying for drug susceptibility (Middlebrook's 7H9–10), or others slightly modified for distinguishing between growth characteristics of strains, have not been found to be adequate for use in phage typing. A series of tests involving most of the components used in these media were carried out in the Atlanta Laboratory to determine the effect on phage activity. Details of these tests are to be found in a publication by Redmond and Ward (1966). Calcium ions have been found to be necessary for lysis. Citrate and phosphates are detrimental because they tend to inactivate or to precipitate calcium. A high concentration of agar and Malachite green, as incorporated in L–J medium, reduced the number of plaques. Also, surface active agents such as Tween-80 were found to be detrimental.

As a result of information obtained from these tests, a new medium for bacteriophage studies on the mycobacteria was formulated (Redmond and Ward, 1966). This is the RVA, RVB medium that has been tested in several laboratories in a number of different countries. It has been recommended by the International Committee on Phage Typing of Mycobacteria associated with the various laboratories involved in the phage typing studies.

Only slight differences other than the addition of agar exist between RVA and RVB media; however, the formulae and procedure for making each are given.

A. RVA-17 agar medium

It has been found necessary to follow certain definite procedures for combining some of the chemical substances in order to prevent undesirable reactions. For best results these procedures should be followed in making the medium.

RVA-17

Solution A

Na$_2$HPO$_4$ (anhydrous)	1·2 g
KH$_2$PO$_4$	0·55 g
NaCl	2·5 g
NH$_4$Cl	0·5 g
Sodium pyruvate	0·75 g
Nutrient broth (Difco)	2·0 g
Proteose peptone No. 3 (Difco)	2·5 g
Casein hydrolysate (enzymatic)	2·0 g
Glycerol	8·0 ml
ZnSO$_4$	0·10 mg
CuSO$_4$·5H$_2$O	0·10 mg
Distilled water	360 ml

Autoclave at 15 lb pressure for 20 min.

Solution B

Glucose (anhydrous)	5·0 g
Distilled water	50 ml

Autoclave at 15 lb pressure for 15 min.

Solution C

MgSO$_4$·7H$_2$O (1 M stock solution) (page 352)	1 ml
CaCl$_2$ (1 M stock solution) (page 352)	1 ml

FeCl$_3$·6H$_2$O	1·0 mg
Agar (certified)	11·0 g
Distilled water	485 ml

Melt agar; add MgSO$_4$ and CaCl$_2$. Autoclave at 15 lb pressure for 20 min. Add 1 ml of Fe3 stock solution (page 352).

Solution D, Sodium glutamate (stock solution) (page 352), 5 ml.

Solution E, Oleic acid–albumin complex.

Albumin (bovine Fraction V)	3·5 g
Distilled water	95 ml
Oleic acid–NaOH mixture†	5 ml
NaOH (2·5 N)	0·35 ml

Sterilise by Seitz filtration. Warm to about 45°C.

Mix solutions A, B, C, and D after each has been sterilised as indicated. Allow to cool to 48°C and add Solution E. Mix gently so as to prevent the formation of bubbles. Pour into Petri dishes, 30 ml/plate.

Allow the medium to solidify. Dry in incubator for 48 h. Store inverted in refrigerator. Do not use when more than three weeks old.

B. RVB-10 liquid medium. Nutrient broth

A liquid medium without agar similar to the above has been found to produce good growth of mycobacteria for phage studies. Tween-80 is added to the RVB medium in order to produce dispersed growth of some mycobacteria.

RVB-10

Solution A

Na$_2$HPO$_4$ (anhydrous)	1·2 g
KH$_2$PO$_4$	1·0 g
NaCl	2·5 g
NH$_4$Cl	0·5 g
Sodium pyruvate	0·75 g
Nutrient broth	4·0 g
Casein hydrolysate (enzymatic)	2·0 g
Glycerol	8·0 ml
ZnSO$_4$	0·10 mg
CuSO$_4$·5H$_2$O	0·10 mg
Tween-80 (stock solution) (page 352)	2 ml
Distilled water	360 ml

† Add 0·12 ml oleic acid to 10 ml 0·05 N NaOH.

Autoclave at 15 lb pressure for 20 min.

Solution B

| Glucose (anhydrous) | 5·0 g |
| Distilled water | 50 ml |

Autoclave at 15 lb pressure for 15 min.

Solution C

$MgSO_4 \cdot 7H_2O$ (1 M stock solution) (page 352)	1 ml
$CaCl_2$ (1 M stock solution) (page 352)	0·5 ml
Fe3 (stock solution) (page 352)	2 ml
Water	485 ml

Autoclave.

Solution D. Sodium glutamate (stock solution) (page 352), 5·0 ml.

Solution E

Bovine albumin (Fraction V)	2·5 g
NaOH (2·5 N)	0·70 ml
Distilled water	100 ml

Sterilise by Seitz filtration.

Mix solutions A, B, C, and D after sterilising. Cool to 47°C, then add solution E. Dispense into tubes, 10 ml/tube.

The preparation of the RVB-10 medium should follow the procedure as given above. Due to the fact that several of the components produce some interaction when autoclaved together this procedure has been found necessary. The stock solutions may be kept in the refrigerator for long periods of time when sterile. Serum may be a satisfactory substitute for the bovine albumin in the RVB-10 medium, but should not be used in the agar medium as serum has an inhibitory effect on phage action.

C. Nutrient broth

Nutrient broth is a satisfactory medium for making and storing suspensions of phage and bacteria. It may be made as follows:

Nutrient broth	8 g
NaCl	5 g
$CaCl_2$ (1 M solution) (page 352)	1·0 ml
Water (distilled)	1 litre

Adjust to pH 6·8–7·0 with 2·5 N NaOH and sterilise by autoclaving at 15 lb pressure for 20 min. Store in refrigerator.

D. Stock solutions and mixtures

To save time in making RVA and RVB media, make, store and use the following stock mixtures and solutions. Use distilled water. Store in refrigerator.

Mixture no. 1: Sodium pyruvate 7·5 g; Na_2HPO_4 (anhydrous), 12·0 g; KH_2PO_4, 5·5 g; NH_4Cl, 5·0 g; NaCl, 25·0 g; water, 1000 ml. Dissolve each compound before adding the next one. Dispense 100 ml amounts into flasks. Sterilise by autoclaving. Use 100 ml for making 1 litre of RVA-17 or RVB-10.

Mixture no. 2: Nutrient broth, 20·0 g; proteose peptone no. 3 (Difco), 25·0 g; casein hydrolysate (enzymatic), 20·0 g; water, 1000 ml. Dissolve by warming and dispense 100 ml amounts into flasks. Sterilise by autoclaving. Use 100 ml for making 1 litre of RVB-10.

Mixture no. 3: Nutrient broth, 40·0 g; casein hydrolysate (enzymatic), 20·0 g; water, 1000 ml. Dissolve by warming and dispense 100 ml amounts into flasks. Sterilise by autoclaving. Use 100 ml/1 litre of RVA-17.

Mixture no. 4: Glucose, 50·0 g; water, 500 ml; dissolve and dispense 55 ml amounts into flasks. Sterilise by autoclaving. Use one flask per 1 litre of RVB-10 or RVA-17.

Stock solutions

$CaCl_2$: 1 M calcium chloride. Dissolve 11·1 g of $CaCl_2$ (anhydrous) in 100 ml water. Dispense 20 ml/tube. Sterilise by autoclaving.

Fe3: ferric chloride. Dissolve 100 mg $FeCl_3 \cdot 6H_2O$ in 100 ml water. Dispense 20 ml/tube. Sterlise by autoclaving.

Mg4: 1 M magnesium sulphate. To 24·65 g $MgSO_4 \cdot 7H_2O$ add distilled water to 100 ml. Dissolve. Dispense 20 ml amounts into tubes. Sterilise by autoclaving.

ZCl: zinc–copper. Dissolve 10 mg $ZnSO_4$ and 10 mg $CuSO_4 \cdot 5H_2O$ in 200 ml water. Dispense 20 ml amounts into tubes. Sterilise by autoclaving.

TW3: Tween-80 (polyoxyethylene sorbitan monooleate). Tween-80, 25 ml; water (distilled) 75 ml. Warm and mix thoroughly. Dispense 20 ml each into tubes. Autoclave for 15 min at 15 lb. Shake tubes while still hot to mix Tween with water.

Sodium glutamate: sodium glutamate 10 g; water 100 ml. Sterilise by Seitz filtration (test for sterility by inoculating 1 ml into 10 ml RVB-10; incubate). Dispense 20 ml amounts into tubes.

To make RVB-10: add 8·0 ml glycerol; 2·0 ml TW3 (stock); 2·0 ml ZCl (stock) to 200 ml water. Autoclave 15 min at 15 lb pressure.

Add 1·0 ml Mg4 (stock); 1·0 ml $CaCl_2$ (stock), 1·0 ml Fe3 (stock) to 450 ml water. Autoclave for 15 min at 15 lb pressure.

Mix the two above suspensions, when cooled add† mixture no 1, 100 ml; mixture no. 2, 100 ml; mixture no. 4, 50 ml; Na glutamate (stock), 5·0 ml; bovine albumin mixture, 100 ml.

To make RVA-17: add 8·0 ml glycerol, 2·0 ml ZCl (stock) to 150 ml water; autoclave.

Dissolve 11·0 g agar in 500 ml water by warming. Add 1 ml Mg4 (stock), 1 ml CaCl$_2$ (1 M) (stock), and 1 ml Fe3 (stock). Autoclave for 20 min at 15 lb pressure. Mix the above two solutions.

Add mixture no. 1, 100 ml; mixture no. 3, 100 ml; mixture no. 4, 50 ml; 5 ml Na glutamate (stock); and oleic acid albumin, 100 ml. Mix all solutions so as to maintain temperature at 46–48°C when oleic acid albumin is added. Do not allow agar to solidify. Dispense into Petri dishes.

III. BACTERIOPHAGE

A. Isolation of phages

Mycobacterial phages appear to be widespread and can be isolated with little difficulty. Soil, human excreta, polluted water and various biopsy specimens, as well as other sources, have yielded phages on each of several occasions. It has been necessary in most instances to enrich soil samples with one or more strains of mycobacteria in order to obtain lytic plaques (Cater and Redmond, 1963; Takeya and Yoshimura, 1957).

The procedure is as follows. To approximately 50 g of soil sample add 50 ml of medium (NB3 or RVB-10 without Tween-80) and 10 ml of culture of a strain of mycobacterium that is likely to serve as host or indicator strain to the phage. ATCC 607 has been used extensively. It is easily grown and readily available. Repeat the addition of 10 ml of the same culture at intervals of one to two times each week for four or five weeks, keeping the flask stoppered with a porous substance and incubate at 37°C. The contents of the flask should be thoroughly mixed at least each time bacteria are added. Organisms washed from solid medium containing Malachite green (or other inhibiting substance) should not be used. Two or more strains of mycobacteria may be used for enrichment and in this case each bacterial strain should be used for testing for phage. About one week after the last enrichment, add 25–50 ml of nutrient broth to the flask, mix thoroughly and centrifuge to eliminate the debris. Pipette off the supernateant liquid and filter through a sintered glass filter or membrane filter with 0·45 µm porosity. Seitz filters retain most of the phage. This phage filtrate may be stored in the refrigerator at 4°C for some length of time for subsequent testing.

† Mix so as to avoid bacterial contamination. Dispense into culture tubes, 7·5 or 10 ml as desired.

Make three ten-fold serial dilutions of the filtrate into nutrient broth. Using fresh cultures of the strain, or strains, of bacteria with which the flasks were enriched make dilutions containing sufficient organisms to produce an easily observable but slight turbidity (about 0·1–0·2 optical density). The concentration will depend on the rate of growth of the organism. Inoculate 2 ml of bacterial suspension with 0·2 ml of each of the four phage dilutions and spread about 0·5 ml on an agar plate (RVA-17) so as to cover the entire surface. The exact amounts will vary depending on the dryness of the plates. Incubate the plates at 36–37°C and examine daily for cleared spots (plaques). On plates of rapidly growing bacteria, such as ATCC 607, the plaques may disappear after one or two days due to the rapid overgrowth of plaques by the bacteria. In other instances, the plaques may continue to enlarge for several days after appearing.

Isolate material from selected plaques with a sterile needle into small amounts of nutrient broth. Make dilutions of each and replate as before using also other strains of mycobacteria to determine lytic activity. Examine each plate carefully for variant plaques and make individual isolations from each type of plaque. Plaque variants may be in the form of large or small, as well as clear or turbid plaques. There may be distinct differences in the margin being either abrupt or gradual, or consisting of a distinct halo. Phage isolated from each plaque type should be tested for lysis on several host strains.

If no plaques are observed on the original plates with the dilutions used test larger samples of the filtrate, e.g. 1 ml of filtrate mixed with 1 ml of bacterial suspension. Also the undiluted filtrate may be spotted on lawns of suitable bacterial strains.

Stool specimens, polluted water and other types of materials that are contaminated with mycobacteria may produce phage when treated in a similar manner. Material containing large numbers of mycobacteria may yield phage without being enriched. However, sputum specimens rarely have given rise to phage.

B. Phage propagation and suspensions

Once a phage strain is established on a susceptible host, it should be propagated and a stock suspension obtained. Make several ten-fold serial dilutions of a good phage suspension and spread each with its host as indicated above to determine the concentration necessary to produce just complete lysis of the host. Using this concentration, inoculate several plates and recover the phage when lysis is complete by flooding each plate with 4 or 5 ml of nutrient broth. Allow to remain at room, or refrigerator, temperature for 2–3 h. Add 2–3 ml of fresh nutrient broth to each plate and agitate and scrape (mix) the lysed bacteria from the surface. Pipette into sterile tubes. Wash the surface with 1 ml of broth and add to the tube. Centrifuge and remove the supernatant liquid with a sterile pipette. Filter using a sintered glass or membrane filter to remove all the bacteria. Store temporarily at 4°C. Make separate suspensions for each phage variant that has been isolated.

C. Phage assay

Each suspension should be assayed to determine the number of plaque forming units (PFU). Make eight ten-fold serial dilutions into nutrient broth. Add 0·2 ml of each of the three highest dilutions to 2 ml of the host bacteria diluted and suspended in nutrient broth. Spread 0·5 ml of each suspension on one RVA-17 plate. Do not remove excess. Incubate at 37°C and observe daily for lytic plaques. When there is optimal development of the plaques, make accurate counts of plates containing approximately 100–200 distinct plaques. Multiply the number by the dilution factor for this plate. This is the number of active phage particles (PFU), per millilitre of the stock suspension. Label, giving the concentration in PFU of each tube, seal and place in refrigerator at 4°C. With these standardised stock suspensions, comparisons can be made with various phages and on different host bacteria and they may be used, after determining the routine test dilution (RTD) in phage typing of mycobacterial strains (see below).

D. Phage mutants and adaptation to new hosts

Enrichment cultures. Although in a few cases phage has been found in soil or other samples without enrichment with mycobacteria, this is a rare occurrence. The logical assumption when phage is isolated from enrichment mixtures is that a few phages were present in the sample and that the number was increased through lysing of the bacteria that were added. Variations in the methods and conditions of the experiments and variations in the activity of the phages isolated make it very difficult to evaluate the factors that may be active in these procedures. The consensus is that adaptation and variation takes place in enrichment cultures (Redmond, 1963).

A procedure somewhat similar to soil enrichment has been used by various workers in attempts to adapt phage to lyse certain bacteria not previously susceptible. Some have been successful, some not. Many workers think that adaptation to new hosts is a very likely means of obtaining specific phage strains for typing purposes (Piguet, 1960; Hnatko, 1953).

This procedure has been used to adapt phage to lyse one of the strains of

M. kansasii. Four phage strains, each with lytic activity on one of four strains of bacteria, were inoculated into a flask containing medium free of Tween-80 that had been inoculated with the four bacterial hosts and *M. kansasii.* On day 5 or 6, subcultures were made and fresh suspensions of bacteria added. This procedure was repeated four or five times following which the mass was transferred to tubes and centrifuged. The supernataut liquid was filtered, using an ultrafine sintered glass filter. The filtrate was diluted and each dilution mixed with a suspension of *M. kansasii* and spread on a RVA-17 plate. Also, small drops of each suspension were placed on a lawn of *M. kansasii.* A few plaques were observed on two of the plates. Phage was isolated from these plaques and plated on *M. kansasii* as host. Repeated isolations and platings resulted in phage AG1 with good lytic action on this bacterium.

Adaptation by heavy inoculation of phage. Another adaptive method that has been used with success is the heavy inoculum technique. An undiluted phage suspension is inoculated in ratios of 0·1 to 10, 0·5 to 10 and 1 to 10 (or more) into a good suspension of the resistant bacteria desired to be lysed. Each mixture is then spread on an agar plate. This procedure may result in non-specific lysis with no phage production. A few clear plaques will indicate true phage lysis, and subsequent isolations of phage from these plaques should produce good lytic activity.

Ultraviolet irradiation and other mutagenic agents. Phages treated with ultraviolet irradiation either before applying to a bacterial host or by irradiation of the infected bacteria is likely to produce variants (mutants) with altered specificity. It has been indicated that irradiation of a resistant bacterium followed by infection with a non-infecting phage is likely to produce a phage with lytic action. In attempts of this type, heavy inoculation of phage is found to produce best results. There are without doubt mutant phages in all heavy suspensions of phage which may produce lysis on a strain of bacteria resistant to the phage mass.

Phage recombinants. Since most bacteria are thought to be lysogenic, most of which are carrying defective prophages not capable of producing lysis, any phage to which the bacterium is resistant but which is capable of entering the bacterial cell may combine with the prophage genome and produce a recombinant phage that may be lytic. Two different phages infecting one bacterium may result in a recombinant phage that differs from each of the parent phages.

To summarise these concepts of adaptation, it is possible that any heavy suspension of phage may contain a few mutant phage particles capable of lysing non-susceptible bacteria. Also many phages enter bacteria but are incapable of producing lysis. These may combine with a phage genome carried by the bacterium and result in a new lytic phage.

IV. EPIDEMIOLOGICAL STUDIES ON MYCOBACTERIA USING PHAGE TYPING

A significant development occurred when Baess (1966) reported the isolation of phage BK_1 from soil taken from poultry runs where the hens were infected with *Mycobacterium avium*. The host bacterium was the F-21 (ATCC 607) strain of *Mycobacterium smegmatis*. Of 82 wild strains of *M. tuberculosis* tested the majority were highly resistant or not lysed by BK_1 at any concentration. This phage was very stable and gave highly reproducible results that were read with ease because the plaque that was produced was large and clear. At that point BK_1 and Redmond's GS4E were the two phages with the greatest potential for subdividing the species, as required for epidemiologic studies.

The following year Bates and Fitzhugh (1967) reported that 92 wild strains of *M. tuberculosis* could be subdivided into three types designated A, B, and C using a battery of phages. The phages used were DS6A, GS4E, and BG_1 isolated by Redmond, and D-34 isolated by Froman. Most strains, 76%, were Type A and were resistant to all but phage DS6A. Fourteen per cent of strains were highly susceptible to lysis by phages DS6A, GS4E, and BG_1 and were designated Type B; and 10% were lysed by all these phages, as well as D-34, and were designated Type C. These data are summarised in Table I.

TABLE I

Phage types of human tubercle bacilli as recognised by Bates and Fitzhugh (1967)

Phage type	AG1[a] (MTPH 1)[b]	DS6A (MTPH 2)	GS4E (MTPH 3)	BG[1] (MTPH 5)	D-34 (MTPH 6)
A	+	+	−	−	−
B	+	+	+	+	−
C	+	+	+	+	+

+ Lysis at the routine test dilution.
− No lysis at the routine test dilution.
[a] AG1 phage included in later studies.
[b] Standard nomenclature for designation of phages; introduced in 1972.

This report was soon followed by a study of Tokunaga *et al.* (1968) who typed human tubercle bacilli using nine phages. The routine test dilution (RTD) spotting technique as used by previous investigators was compared to the more laborious plaque counting method and the results obtained by these two methods were highly comparable. Of 37 strains tested all were lysed to the same degree by phages DS6A, D29, C3, and L1,

Of some importance was their observation that GS4E, BK_1, and B1 gave nearly identical lysis patterns except for two strains which showed an intermediate degree of susceptibility to GS4E and BK_1, but were highly susceptible to phage B1. They reported additional results of considerable significance; the phage patterns of the drug-resistant mutants were similar to those of the original drug-susceptible strains. Of seven strains isolated in India, three were lower in virulence for guinea-pigs than the remaining four, but there was no correlation between virulence and phage type in this small sample. These workers included, among the 37 strains tested, five obtained from infants suspected of having been infected from a single source in an institution. All five showed similar lysis patterns. In contrast, six strains obtained from children whose sources of infection were thought to be different showed different lytic patterns in five cases.

Subsequently, Bates and Mitchison (1969) used phage typing of tubercle bacilli as a means to compare this type of marker with biochemical or virulence characteristics of strains collected from widely separated locations throughout the world. Previous epidemiological studies had shown that tubercle bacilli isolated from patients in southern India were generally less virulent than organisms isolated in Britain (Frimodt-Moller, 1975; Mitchison *et al.*, 1960; Bhatia *et al.*, 1961) and were more susceptible to the bactericidal action of hydrogen peroxide (Subbaiah *et al.*, 1960) and more resistant to thiacetazone and para-aminosalicylic acid (Selkon *et al.*, 1960; Joseph *et al.*, 1964) than were British strains. Strains from Hong Kong were more resistant to thiacetazone than British strains, but were similar to British strains in their virulence for guinea-pigs (Dickinson *et al.*, 1963; Mitchison and Lloyd, 1964). To search for a possible relationship between phage type and any of these phenotypic characteristics, 225 strains of tubercle bacilli obtained from patients in Britain, Hong Kong, Rhodesia, and southern India were phage typed. The strains were isolated before drug therapy had been given, but from 28 of these same patients additional isolates were collected at intervals during treatment. The latter strains had become drug resistant during treatment and this change allowed for a study of phage type and drug resistance emerging *in vivo*.

Several items of epidemiological importance emerged from this study. First, an unexpected phage type designated as "intermediate" was observed to be harboured by patients from southern India. The term "intermediate" was given for these strains because they showed a pattern intermediate between types A and B in regard to their susceptibility to lysis by the phages GS4E, BG_1, and BK_1. Previously all strains tested from other geographic areas had shown a consistent pattern of either total susceptibility or resistance to these three phages. Many of the strains from southern India did not show this homogeneity and therefore were not clearly type A or B. (In a study of seven strains from India, Tokunaga *et al.* (1968) reported two which showed intermediate degrees of susceptibility to GS4E and BK₁.) Second, it was found that tubercle bacilli were not evenly distributed geographically according to phage type. Type B strains were more common in Britain than in Hong Kong or southern India, the Type A strains predominated in Hong Kong, and the "intermediate" strains together with the Type A strains were observed in southern India. Type C strains were not observed among any of the strains in this study. Third, it was found that the phage type did not change despite *in vivo* changes in drug susceptibility, animal virulence, or catalase production. Fourth, it was found that phage type did not correlate with drug resistance when Isoniazid, Streptomycin, and para-aminosalicylic acid were the drugs studied. Fifth, it was observed that the southern Indian Isoniazid-sensitive strains of decreased virulence for guinea-pigs were randomly distributed between phage Type A and the "intermediate" type.

Another large epidemiological study was reported by Baess (1969) who evaluated 230 strains obtained from Danish patients who were linked epidemiologically. Phage BK₁ divided 205 human strains into 62 phage-susceptible and 143 phage-resistant strains. (Of 88 groups of patients with a presumed epidemiological relation within each group, 22 groups harboured only phage-susceptible and 61 only phage-resistant strains.) Five patient groups harboured both phage-resistant and phage-susceptible strains. Baess found only a few strains lysed by D34, observed that the plaques produced by it were turbid, and expressed concern that D34 would not be a useful phage for epidemiological studies. An additional five phages were used by Baess and none served to subdivide the tubercle bacilli into more types than the two shown by BK₁.

Thereafter several additional epidemiological studies were reported by investigators who continued using the techniques developed by Redmond *et al.* (1963) or by Tokunaga and Morohashi (1963). Stead and Bates (1969) used phage typing to support other epidemiological data which indicated that a young soldier had become infected while on military assignement in the Far East and upon returning to the United States had infected two persons. Steiner *et al.* (1970) studied an outbreak of primary tuberculosis where the source case was excreting organisms resistant to Isoniazid, Streptomycin, and para-aminosalicylic acid. The phage types of the epidemiologically related strains were identical in each instance, and when correlated with the drug susceptibility pattern for the strains, confirmed that this Isoniazid-resistant organism was highly infectious and virulent. Stead and Bates (1971) used phage typing to study the pathogenesis of tuberculosis by typing tubercle bacilli isolated from a chest wall abscess which had developed following local trauma during an episode of primary tuberculosis. The phage type was identical to that of a strain obtained from the patient's close associate. This study was the first of several showing that phage typing could be used to determine relationships in tuberculosis epidemiology.

Kitahara (1973) studied the geographic distribution of phage types of tubercle bacilli in Japan and Kenya, and found that in Japan about 20% of strains were susceptible to lysis by BK_1, but in Kenya only 5% were of this type. Of considerable interest and importance was the observation that 32% of the Kenyan strains were susceptible to lysis by D34 whereas only a very few such strains were found in Japan. The increased frequency of D34-susceptible strains obtained from Kenyan patients coincides with unpublished data obtained by Bates and Rado (1970), who have found that such strains are also frequent among Egyptian and Bolivian patients. In this same report Kitahara used phage typing to study the intrafamilial spread of tuberculosis in Japan.

Additional geographic studies on the phage type of tubercle bacilli were reported by Mizuguchi *et al.* (1973) who studied the phage susceptibility of strains isolated in Japan, the Netherlands, and Ceylon. A number of phages not used extensively prior to this report were included in the phage testing. Phages DNA III 8, Clark, Legendre, and Sedge isolated by Mankiewicz (1973) were included, as was phage pH isolated by Sushida and Hirano (1971). The phage susceptibilities of the 54 Ceylonese strains were notable in that 87% were susceptible to BK_1, 68% to pH, and 8% to D34. The Japanese and Dutch strains were generally more resistant to these phages than previously reported.

Among the several investigations on mycobacteriophages many variations in technique were used. The designations for phages, phage types, and mycobacterial host strains were diverse and confusing. To correct this problem, the World Health Organisation gave support to a working group studying the development of phage typing for mycobacteria. The efforts of this group have been directed towards formulating a more uniform methodology and nomenclature. An important report from this group appeared in 1973 which suggested that uniform techniques could be applied world-wide and that the technology required to achieve uniform results was within reach of investigators working in all developed countries (Sula *et al.*, 1973). A subsequent report described in detail standard methodology for phage typing *M. tuberculosis* using techniques and media widely available throughout the world (Rado *et al.*, 1975). In addition, the phages used for typing, together with their host strains, were given lettered and numbered designations. This permitted the development of a systematic method for assigning a phage type to a tubercle bacillus and allowed for expansion and modification of the system as new information might require. The phage and host strain nomenclature and the recognised phage types are shown in Tables II and III. When these methods are followed meticulously, the data obtained are highly reproducible among different laboratories.

Subsequent studies have now appeared in which were used the methodology and the typing phages recommended by the World Health

TABLE II

Mycobacteriophages used for phage typing *Mycobacterium tuberculosis*

Present designation	Original designation and $1 \times RTD^a$		Host	Source
MTPH[b] 1	AG1	1×10^5	AT7 M. kansasii	W. B. Redmond, Atlanta, U.S.A.
MTPH 2	DS6A	1×10^6	M. tuberculosis, H37Rv	W. B. Redmond, Atlanta, U.S.A.
MTPH 3	GS4E	1×10^5	M. tuberculosis, H37Rv	W. B. Redmond, Atlanta, U.S.A.
MTPH 4	BK1	1×10^6	M. smegmatis, ATCC 607	I. Baess, Copenhagen, Denmark
MTPH 5	BG1	4×10^3	M. intracellulare[c] P-17 (Runyon)	W. B. Redmond, Atlanta, U.S.A.
MTPH 6	D-34	1×10^7	Mycobacterium not speciated, Froman's F-130	S. Froman, Los Angeles, U.S.A.
MTPH 7	DNA III 8	1×10^5	M. tuberculosis, H37Rv	E. Mankiewicz, Montreal, Canada
MTPH 8	X-20	1×10^6	M. tuberculosis, H37Rv	J. H. Bates, Little Rock, U.S.A.
MTPH 9	PH	1×10^5	M. tuberculosis, H37Rv	K. Sushida, Tokyo, Japan
MTPH 10	Clark	1×10^6	M. smegmatis, ATCC 607	E. Mankiewicz, Montreal, Canada
MTPH 11	Sedge	1×10^6	M. smegmatis, ATCC 607	E. Mankiewicz, Montreal, Canada
MTPH 12	Legendre	1×10^6	M. smegmatis, ATCC 607	E. Mankiewicz, Montreal, Canada

[a] Minimal concentration of virus (PFU/ml) that gave almost confluent lysis when 0·01 ml of stock was "spotted" on propagating host strain. RTD = routine test dilution.

[b] MTPH = mycobacterial typing phage, human.

[c] This virus was titrated on ATCC 607.

This Table is reproduced by permission from the *American Review of Respiratory Diseases* **111**, 460.

TABLE III

The major phage types of *Mycobacterium tuberculosis* determined
by susceptibility to subdividing mycobacteriophages

Phage type	Lysis by mycobacterial typing phage, human				
	3	4	5	6	9
A	−	−	−	−	−
B	+	+	+	−	+
C	+	+	+	+	+

Reproduced by permission of the *American Review of Respiratory Diseases* **111**, 461.

Organisation working group. Mankiewicz and Liivak (1975) studied individual colonies of *M. tuberculosis* isolated from Canadian Eskimos and from Canadians of European heritage. Sputum samples were evaluated from 233 Eskimos and 150 patients of European heritage, and three individual colonies from each patient were selected for phage typing. The Eskimo patients had experienced repeated hospital admissions. Multiple drug therapy regimes had been prescribed and treatment failures were common. In contrast, the patients of European heritage had few treatment failures and multiple treatment regimens were rare. In 33 instances, the sputum from Eskimos exhibited the presence in the same specimens of tubercle bacilli having different phage types. This remarkable finding was observed only among the Eskimo patients. Since only three separate colonies from each sputum were selected for typing, the finding of multiple phage types in the same specimen might well have been greater had a larger number of colonies been studied from each specimen.

The possibility of a patient being infected with two or more phage types and its impact on epidemiologic data useful for planning a tuberculosis control programme have been reviewed by Raleigh *et al.* (1975). In a study of 26 patients who had relapsed following initial treatment for pulmonary tuberculosis, these investigators found that in nine patients the phage type of the pre-treatment and last-positive isolate prior to becoming culture negative were different from the phage type of the isolate obtained after bacteriologic relapse. It was suggested that both phage types were present in the pre-treatment sample but one phage type predominated and thus was selected as the isolate for the phage typing.

These two reports require further study and documentation, but the epidemiological implications are apparent. If a person can be infected by two or more phage types, then the effectiveness of the immunity induced by the initial infection may be less than has been thought previously.

Indeed, the frequency of dual infection may be greater than these studies would suggest even in developed countries where tuberculosis is coming under control.

Further data and comment regarding dual infection in man have been supplied by Bates *et al.* (1975) who studied isolates of tubercle bacilli obtained from patients who were infected in two or more different anatomic sites. In this study single colony isolates were not evaluated but instead numerous colonies from each culture were mixed together and subcultured for typing. This technical point is noteworthy because of the overshadowing effect of resistant bacteria in any study involving resistant and susceptible strains. Since Type A is resistant to all phages except MTPH 1 and MTPH 2 (see Table I) if this strain is present together with bacilli of any other type, the lytic pattern of the latter will not be apparent because the potential zones of lysis will be overgrown by the Type A phage-resistant bacilli present on the medium.

The point is obvious. If more than one phage type is present in a culture, present typing methods will not detect the type which is more susceptible to phage lysis. Despite this limitation, the results showed that two of the 88 patients with concurrent infection at two or more sites were infected with tubercle bacilli of different phage type. Thus despite inadequate techniques, dual infection was recognised with unexpected frequency among patients in the United States. These reports suggest that repeated exogenous infection with tubercle bacilli does occur in developed countries, and it can be speculated that exogenous reinfection is much more frequent among persons living in locations where tuberculosis continues in epidemic proportions.

Efforts to phage type *Mycobacterium bovis* have been unproductive and no satisfactory technique for recognition of different types within this species has emerged. Baess (1969) studied 25 *M. bovis* strains for their susceptibility to nine phages and found all strains resistant to lysis. Caroli *et al.* (1967) reported the phage typing of 75 strains of *M. bovis* isolated from cattle in Italy and observed seven phage types using 14 different phages. Redmond (1973) adapted seven phages to lyse *M. bovis* and used these adapted phages to type 53 cultures of bovine tubercle bacilli isolated from cattle and other animals in widely separated areas of the United States. Prior to adaptation, the phages were inactive for bovine mycobacteria. Ten types were observed with only three cultures being resistant to all phages. To date, no other investigations have been reported.

The epidemiology of *M. kansasii* infection is poorly understood. Clinical evidence indicates that it is not transmitted from man to man and yet a reservoir in nature has not been clearly identified. Bailey *et al.* (1970) and Kaustova *et al.* (1974) reported isolation of *M. kansasii* from tap water, and

Leclerc *et al.* (1971) grew this organism from sewage. Many other attempts to discover *M. kansasii* in nature have failed. Wayne (1962) reported two varieties of *M. kansasii* according to their catalase activity, with the low catalase strains being non-pathogenic and the high catalase strains being pathogens for man. Their observations were confirmed by Gruft (1972).

The subdivision of the species *M. kansasii* by phage susceptibility testing was first reported by Hobby *et al.* (1967). Engel (1975) phage typed 209 strains of *M. kansasii* isolated from the Netherlands using 14 phages and was able to recognise 13 types. Among the organisms tested, 124 were from Rotterdam and of these 96 showed a single phage type. Only nine strains of this phage type were clinically significant. However, three of seven and five of 21 strains of other phage types were human pathogens. No correlation was found between catalase activity and phage type. Thus it appears that phage typing may permit the subdivision of *M. kansasii* into two large groups, one group having a high probability of being a human pathogen and the second having a high probability of being a saprophyte.

Gunnels and Bates (1972) reported the isolation of three phages useful for the subdivision of 29 strains of *Mycobacterium xenopi* into four types, but the types did not correlate with drug susceptibility, pathogenicity, or geographic location.

Phage typing and/or subdivision of other species of mycobacteria which are pathogenic for man such as *M. intracellulare* and *M. scrofulaceum* have not been reported.

V. PRESERVATION AND STORAGE; SHIPPING

A. Bacteriophage

Knowledge of the best methods and conditions for preservation of phages and of their bacterial hosts is essential for anyone involved in phage work. Stocks of phages or bacteria which are unstable require frequent propagation or subculturing. The main properties, such as the plaque morphology and lytic spectrum of the phages and host range of the bacteria, must be checked at frequent intervals for each preparation to detect possible interfering mutations or other changes.

In general, the stability of phages depends on the suspending substrate. In media containing small amounts of proteins (e.g. bovine serum albumin, fraction V, or gelatin) or in broth (e.g. nutrient broth, Difco) most mycobacteriophages remain relatively stable (Redmond, 1963). In salt solutions, the presence of small amounts (10^{-3} mol) of divalent cations such as Mg^{2+} or Ca^{2+} increases stability (Sellers *et al.*, 1962). Detergents, heavy metals and other agents may affect the stability of phage suspensions. A good review of the effects of various substances on phages in general is presented by Adams (1959). Tween-80 in concentrations normally used in myco-bacterial culture media has no effect on the stability of the mycobacterio-phages, but interferes with adsorption to the hosts (Bowman, 1958; Sellers et al., 1962). Low temperatures favour stability. In a good suspend-ing medium, most mycobacteriophages remain stable for months or years at 4°C, although titres will decrease gradually. At room temperature, many phages lose their activity within a few days, although stability may be increased by keeping the phages on small agar plugs sealed in vials (Red-mond and Ward, 1966). Thus storage at 4°C in broth appears to be the most simple method for short-term preservation of phages in most in-stances, provided that viability and other important characters are regularly checked.

For long-term preservation (for two years or more), drying from the unfrozen or frozen state (freeze-drying) and freezing at ultra-low tempera-tures (near −196°C) have been applied. Although drying of phages on filter paper discs from the unfrozen state was reported as quite unsuccessful by Clark (1962), rapid drying of mycobacteriophages in small volumes of heart infusion broth gave satisfactory results according to Will et al. (1961). The mycobacteriophages survived well and retained all their characteristics upon reconstitution after storage in the dark, or even at room temperature. It is well known that great differences in sensitivity to drying and freezing do exist among various phages. Campbell-Renton (1941) even proposed using these differences as a basis for the classification of phages. Freeze-drying of phages has been performed by a number of workers with more or less satisfactory results. Although Clark (1962) reported high losses on freeze drying of various phages, he later mentioned the routine use of this technique for distribution of bacteriophages by the American Type Culture Collection (Clark, 1970). Successful freeze-drying depends on such factors as suspending medium, cooling rate and drying temperature as demonstrated by Davies and Kelly (1969), who developed a satisfactory freeze-drying method for a corynebacteriophage that was extremely labile in suspension at 4°C and even at −25°C. The method, using a solution containing 20% peptone, 10% sucrose, and 2% sodium glutamate as a protective substrate, has been successfully applied to a series of other corynebacteriophages by Carne and Greaves (1974). Muro-hashi et al. (1960, 1969) freeze-dried high concentrations (10^9–10^{11} PFU/ml) of eight mycobacteriophages in their original propagation medium (trypti-case-soy-broth) without any additional protecting substances. Inactivation by freeze-drying varied from 10–50%. Inactivation during storage of the freeze-dried phages at 5°C for more than six years caused no more than a three-fold reduction in active particles. Engel et al. (1974), comparing various protecting substrates, reported on satisfactory freeze-drying of 52 of 53 different mycobacteriophages in a medium of 5% sodium glutamate and 0·5% gelatin. After propagation of the bacteriophages on their respective hosts on solid medium, the phage lysates were washed from the surface of each plate with sterile distilled water in order to keep the salt concentration as low as possible. (Salts decrease the freezing point of the phage suspension thus causing foaming during the freeze-drying process.) The lysates were filtered through a 0·45 μm membrane filter.

The filtered lysates were mixed with equal amounts of a solution of 10% sodium glutamate (British Drug House Chemicals) and 1% gelatin (Difco). One millilitre samples were dispensed into 2 ml glass ampoules, which were subsequently aseptically attached to the ampoule holder of the freeze drying apparatus. The holders were carefully rotated by hand in an ethanol bath kept at $-40°$C by means of refrigeration to obtain a uniform coating of frozen phage suspension on the walls of the ampoules. Subsequently the holders were attached to the freeze drying apparatus.

The drying process was a two-step procedure. The preliminary drying was carried out by application of vacuum for 24 h until a vapour pressure of 8 μm Hg was obtained. The water vapour was transported to a low temperature condenser ($-60°$C). Uncondensed vapours were drawn off through F.G. 50 filters (American Air Filter Company, Louisville, Kentucky). In order to prevent the phage suspensions reaching the melting point too early during the drying process, thus causing foaming, the ampoules were refrigerated for the first 4–5 h by means of a CO_2-ethanol mixture.

The first step was terminated by sterile aeration of the ampoules with filtered air, after which they were removed and their necks flame-heated and drawn to capillaries.

For the second step the ampoules were again attached. In order to obtain a pressure of less than 1×10^{-2} μm Hg, an oil diffusion pump in combination with P_2O_5 absorption was used. This pressure was reached after 4–5h and maintained for another 18 h.

The second step was completed by flame-sealing of the ampoules at the capillary part keeping them still under vacuum. All ampoules were finally checked for vacuum by means of an Edwards high frequency tester (Model T2).

Titres determined after processing generally did not show more than a ten-fold reduction. No changes in plaque morphology or lytic spectrum were noted after reconstitution when compared with the original phage preparation. The freeze-dried preparations could be stored in the dark at room temperature for as long as four years or even longer (unpublished results). Dubina *et al.* (1974) were able to confirm these results in work on mycobacteriophage AG1.

The value of storage at temperatures of liquid nitrogen has been demonstrated by Clark et al. (1962), Clark and Klein (1966) and Carne and Greaves (1974). Titres were reported constant for as long as five years (Clark and Klein, 1966). The phages were frozen at controlled rate in three steps to −196°C and quickly thawed for use. As a protective additive 10% glycerol was added to the lysates (Clark et al., 1962). Storage at −196°C, however, is limited by refrigeration space and shipping possibilities. Freeze-drying permits storage and shipping at room temperature and therefore seems the most useful method for long-term preservation of phages.

B. Mycobacteria

For storage and shipping of the bacterial hosts similar methods are effective. Most mycobacterial strains keep well on slants at 4°C for several months or years. This means of preservation is satisfactory for short-term storage, but regular checks on viability and phage susceptibility remain necessary. Other important characteristics (e.g. biochemical, immunological, virulence) must be checked to minimise selection of undesirable mutants in subcultures. For long-term preservation, drying from the unfrozen state was reported to provide stable stocks (Will et al., 1961). Storage at −70°C in Middlebrook 7H-9 liquid medium provided stocks with 100% viability and stability of genetic characteristics for long periods (Kim and Kubica, 1973). Freeze-drying of bacteria in various suspending media (e.g. skimmed milk, 7·5% glucose, broth) is commonly used with satisfactory results and is extensively reviewed by Lapage et al. (1970). It may result in an initial kill of 40–50% as has been demonstrated for some BCG preparations (Lind, 1967; Ungar, 1949). Storage and shipping without refrigeration are definite advantages of this method. The choice of method will depend on local circumstances and requirements.

VI. PHAGE TYPING PROCEDURES ADAPTED TO THE MYCOBACTERIA

A. The routine test dilution (RTD)

The routine test dilution (RTD) has been defined as the highest dilution of phage that will produce complete lysis of the propagating strain of bacteria. For typing M. tuberculosis a given prototype strain (see page 361) is used to determine the RTD of each phage.

Despite some problems, the RTD method of phage typing appears to be the most satisfactory procedure. The major difficulties encountered are due to variations in susceptibility to the typing phages of bacterial strains from different sources. Variations in the degree of lysis often make comparisons between strains somewhat questionable. This problem has been solved to a certain extent by using two additional phage dilutions along with the RTD; for example, using 0·1 RTD, 1 RTD and ten-fold RTD for spotting on each isolate to be typed. This requires more work, but its often valuable and is recommended when questionable results are obtained.

The RTD is determined as follows: prepare a suspension of the bacterial culture to be used (the propagating or the prototype strain) that will produce a good, but not heavy lawn on the agar plate within one to three days. For this purpose log-phase bacteria from broth culture (RVB-10) are preferable. Dilute the bacterial culture into sterile saline, water or nutrient broth to an optical density of about 0·2 (much more dilute for rapidly growing strains). The concentration of the suspension will vary in indirect ratio to the rate of growth of the individual strain. Determine this by previous experiment. Inoculate an RVA-17 agar plate by flooding with 1–1·5 ml of the suspension. Spread the suspension uniformly over the entire surface and tilt the plate to drain the surplus inoculum to one side. Pipette off the surplus and allow the plate to dry in an incubator for 2–3 h (24 h in the case of slowly growing organisms). Make eight (more if necessary) ten-fold serial dilutions of the phage suspension into NB3 using a different pipette (1 ml or 0·1 ml) for each transfer. Using the six highest dilutions, i.e. 10^{-3}–10^{-8}, spot 0·01 ml on marked areas of the inoculated plate. Incubate the plate and examine daily for lysis. When maximal lysis has been attained determine the highest dilution which has produced complete lysis. Because of the slowness of growth and lysis this may require three or four days. This dilution is the RTD. Phage dilutions with smaller increments may be used if necessary, e.g. if the 10^{-6} dilution produces a clear spot and the 10^{-7} dilution produces a slightly turbid spot, an intermediate dilution of 5×10^{-7} may be tested and used as the RTD, if it produces complete lysis. Keep stock phage suspensions in the refrigerator and make fresh RTD dilutions each day tests are made on unknown strains.

B. Phage and bacterial sources

Typing phages and their respective propagating strains of mycobacteria may be obtained from the Rijks Instituut Voor de Volksgezondhera, Bilthoven, Netherlands, or they may be isolated and propagated locally as indicated in Section III.

C. Spotting with phage suspensions

With a small syringe and small gauge needle, or the Accu-drop assembly sold by the Sylvania Company, Milburn, New Jersey, place a fraction of a drop, $0·5–1 \times 10^{-2}$, of each typing phage suspension diluted to the RTD on each plate. Each plate should be marked to identify the position of the droplets, or each plate should be placed on top of a circle of paper marked to designate the positions of the droplets. Handle the plate carefully so as to avoid spreading and coalescence of the droplets. Place in the incubator at 36°C. Read results after 24 h and daily thereafter for as long as necessary, up to 7 days. The final reading on rapidly growing bacteria can be made after one or two days; for slowly growing pathogenic bacteria five to seven days are necessary for full development of the phage activity.

D. Reading and reporting of the results

The results of the phage activity should be recorded daily. The following designations are recommended. Complete lysis + + + + ; nearly complete lysis + + + ; lysis with slight turbidity + + ; individual plaques + (number may be indicated as +(16)); no lysis −. For routine reporting of results of phage typing a simpler form is frequently used as follows; complete or nearly complete lysis + + + ; partial lysis + + ; few plaques + ; no lysis −.

E. Methods for typing *M. tuberculosis*

Slight modifications of the techniques have been found beneficial in subtyping strains of *M. tuberculosis*. The media and methods for growth of the bacterial cultures and the propagation of the phages are the same as indicated in the previous Section. In order to obtain homogeneous growth of the cultures to be tested, each strain should be transferred two or more times in RVB-10 broth until a well-dispersed suspension is obtained. When the cultures have reached readily visible and dispersed growth, 10 ml should be centrifuged and resuspended in NB3 so as to give a turbidity corresponding to the number 3 or 4 tube on the McFarland Scale. Spread 1 ml amounts on the surface of RVA-17 plates and allow to dry, with the cover on, in the incubator for 24 h. Spot with 0·01 ml of each phage using three dilutions, 0·10 × RTD, 1 × RTD and 10 × RTD. Titration of the stock phages should be made periodically and not more than 14 days before their use.

TABLE IV

Proposed division of phage type A using auxiliary phages

Subtype	
A₀	Type A lysed by not more than one auxiliary phage
Aₓ	Type A lysed by two or more auxiliary phages

Adapted from Rado *et al.* (1975).

13

Prototype strains for each subtype of *M. tuberculosis* have been selected and are available. These will be distributed with the phages.

Plates should be checked daily beginning 24 h after spotting and records kept of lysis for each dilution for 10 days. The final results to be reported should be the maximal lysis of each phage for each bacterial strain. This should be based on the scheme shown in Fig. 1.

Twelve typing phages are available. These are designated MTPH 1 to 12, as found in Table II (Rado *et al.*, 1975). They fall into three categories, thus: general, MTPH 1 and 2; subdividing, MTPH 3–6 and MTPH 9; and auxiliary phages MTPH 7, 8 and 10–12. The general phages lyse all *M. tuberculosis* strains. The subdividing phages divide the strains into three basic types A, B, and C as shown in Table III. The auxiliary phages show variable patterns on strains of Type A, but lyse all strains of Types B and C. Only preliminary tests and results are available for the auxiliary phages, the scheme of classification being given in Table IV. This proposed method allows for expansion and inclusion of additional phages and subtypes as may be needed in future work.

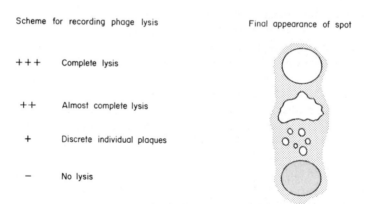

Scheme for recording phage lysis		Final appearance of spot
+++	Complete lysis	
++	Almost complete lysis	
+	Discrete individual plaques	
−	No lysis	

FIG. 1. Method of recording phage lysis and schematic appearance of spots on bacterial lawn following lysis in phage droplets. Reproduced by permission from the *American Review of Respiratory Diseases* **111**, 461.

At the last meeting of the phage typing study group on October 13–15, 1976, results of studies were presented (W. D. Jones, pers. comm.) which showed much clearer differentiation of lytic spots, especially with the auxiliary phages. The method utilises an adaptation of the soft agar overlay technique that can be used to obtain improved results with *M. tuberculosis*.

F. The soft-agar overlay method for phage typing *M. tuberculosis*

1. *7H9 broth medium*

Add 4·7 g of Middlebrook 7H9 broth base (Difco) and 0·5 ml of Tween-80 to 900 ml of distilled water and stir until each is in solution. Sterilise by autoclaving at 15 lb (121°C) pressure for 15 min. Allow to cool and add aseptically, 100 ml of sterile Middlebrook ADC enrichment (BBL). Dispense in 5 ml amounts into sterile 20 × 150 mm screw-cap tubes.

2. *Soft agar overlay*

1. Dubos broth base without Tween-80 (Difco)   6·5 g
2. Proteose peptone No. 3 (Difco)   10·0 g
3. Glycerol   10·0 g
4. Agar   7·5 g
5. Distilled water   1000 ml

Dissolve ingredients 1, 2, and 3 in water then add agar. Heat to dissolve agar. Stir frequently while heating to dissolve the agar in order to prevent scorching. Dispense in 3·5 ml amounts into sterile 16 × 125 screw-cap tubes. Sterilise medium after tubing by autoclaving at 15 lb (121°C) pressure for 15 min.

3. *Hard basal medium (oleic acid–albumin agar)*

1. Dubos Oleic Agar Base (Difco)   20·0 g
2. Bacto-Dubos Oleic Albumin Complex (Difco)   100 g
3. Distilled water   900 ml

Dissolve agar base in water with frequent stirring to prevent scorching. Sterilise by autoclaving at 15 lb (121°C) for 15 min. Temper medium to 52°C then add aseptically, 100 ml of sterile Bacto-Dubos oleic albumin complex (Difco). Mix contents of flask by gentle swirling to prevent formation of bubbles and dispense into sterile Petri dishes.

All supplements are available commercially in sterile amounts of 20 ml/tube. Smaller quantities of each medium may be prepared by using 10 ml supplement for each 90 ml of basal medium.

Inoculate 7H9 broth with the isolates of *M. tuberculosis* that are to be phage typed. Incubate the 7H9 broth cultures at 37°C for ten days. Shake cultures daily to maintain dispersed growth. Just prior to use, shake cultures and allow to stand undisturbed for 5 min to permit the larger clumps to settle.

Melt the soft-agar overlays in a boiling water bath. After melting the agar temper the soft-agar to 52°C. Pipette 0·5 ml of the 7H9 broth culture into the tube with the tempered soft-agar, shake the tube gently to mix the bacteria and agar, and pour the mixture on to the hard basal oleic albumin agar. Gently rotate the Petri dish to ensure even distribution of the soft-agar over the surface of the hard basal medium. Place the Petri dishes on a flat level surface to allow the soft-agar to cool and harden. After allowing the soft-agar to thoroughly harden, place the inoculated plates in a 37°C incubator, in the *inverted* position, and incubate overnight. This incubation period allows the surface of the medium to dry sufficiently to prevent "running" of the drops of the phage lysates. The dried plates are then spotted with the RTDs of the respective test phages as recommended.

The 7H9 broth cultures of *M. tuberculosis* may be used between the tenth and thirteenth days of incubation. All isolates of *M. tuberculosis* utilise Tween-80, but very slowly and with variation among strains; therefore, 7H9 broth cultures less than ten days old should not be used. This procedure will circumvent the occasional problem with the interference of Tween-80 with the adsorption of phage which may be encountered if younger cultures are used.

G. Reproducibility of phage typing results

Statistical evaluations of results reported by five of the WHO Study Group Laboratories have been made and reported by Rado *et al.* (1975). In this study 100 strains of *M. tuberculosis* were typed, the same phages and techniques being used in each laboratory. The tests were repeated in a second study one year later.

It was found that the strains could be assigned to phage types A and B with better than 98% reliability. In both studies there was 100% agreement among all laboratories with one phage, MTPH 2. Although the reproducibility of type C strains (total less than ten strains) was below 50%, improvements in methods appeared to justify continuation of this as a major type.

Subdivision of type A strains by the auxiliary phages showed less agreement among the five laboratories. The results of the final study and analysis on the 100 strains were in agreement 70·4% of the time. It is apparent that phages with greater specificity will make it possible to increase the reproducibility in this group.

VII. LYSOGENY

Lysogeny in the mycobacteria appears to be very wide-spread, if not universal. Recent reports by Šula *et al.* (1973), Jones (1973), Mankiewicz (1961), Buraczewska *et al.* (1971), Murohashi *et al.* (1959) and others have revealed phage in bovine, atypical group III, and *M. fortuitum* strains isolated from human or animal sources. The lysogenic condition affects the response of bacteria to other strains of phage, frequently inhibiting lysis, or modifying the phage typing pattern.

It would be good policy to check for temperate phage in all strains of mycobacteria that are resistant to phages that lyse closely related strains. This is easily accomplished by plating such strains on indicator bacteria susceptible to many of the phages used in typing. The use of ultraviolet light, or chemical inducers such as Mitomycin C, azaguanine or nitroso-guanidine, will increase the possibility of isolating phage. A defective prophage may be indicated when bacteria are lysed with a virulent phage following which a large number of variant phages are found (Mankiewicz and Redmond, 1968).

REFERENCES

Adams, M. H. (1959). "*Bacteriophages*". Interscience, New York.
Baess, I. (1966). *Am. Rev. resp. Dis.* **93**, 622–623.
Baess, I. (1969). *Acta path. microbiol. scand.* **76**, 464–474.
Bailey, R. K., Wyles, S., Dingley, M., Hesse, F. and Kent, G. W. (1970). *Am. Rev. resp. Dis.* **101**, 430–431.
Bates, J. H. and Fitzhugh, J. K. (1967). *Am. Rev. resp. Dis.* **96**, 7–10.
Bates, J. H. and Mitchison, D. A. (1969). *Am. Rev. resp. Dis.* **100**, 189–193.
Bhatia, A. L., Csillag, M., Mitchison, D. A., Selkon, J. B., Somasundaram, P. R. and Subbaiah, T. V. (1961). *Bull. Wld Hlth Org.* **25**, 313–322.
Bowman, B. U. (1958). *J. Bact.* **76**, 52–62.
Bowman, B. U. and Redmond, W. B. (1959). *Am. Rev. resp. Dis.* **80**, 232–239.
Buraczewska, M., Manowska, W. and Rdultowska, H. (1971). *Am. Rev. resp. Dis.* **104**, 760–762.
Buraczewska, M., Kwiatkowski, B., Manowska, W. and Rdultowska, H. (1972). *Am. Rev. resp. Dis.* **105**, 22–29.
Campbell-Renton, M. L. (1941). *J. Path. Bact.* **53**, 371–384.
Carne, H. R. and Greaves, R. I. N. (1974). *J. Hyg. (Camb.)* **72**, 467–470.
Caroli, G., Mazzarone, R. and Lauro, P. (1967). W.H.O.
Cater, J. C. and Redmond, W. B. (1961). Symposium on Bacteriophage of Macobacteria, Bilthoren, the Netherlands, pp. 32–34.
Clark, W. A. (1962). *Appl. Microbiol.* **10**, 466–471.
Clark, W. A. (1970). *In* "Culture Collections of Micro-organisms" (H. Tizuka and T. Hasegawa, Eds), pp. 309–318.
Clark, W. A., Horneland, W. and Klein, A. G. (1962). *Appl. Microbiol.* **10**, 463–465.
Clark, W. A. and Klein, A. (1966). *Cryobiology* **3**, 68–75.
Davies, J. D. and Kelly, M. J. (1969). *J. Hyg. (Camb.)* **67**, 573–583.
Dickinson, J. M., Lefford, M. J., Lloyd, J. and Mitchison, D. A. (1963). *Tubercle* **44**, 446–451.
Dubina, J., Šula, L. and Slosarek, M. (1974). *Stud. Pneumol. Phtisiol. Cechoslov.* **34**, 617–620.
Dubos, R. (1954). *Am. Rev. Tuberc.* **70**, 391–401.
Edson, N. L. (1951). *Bact. Rev.* **15**, 147–182.
Engel, H. W. B., Smith, L. and Berwald, L. G. (1974). *Am. Rev. resp. Dis.* **109**, 561–566.

Frimodt-Moller, J. (1975). *Indian Coun. med. Res.*, Tech. Report of the Sci. Advis. Board, New Delhi, p. 153.

Froman, S., Will, D. W. and Bogen, E. (1954). *Am. J. publ. Hlth* **44**, 1326–1333.

Gardner, G. M. and Weiser, R. S. (1947). *Proc. Soc. exp. Biol. Med.* **66**, 205–206.

Gruft, H. (1972). *Am. Rev. resp. Dis.* **106**, 119–120.

Gunnels, J. J. and Bates, J. H. (1972). *Am. Rev. resp. Dis.* **105**, 388–392.

Hauduroy, P. and Rosset, W. (1948). *C. r. hebd. Séanc. Acad. Sci.*, Paris **227**, 917–918.

Hobby, G. L., Redmond, W. B., Runyon, E. H., Schaefer, W. B., Wayne, L. G. and Wichelhausen, R. H. (1967). *Am. Res. resp. Dis.* **95**, 954–971.

Hnatko, S. I. (1953). *Can. J. med. Sci.* **31**, 462–473.

Jones, W. D. (1973). *Am. Rev. resp. Dis.* **108**, 1438–1441.

Jones, W. D. and White, A. (1968). *Can. J. Microbiol.* **14**, 551–555.

Joseph, S., Mitchison, D. A., Ramachandran, K., Selkon, J. B. and Subbaiah, T. V. (1964). *Tubercle* **45**, 354–359.

Kaustova, J., Kandus, J. and Cechova, A. (1974). *Stud. Pneumol. Phtisiol. Cechoslov.* **34**, 219.

Kim, T. H. and Kubica, G. P. (1973). *Appl. Microbiol.* **25**, 956–960.

Kitahara, K. (1973). *Kekkaku* **48**, 61–69.

Lapage, S. P., Shelton, J. E. and Mitchell, T. G. (1970). "Methods in Microbiology", Vol. 3A, pp. 1–133. Academic Press, London.

Leclerc, H., Nguematcha, T., Debruyne, J. and Tacquet, A. (1971). *Ann. Inst. Pasteur* **22**, 177–188.

Lind, A. (1967). *Scand. J. resp. Dis.* **48**, 343–347.

Mankiewicz, E. (1961). *Nature, Lond.* **191**, 1416–1419.

Mankiewicz, E. (1972). *Can. J. publ. Hlth* **63**, 342–354.

Mankiewicz, E. (1973). Proc. 8th Wld Hlth Org. Committee Symp. on Phage Typing of Mycobacteria, Montreal, pp. 121–143.

Mankiewicz, E. and Liivak, M. (1975). *Am. Rev. resp. Dis.* **111**, 307–312.

Mankiewicz, E. and Redmond, W. B. (1968). *Am. Rev. resp. Dis.* **98**, 41–46.

Mitchison, D. A. and Lloyd, J. (1964). *Tubercle* **45**, 360–369.

Mitchison, D. A., Wallace, J. G., Bhatia, A. L., Selkon, J. B., Subbaiah, T. V. and Lancaster, M. C. (1960). *Tubercle* **41**, 1–22.

Mitzuguchi, Y., Maruyama, Y., Suga, K. and Murohashi, T. (1973). *Kekkaku* **48**, 219–225.

Murohashi, T., Tokunaga, T., Maruyama, Y. and Mizuguchi, Y. (1969). *Proc. Ann. Meeting Freeze Drying Study Gcoup*, **15**, 117–119.

Murohashi, T., Tokunaga, T., Mizuguchi, Y. and Maruyama, Y. (1963). *Am. Rev. resp. Dis.* **88**, 664–669.

Murohashi, T., Tokunaga, T. and Seki, M. (1959). *Med. Biol.* **53**, 242–246.

Murohashi, T., Tokunaga, T. and Seki, M. (1960). *Med. Biol.* **54**, 214–218.

Piguet, J.-D. (1960). Contribution à l'etude des mycobacteriophages, pp. 1–136. Thesis, University of Lausanne (Faculté des Sciences).

Rado, T., Bates, J. H., Engel, H. W. B., Mankiewicz, E., Murohashi, T., Mizuguchi, Y. and Šula, L. (1975). *Am. Rev. resp. Dis.* **111**, 459–468.

Raleigh, J. W., Wichelhausen, R. H., Rado, T. A. and Bates, J. H. (1975). *Am. Rev. resp. Dis.* **112**, 497–503.

Redmond, W. B. (1963). *Adv. tuberc. Res.* **12**, 191–229.

Redmond, W. B. (1973). Proc. 7th and 8th Symp. on Isolation, Classification and World Wide Distribution of Mycobacterial, Montrea, pp. 169–175.

Redmond, W. B., Cater, J. C. and Ward, D. M. (1963). *Am. Rev. resp. Dis.* **87**, 257–263.
Redmond, W. B. and Ward, D. M. (1966). *Bull. Wld Hlth Org.* **35**, 563–568.
Rieber, M. and Imaeda, T. (1969). *J. Virol.* **4**, 542–544.
Segawa, J., Takeya, K. and Sasaki, M. (1960). *Am. Rev. resp. Dis.* **81**, 419–420.
Selkon, J. B., Subbaiah, T. V., Bhatia, A. L., Radhakrishna, S. and Mitchison, D. A. (1960). *Bull. Wld Hlth Org.* **23**, 599–611.
Sellers, M. I., Baxter, W. L. and Runnals, H. R. (1962). *Can. J. Microbiol.* **8**, 389–399.
Stead, W. W. and Bates, J. H. (1969). *Ann. int. Med.* **70**, 707–711.
Stead, W. W. and Bates, J. H. (1971). *Ann. int. Med.* **74**, 559–561.
Steiner, M., Chaves, A. D., Lyons, H. A., Steiner, P. and Portugaleza, C. (1970). *New Engl. J. Med.* **283**, 1353–1358.
Subbaiah, T. V., Mitchison, D. A. and Selkon, J. B. (1960). *Tubercle* **41**, 323–333.
Šula, L. and Mohelska, H. (1965). Proc. 2nd Symp. on Isolation, Classification and World Wide Distribution of Mycobacteria, Prague.
Šula, L., Redmond, W. B., Coster, J. F., Baess, I., Bates, J. H., Caroli, G., Mankiewicz, E., Murohashi, T. and Vandra, E. (1973). *Bull. Wld Hlth Org.* **48**, 57–63.
Šula, L., Šulova, J. and Spurná, M. (1973). *Zentbl. Bakt. Hyg. Abt. I* **223**, 520–532.
Sushida, K. and Hirano, N. (1971). *Am. Rev. resp. Dis.* **106**, 269–271.
Takeya, K. and Yoshimura, T. (1957). *J. Bact.* **74**, 540–541.
Takeya, K., Yoshimura, T., Yamaura, K. and Tadao, T. (1959). *Am. Rev. resp. Dis.* **80**, 543–553.
Tokunaga, T., Maruyama, Y. and Murohashi, T. (1968). *Am. Rev. resp. Dis.* **97**, 469–471.
Tokunaga, T. and Murohashi, T. (1963). *Jap. J. Med. Biol.* **16**, 21–30.
Ungar, J. (1949). *Tubercle* **30**, 2–4.
Wayne, L. G. (1962). *Am. Rev. resp. Dis.* **86**, 651–656.
Will, D. W., Froman, S., Akiyama, Y. and Scammon, L. (1961). *Am. Rev. resp. Dis.* **84**, 739–743.

CHAPTER IX

Identification of Mycoplasmas

E. A. Freundt, H. Ernø

FAO/WHO Collaborating Centre for Animal Mycoplasmas, Institute of Medical Microbiology, University of Aarhus, Denmark

and Ruth M. Lemcke†

The Lister Institute of Preventive Medicine, University of London, England

† Present address: Agricultural Research Council, Institute for Research on Animal Diseases, Compton, Newbury, Berks, England.

I. INTRODUCTION

The primary objective of this Chapter, which may be regarded as a continuation of the chapter by Fallon and Whittlestone (1969) on the "Isolation, cultivation and maintenance of mycoplasmas", published in Volume 3B of this Series, is to describe in detail the practical aspects of methods used for the identification of mycoplasmas.

Identification of an unknown organism aims at defining its taxonomic position at the species or subspecies level. In order to provide a background for the presentation of the data required for this purpose, an outline is given in Section II of the taxonomy of the mycoplasmas. In this connection, the biochemical and more particularly the serological methods used in species classification and identification are discussed in general terms.

Section III provides an outline of standard procedures that, step by step, will lead to final identification.

Technical details of biochemical and serological methods used to identify mycoplasmas are presented in Sections IV and V, respectively.

The presentation and the views expressed in this Chapter are based to a great extent on the experience gained by the authors through several years of work on identification problems. Also, the methods and techniques currently adhered to in our respective laboratories provide the primary basis of the descriptions of recommended standard methods. In some cases, useful alternative methods are briefly described or selected references are given.

For a more general orientation in the field of mycoplasmology, and in particular for information about the significance of mycoplasmas as aetiological agents of diseases in man, animals and plants, the reader is referred to recent books (Hayflick, 1969; Sharp, 1970; Barile *et al.*, 1979), symposia (Elliott and Birch, 1972; Maramorosch, 1973; Bové and Duplan, 1974) and reviews (Razin, 1969, 1978; Freundt, 1974a).

II. TAXONOMY OF THE MYCOPLASMAS

A. Class and order

Following a recommendation made by the Subcommittee on the Taxonomy of the *Mycoplasmatales* (1967), Edward and Freundt in 1967 formally proposed that the order *Mycoplasmatales* (vernacular or trivial name "mycoplasmas") be assigned to a new microbial class, the Mollicutes (Table I). The name of this new class, which means "soft skin", refers to a most striking fundamental property of the mycoplasmas: their lack of a rigid cell wall and inability to synthesise basic constituents of the bacterial cell wall such as muramic acid and diaminopimelic acid. The last part of

TABLE I

Taxonomy of class Mollicutes

Class: Mollicutes
  Order I: *Mycoplasmatales*
    Family I: *Mycoplasmataceae*
      Genus I: *Mycoplasma*
          1. Sterol required for growth
          2. Sensitive to digitonin
          3. Genome size $4 \cdot 5 \times 10^8$ daltons
      Genus II: *Ureaplasma*
          1. Sterol required for growth
          2. Sensitive to digitonin
          3. Urea catabolised
          4. Genome size $4 \cdot 5 \times 10^8$ daltons
    Family II: *Acholeplasmataceae*
      Genus I: *Acholeplasma*
          1. Sterol not required for growth
          2. Resistant to digitonin
          3. Genome size $1 \cdot 0 \times 10^9$ daltons
    Family III: *Spiroplasmataceae*
      Genus I: *Spiroplasma*
          1. Sterol required for growth
          2. Sensitive to digitonin
          3. Helical morphology, rotational and undulating motility
          4. Genome size $1 \cdot 0 \times 10^9$ daltons

Genera of uncertain affiliation
A. *Anaeroplasma*
          1. Some strains require sterols, some do not
          2. Some strains are resistant to digitonin, some are sensitive
          3. Strict anaerobes
          4. Genome size: not determined
B. *Thermoplasma*
          1. Sterol not required for growth
          2. Sensitivity to digitonin: not determined
          3. Optimum temperature about 59 °C and optimum pH 1–2
          4. Genome size $1 \cdot 0 \times 10^9$ daltons

---

the definition is intended to exclude from the Mollicutes the L-phase variants and similar aberrant bacterial forms which are also characterised by the absence of a complete cell wall and by a number of other properties derived from this basic feature (Edward, 1967; Freundt, 1973). On the practical level, classification of an organism as a member of the order *Mycoplasmatales*, class Mollicutes, is based primarily on the following criteria: absence of a cell wall, typical "fried egg" appearance of the minute colonies, filterability through a membrane filter of 450 nm pore diameter, and absence of reversion of fresh isolates to bacteria under appropriate conditions (Subcommittee on the Taxonomy of *Mycoplasmatales*, 1972).

B. Families and genera

Two families are currently recognised in the order *Mycoplasmatales*. Family I, the *Mycoplasmataceae*, includes two genera, *Mycoplasma* and *Ureaplasma*, both of which are distinguished by requiring cholesterol or other sterols for growth. Associated with this property is their sensitivity to 1·5% digitonin (Ernø and Stipkovits, 1973; Freundt *et al.*, 1973a). The genome size is about 5×10^8 daltons (Bak *et al.*, 1969). Members of the genus *Ureaplasma*, previously known as T-mycoplasmas, differ from those of the genus *Mycoplasma* in their ability to catabolise urea (Shepard *et al.*, 1974). Almost all the mycoplasmas of proven pathogenicity to man and animals are found within the genus *Mycoplasma*.

Family II, the *Acholeplasmataceae*, is characterised by not depending on sterols for growth and by being resistant to 1·5% digitonin. The genome size is $1·0 \times 10^9$ daltons (Bak *et al.*, 1969). Only one genus, *Acholeplasma*, is recognised.

A new genus *Anaeroplasma*, that will neither fit into *Mycoplasmataceae* nor *Acholeplasmataceae*, because it includes sterol-requiring as well as sterol-non-requiring organisms, was recently established by Robinson *et al.* (1975). Species of this genus are strict anaerobes and have been isolated hitherto only from the rumens of cattle and sheep.

Recently Skripal (1974) suggested the establishment within the order *Mycoplasmatales* of a third family, the *Spiroplasmataceae*, which was later accepted by the Subcommittee on Taxonomy of *Mycoplasmatales* (1977). The proposed new family, which contains only one genus, *Spiroplasma* (Saglio *et al.*, 1973), differs from the two other families in exhibiting helical morphology and rotational and undulating motility. Sterols are required for growth, and sensitivity to 1·5% digitonin has been demonstrated. The genome size is $1·0 \times 10^9$ daltons. The discovery in 1973 of the first member of the new taxon, *S. citri*, an organism associated with and probably causing "stubborn disease" of citrus plants, very obviously opened up an exciting and most promising new field in mycoplasmology. In addition to important plant and insect pathogens, the *Spiroplasmataceae* includes an organism, the suckling mouse cataract spiroplasma, that is pathogenic to vertebrates.

Thermoplasma (Darland *et al.*, 1970) is yet another genus that has been accepted by the Subcommittee for inclusion in the class Mollicutes, although its allocation to order and family has been postponed until more information is available (Subcommittee on the Taxonomy of *Mycoplasmatales* 1975; Edward, 1974). The one species described within this genus, *T. acidophilum*, is distinguished by a temperature optimum for growth of about 59°C and a pH optimum of about 1–2. Sterol is not required for growth. The genome size is 1.0×10^9 daltons (Christiansen *et al.*, 1975). It has been recovered up till now only from burning coal refuse piles.

Note. The description in Sections III–V of the biochemical and serological methods in the identification of mycoplasmas is prepared with special reference to those species that are classified within the genera *Mycoplasma*, *Ureaplasma* and *Acholeplasma*, which together comprise the vast majority of the species that are of interest to the laboratory worker concerned with mycoplasmas of human and animal provenance. In all essentials, the procedures used to identify these mycoplasmas will also be applicable to organisms belonging to the genus *Spiroplasma*. The growth conditions and hence the techniques to be used for the isolation and propagation of *Anaeroplasma* species, and the methods used for their characterisation and identification, do, on the other hand, differ rather profoundly from those suitable for the other mycoplasmas. Since moreover the authors have as yet no personal experience with these organisms, no further mention will be made of them in this Chapter. Readers having a special interest in *Anaeroplasma* species are referred to the papers by Robinson and Hungate (1973), Robinson *et al.* (1975), and Robinson and Allison (1975). Similar considerations hold true for *Thermoplasma* for which reference should be made to Darland *et al.* (1970) and Belly *et al.* (1973).

C. Species

1. *The species concept*

The basic unit of biological relatedness is the species. In organisms that reproduce sexually species are defined by the ability of their members to breed with one another. In organisms that reproduce asexually the species concept represents a major problem that has never attained a fully satisfactory solution. Whereas in traditional microbial taxonomy the grouping of supposedly related strains in the category of a species is based mainly on phenotypic characters, the burst of nucleic acid homology studies which has occurred in recent years does in fact represent a genetic approach to the classification of micro-organisms. For obvious reasons nucleic acid hybridisation techniques do not offer themselves as a means of routine identification and classification. Nevertheless, nucleic acid homology data may, at least under ideal circumstances, provide a rational and meaningful basis for species classification by serving as a means of selecting those phenotypic characteristics that may be particularly useful in identifying an organism as belonging to a certain species (Johnson, 1973). As a matter of fact, a very remarkable agreement has been found between the results of hybridisation experiments and the present species classification that has been based, throughout the years, on the combination of biological properties and serological data.

It is only logical, therefore, that the Subcommittee on the Taxonomy of *Mycoplasmatales* (1972) should strongly recommend the inclusion of cultural and biochemical as well as antigenic characters among the properties to be determined when a new species is being proposed. It follows that the identification of an unknown strain as belonging to one or other of the mycoplasma species already established must be based on similar criteria. Since, as mentioned already, the mycoplasmas unfortunately possess relatively few distinctive morphological, cultural and biochemical properties, the identification of an unknown strain depends ultimately on serology. Although it is indisputable that high priority must be given to the use of antigenic markers in identification there is still doubt as to which serological methods are the most suitable.

In the following Sections, the serological methods that are most frequently used in species classification and thereby also for the identification of mycoplasmas will be mentioned. The theoretical basis of the methods will be discussed only for those techniques that are specifically used in mycoplasmology, whereas an attempt will be made to evaluate the specificity and sensitivity of each test.

Although specificity is, of course, a relative term, a specific test is defined here as a test that will identify an organism as belonging to a certain species without revealing a significant degree of cross-reaction with representatives of any other species. Thereby, a working definition of the species concept for mycoplasmas becomes a prerequisite for a further meaningful discussion of what is to be understood by the specificity of a test. On empirical grounds, a definition of the mycoplasma species may be deduced from experience gained with the two tests that have proved most useful and to which reference will be made frequently in subsequent sections, i.e. the growth inhibition and metabolism inhibition tests. In terms of the methods that are routinely used to identify an organism at species level, a mycoplasma species may be defined as a group of strains which are so closely related that antibody produced against one strain will prevent replication and metabolism of all the other strains within that group. The specificity of any other method has thus to be evaluated on the basis of the extent of agreement it shows with the results obtained by the growth and metabolism inhibition tests: a test is specific if it results in the same grouping as do these tests, but it is less specific if it gives a broader grouping or shows more cross-reactivity between species established by these tests.

The term "sensitivity" is in fact even more ill-defined than "specificity" and is often used in differing senses by different authors. In the following Sections a test is described as sensitive when a positive reaction can be obtained with a small amount of antibody.

2. Serology as a basis for species classification

(a) *Growth inhibition.* The growth inhibition (GI) test was introduced by Nicol and Edward (1953) who observed that antiserum incorporated in solid agar medium inhibited the growth of colonies of homologous and closely related strains. Inhibition of growth by specific antiserum may also occur in liquid medium and can be measured either directly, by turbidimetry or by subculturing on to solid medium, or indirectly, by the metabolism inhibition test (see below). Direct methods of demonstrating growth inhibition in liquid medium have gained no common usage and the term GI test now refers only to growth inhibition carried out on solid medium.

Although the nature of the growth inhibiting effect of specific antibody is not fully understood two mechanisms may be operating, a blocking of the transport systems of the cell membrane, possibly because of the lack of a true cell wall (Edward and Fitzgerald, 1954), and a complement dependent lysis of the mycoplasma cell. Although growth inhibition on solid medium is generally believed to be independent of heat-labile accessory factors, some recent reports (Roberts, 1971; Roberts and Pijoan, 1971) suggest that such factors may be involved. It should also be remembered that some complement is usually present in the growth medium used for the test, being derived from unheated horse serum and/or from the antiserum if this is used in the non-inactivated state. Further studies, including, *inter alia*, electron microscopic visualisation of the interaction between antibody and specific components of the cell membrane, are needed to elucidate the theoretical basis of the growth inhibition phenomenon.

In practical terms, the GI test is based on the fact that the development of colonies is inhibited either to the extent that they do not develop at all or that their number and/or size are significantly reduced. The original version of the test as described by Edward and his colleagues was soon replaced by techniques requiring smaller amounts of antiserum, viz. the disc-, agar well- and serum-drop methods. The principles of these modifications of the GI test will be described in Section V.B.

That the GI test shows, overall, a satisfactory degree of specificity is attested by the experience of its use over many years. On the other hand, some antisera may occasionally exhibit species non-specific inhibitory activity. For example, antisera produced against strains of the bovine "serogroup 7" (Leach, 1967) have been observerved in various laboratories to inhibit the growth of *M. bovigenitalium*, although neither biochemical nor other serological data suggest a relationship on the species level between these two taxa (Ernø and Jurmanová, 1973). Mention may also be made of the cross-reactions as yet unexplained, which are occasionally seen between different species of the genus *Acholeplasma*.

A major disadvantage of the GI test is its low degree of sensitivity. Another inconvenience, that may or may not be associated with the relative lack of sensitivity of this test, is the difficulty sometimes experienced in obtaining satisfactory inhibition of growth on solid medium, even with antisera that may prove highly potent in the metabolism inhibition test. This phenomenon tends to be more pronounced with some species than with others. It is encountered particularly frequently within the genus *Acholeplasma*, but may be seen also with rapidly growing strains of the genus *Mycoplasma*, such as *M. mycoides* subsp. *capri*. As described in more detail later, several modifications have been developed with the purpose of enhancing the sensitivity of the GI test. The following parameters have a decisive influence on the test results: the size of the inoculum of mycoplasmas, medium composition, incubation temperature, amount of antiserum, and the relative proportions of the different classes of immunoglobulin. As to the size of the inoculum, the density of colonies produced by an inoculum containing 10^4–10^5 colony forming units per ml (CFU/ml) is about optimal. A lower density of colonies will detract from the accuracy of measuring the size of the inhibitory zone or even make the reading of the test unreliable. A higher density will diminish the zone of inhibition. Variations in the medium composition, incubation temperature, and the amount of antiserum used are made in order to create optimal conditions for the antigen–antibody reaction, by allowing a larger amount of antibody to react with a smaller number of mycoplasma cells within a given period of time.

Very little is known about the involvement of different classes of immunoglobulins in growth inhibition, but some observations suggest an association of the inhibitory effect with IgM and IgG, though perhaps to a greater extent with the latter.

(b) *Metabolism inhibition*. The MI test utilises the ability of antibody to inhibit metabolic activities such as glucose fermentation (Taylor-Robinson *et al.*, 1966), arginine catabolism (Purcell *et al.*, 1966a), hydrolysis of urea (Purcell *et al.*, 1966b) or reduction of 2,3,5-triphenyl tetrazolium chloride (Senterfit and Jensen, 1966).

The inhibitory activity is demonstrated by the prevention of a change of pH (a decrease in the case of glucose-fermenting organisms and an increase with organisms hydrolysing arginine or urea), or by the absence of the change to a red colour resulting from the reduction of tetrazolium. The MI test may occasionally be performed even on the basis of a metabolic activity whose exact nature is not yet known. Thus, the acidity produced during the growth of *M. bovigenitalium* and *M. bovis* has been utilised for determining the metabolism inhibitory activity of antisera against these organisms, although it is not a result of glucose degradation and its origin is obscure.

Although there is no consistent correlation between the GI and MI titres of an antiserum, convincing evidence is available to support the view that the MI test is essentially a growth inhibition technique carried out in liquid medium (Purcell *et al.*, 1967; Taylor-Robinson and Berry, 1969; Woode and McMartin, 1973). The specific inhibitory activity of antiserum is believed to be related to the adsorption of antibody to the mycoplasma cell membrane with a consequent depression of cell metabolism and growth (Woode and McMartin, 1973), possibly associated under appropriate conditions with a mycoplasmacidal process.

The specificity and sensitivity of the MI test are both at a very high level. Cross-reactions between different species are seen only occasionally and at rather low titres. Homologous titres of hyperimmune sera exceeding 1:20 000 are by no means unusual. It should be pointed out that a great deal of care and experience is required in performing the test. The time of reading the test is often particularly critical, and for that and other reasons a proper standardisation of the MI test is difficult. In consequence, MI titrations of one and the same serum carried out in different laboratories may well result in titres differing by a factor of ten.

It follows from this brief review of the MI test that this method is extremely useful for taxonomic studies, for which purpose the time and special precautions needed for its performance are justified. It may be found less suitable, on the other hand, for the routine identification of mycoplasma isolates. In certain situations, however, the MI test may be used with advantage in the identification procedure, for example, where an identification obtained by the immunofluorescence test cannot be confirmed by the GI test.

(c) *Immunofluorescence.* The fluorescent antibody technique was introduced in mycoplasmology by Liu (1957) for localisation of the important human pathogen, *M. pneumoniae* (the "Eaton agent") in chick embryos and was used also for demonstration or identification of this organism by Clyde (1961), Goodburn and Marmion (1962) and Chanock *et al.* (1962). As antigen, Chanock and co-workers used agar-grown colonies transferred to microscope glass slides. This technique, which was further modified by Lind (1970) and Rosendal and Black (1972), is relatively laborious and time-consuming. An approach to a rapid laboratory diagnosis of *M. pneumoniae* pneumonia was subsequently made by Hers (1963) using immunofluorescence to identify *M. pneumoniae* associated with respiratory tract cells excreted in the sputum of patients suffering from primary atypical pneumonia. This method apparently never came into common use. A major advance in the utilisation of the immunofluorescence technique for the identification of mycoplasmas was made in 1967 by Del Giudice *et al.*, who developed the epi-immunofluorescence test. This allowed the application of the fluorescent antibody staining method to mycoplasma colonies grown on solid medium and the direct examination of the stained colonies using a microscope with an attachment for incident illumination. Whereas the direct fluorescent antibody staining technique was employed by Del Giudice *et al.*, definite advantages were demonstrated by Rosendal and Black (1972) in using the indirect staining technique.

As is the case with the GI and MI tests the antigens reacting with antibody in the immunofluorescence test are mainly associated with the cell membrane. The testing of living unfixed colonies *in situ* on the agar surface, i.e. the avoidance of any manipulation with the mycoplasma cells prior to antibody-labelling, implies that the native structure of the antigens involved in the reaction is maintained.

The specificity of the epi-immunofluorescence test is equal to that of the GI test and its sensitivity is much higher. The epi-immunofluorescence test has the additional advantage over the GI test of producing a quantitative result, since sera can be titrated and endpoints of 1:1000 or more are usually obtained. Since, moreover, the epi-immunofluorescence test is easy to standardise, a high priority must obviously be given to this test for the routine identification of mycoplasmas. The only real drawback of the test is the tendency of strains of some species to autofluoresce.

It should further be pointed out that the unexpected cross-reactions between different species of the genus *Acholeplasma* that were previously mentioned when discussing the GI test may also be seen with the immunofluorescence test. Collectively, these observations suggest that species differentiation within the genus *Acholeplasma* is perhaps less well-founded and not as clear-cut as in the genus *Mycoplasma*.

(d) *Direct agglutination*. Agglutination of mycoplasma cells by antiserum was among the first tests used for serological classification. Although in the hands of experienced investigators (Klieneberger, 1938; Edward, 1950; Edward and Fitzgerald, 1951, 1954; Nicol and Edward, 1953; Edward and

Kanarek, 1960) it appears to give very satisfactory results, it is not used much at the present time for the identification of mycoplasmas. In the field of avian mycoplasmology, various modifications of direct agglutination are, however, used quite extensively for the demonstration of antibodies in serum specimens in diagnostic and epidemiological studies of diseases caused by important avian pathogens such as *M. gallisepticum* and *M. synoviae* (Adler, 1954, 1958; Adler and Yamamoto, 1956; Adler and Da-Massa, 1964, 1967; Roberts and Olesiuk, 1967; Hromatka and Adler, 1969). The test seems to be highly specific to the extent that it may even distinguish between groups of strains within species that show some degree of heterogeneity, for example *M. hominis*. The original subdivision of *A. laidlawii* into three "types", A, B and C (Laidlaw and Elford, 1936) was also based on the agglutination test, although at least "types" A and B were later shown by other methods to be so very closely related that this distinction is no longer maintained. The sensitivity of direct agglutination tests appears to depend on a number of variables (Section V. E, 2).

(e) *Indirect agglutination.* This technique is employed most widely in the form of the indirect haemagglutination (IHA) test, using tanned fresh or either formalinised or glutaraldehyde treated red blood cells coated with mycoplasma antigen as the agglutinable particles (Ross and Switzer, 1963; Tully, 1963; Dowdle and Robinson, 1964; Taylor-Robinson *et al.*, 1964, 1965a, b; Adler and DaMassa, 1967b; Lind, 1968; Krogsgaard-Jensen, 1971; Freundt *et al.*, 1973; Holmgren, 1973; Lam and Morton, 1974; Cho *et al.*, 1976). Latex particles have also been used as antigen carriers (Morton, 1966; Kende, 1969).

The potential suitability of the IHA test for species classification was particularly borne out by the studies of Tully (1963) and by Taylor-Robinson *et al.* (1964, 1965a) who obtained clear-cut serological distinctions between all of the then established human *Mycoplasma* species. In the study by Tully, the specificity was equal to that of the simple agglutination and fluorescence antibody techniques and in those by Taylor-Robinson *et al.* it was equivalent to or above that of the complement fixation test. The IHA technique used by the latter authors obviously resulted in a high degree of sensitivity. Although at the same time cross-reactions between species frequently occurred, they were always at low titre and negligible compared with the exceedingly high homologous titres. Results reported by Freundt *et al.* (1973b) from an assay of a considerable number of serum reference reagents for a variety of *Mycoplasma* and *Acholeplasma* species were in general in line with those of Taylor-Robinson *et al.*, although in some cases heterologous cross-reactivity at fairly high titre was observed when fresh tanned red blood cells were used. The specificity of the test was, however, considerably improved when formalinised rather than fresh red blood cells were used.

Because of the rather laborious technique and the difficulties involved in standardising the IHA method, this test is of only limited practical value for classification and identification purposes. The high sensitivity of the test does, on the other hand, make it particularly useful for sero-epidemiological studies.

(f) *Complement fixation.* The complement fixation (CF) test has been used widely in the classification of mycoplasmas (Edward, 1950; Edward and Fitzgerald, 1951; Huijsmans-Evers and Ruys, 1956; Card, 1959; Taylor-Robinson *et al.*, 1963, 1965a; Lemcke, 1964; Clyde, 1964; Fox *et al.*, 1969), although it is currently less popular for routine identification than, for example, GI and MI tests. Nevertheless, the fact that the CF test is used extensively in the serological diagnosis of mycoplasma infections of man and animals and in sero-epidemiological surveys should encourage the continued use of this test in the classification of mycoplasmas.

Although considerable interspecies cross-reactivity has been reported with this test, quantitative differences between homologous and heterologous titres are usually sufficient to allow clear taxonomic distinctions to be made. The high levels of cross-reactivity noted in some reports may in some cases be attributed to contamination of both test and immunising antigens with antigenic material from the medium, in particular adsorbed serum proteins, when the immunising antigen is grown in the presence of serum foreign to the animal being immunised. Apart from this, the use of antigens at dilutions that tend to be anticomplementary and the practice of comparing antigens at a fixed unitage or at a standard opacity have probably magnified the cross-reactions between species. Heterologous reactions can be much reduced or even eliminated by testing each antigen at a previously determined optimal concentration (Section V. F, 2 (h)).

Nevertheless, when all technical precautions have been taken to avoid non-specific cross-reactions and thus to increase the specificity of the test, low-level cross-reactions (at 2% or less of the homologous titre) still occur between some species. These may be due to the presence of common or related antigens. In CF reactions it appears that cytoplasmic antigens as well as membrane antigens are involved (Hollingdale and Lemcke, 1969; Pollack *et al.*, 1970). In this respect, CF differs from tests such as direct agglutination, GI, MI or immunofluorescence. Cross-reactions revealed by CF but not by GI, MI or immunofluorescence tests may therefore be due to cytoplasmic antigens, especially in species which share a common metabolic pathway and in which serologically related enzymes have been demonstrated (Thirkill and Kenny, 1974). In this sense, CF tests can be said to be less specific than GI, MI or immunofluorescence tests. On the other hand, the CF test is useful at the intraspecies level, since it shows more homogeneity between strains of the same species than the GI and MI tests, which can apparently reflect small differences in the surface antigens (Hollingdale and Lemcke, 1970; Forshaw and Fallon, 1972). Thus in an identification procedure, CF may in some cases be useful in confirming the relationship of an unknown to a recognised species when other more specific tests have shown only a limited degree of cross-reactivity. The sensitivity of the CF test is greater than that of GI and double immunodiffusion tests, probably equal to that of direct agglutination and immunofluorescence tests, but less than that of the MI test.

(g) *Double immunodiffusion.* Double immunodiffusion has been used quite extensively in the classification of mycoplasmas (Taylor-Robinson *et al.*, 1964, 1965a; Lemcke, 1965, 1973; Kenny, 1969, 1972, 1973; Ernø and Jurmanová, 1973; Rosendal, 1974b). The sensitivity of this method is much lower than that of MI, immunofluorescence or CF tests in that antisera usually have to be used undiluted or at very low dilutions and high concentrations of antigen are required. However, the method has the advantage that reactions of individual determinants or antigens are observable. Other serological methods show the degree of cross-reactivity between strains; immunodiffusion tests can indicate the basis of the cross-reactions in terms of the number of determinants involved and the relatedness of the cross-reacting components. Double immunodiffusion is therefore unlikely to be used as a first step in identification, but rather to investigate or substantiate relationships which have been suggested by GI, MI or immunofluorescence tests. For example, the close relationship of a group of equine mycoplasmas to *M. felis* that was suggested by GI and MI tests was confirmed by immunodiffusion tests (Lemcke and Allam, 1974; Allam and Lemcke, 1975).

In regard to classification, analyses of the antigenic patterns of mycoplasmas by the double immunodiffusion test, or by the more sensitive two-dimensional immunoelectrophoresis technique, have revealed the existence of common antigens within groups of glycolytic *Mycoplasma* or *Acholeplasma* species (Kenny, 1969, 1972, 1973) as well as among arginine-utilising *Mycoplasma* species (Taylor-Robinson *et al.*, 1963; Lemcke, 1965; Fox *et al.*, 1969; Kenny, 1972, 1973; Thirkill and Kenny, 1974). No antigenic overlapping is, however, seen between these two biochemically distinct groups (Fox *et al.*, 1969; Ernø and Jurmanová, 1973; Kenny, 1973).

Information about the presence or absence of antigenic relationships between mycoplasma species found in a particular animal host may assist the interpretation of results obtained in serodiagnostic or sero-epidemiological surveys by other methods such as CF and IHA tests. For example, the fact that the CF test has proved specific for diagnosing *M. pneumoniae* pneumonia correlates with its uniqueness as shown by immunodiffusion tests; this species apparently shares no common antigens with other mycoplasmas from man.

Double immunodiffusion tests as usually carried out are less species-specific than GI, MI or immunofluorescence, and this is probably because cytoplasmic antigens are involved. In tests employing ultrasonically treated or frozen-thawed suspensions and adjuvant-produced sera, reactions seem to be due mainly to cytoplasmic antigens, although surface components which are easily released from the membrane may be involved. When suspensions are treated with detergents, at least two additional precipitin bands released by lysis of the membrane are observed. It appears that cytoplasmic antigens are less species-specific than membrane antigens. Within a single species, it is antigens in the membrane rather than the cytoplasmic fraction that, in immunodiffusion tests, reveal differences between strains (Hollingdale and Lemcke, 1970).

D. Subspecies

At present only one species is subdivided into the lower category of subspecies, viz. *M. mycoides*. It has been a matter of discussion, however, for a long time whether the subspecies concerned might not rather deserve the rank of species. Whatever the final decision may be in this case, a subdivision of some other species may well be necessary in the future since an increasing number of isolates are being analysed and biochemical methods are being refined and standardised. A further subdivision based on the infrasubspecific terms serotype, chemotype and biotype (equivalent to the recently proposed terms "serovar", "chemovar" and "biovar" (Lepage *et al.*, 1973, Appendix 10)) might then be taken into consideration.

E. Present status of classification into species and subspecies

In Table II are listed the species and subspecies of the genus *Mycoplasma* recognised at present, together with their type strains, a medium found suitable for the growth of each species, and some biochemical characteristics. The number of strains examined for each species or subspecies is in some cases so small that future investigations may well reveal biochemical variants. Taxonomic references are given for those species which are not included in the 8th edition of "Bergey's Manual of Determinative Bacteriology" (Freundt, 1974c).

In Table III are tabulated the corresponding data for the genus *Acholeplasma*. The results of the aesculin and arbutin tests are recorded specifically for this genus because these tests appear to be of some use for the identification of species of this taxon.

III. IDENTIFICATION
AN OUTLINE AND DISCUSSION OF STANDARD PROCEDURES

It follows from the consideration of the general principles of mycoplasma classification presented in the previous Section, that identification procedures usually include biochemical as well as serological methods, and that final identification has to be based on serology. Because of the fairly large and constantly growing number of known mycoplasma species, the battery of test sera to be used must necessarily be reduced as much as possible. The results obtained by a preliminary determination of distinguishing biochemical characters will in itself provide a guideline for a rational selection of the test sera needed. Information about the source of the isolate may further reduce the number of test sera that are required for a first attempt at identification. If, for example, the isolation is made from the genital tract of a bull, the origin of the strain and the result of the digitonin test may together imply that testing with a single antiserum will provide the diagnosis. This is because the two species *M. bovigenitalium* and *A. laidlawii* comprise a high percentage of the mycoplasma flora of the genital tract of bulls. In effect, knowing from which host the strain derives will by itself provide a useful indication as to which species should be looked for first. Although it should be remembered here that the host specificity of mycoplasmas is less rigid than hitherto believed (Freundt, 1974a). The task is more difficult when the strain to be identified is recovered from a cell culture where possible sources of contamination are numerous. The outline given below of the standard identification procedure currently adhered to by the FAO/WHO Collaborating Centre for Animal Mycoplasmas takes into consideration in the first instance the situation which exists when no information is available about the origin of the test strain. The number of tests required to achieve the final identification may then be reduced according to the amount of information about the origin or source of the strain and its biological properties.

Prior to any detailed examination, the strain is subjected to a simplified cloning procedure. An isolated colony on solid medium is picked by suction with a Pasteur pipette. The agar plug with the colony is transferred to 1·7 ml of broth and crushed against the wall of the test-tube. The culture is incubated at 37°C for four days and used as a stock culture for an

TABLE II

Species of genus *Mycoplasma*

Species	Type strain	Fermentation of glucose	Hydrolysis of arginine	Phosphatase activity	Growth medium	Reference
M. agalactiae	PG2	−	−	+	B	
M. alkalescens	D12 (PG51)	−	+	+	B	
M. alvi	Ilsley	+	+	n.d.	B	Gourlay *et al.* (1977)
M. anatis	1340	+	−	+	N	
M. arginini	G230	−	+	−	B	
M. arthritidis	PG6	−	+	+	B	
M. bovigenitalium	PG11	−	−	+/−	N	
M. bovirhinis	PG43	+	−	−	B	
M. bovis	Donetta (PG45)	−	−	+	B	Askaa and Ernø (1976)
M. bovoculi	M165/69	+	−	−	B	Langford and Leach (1973)
M. buccale	CH20247	−	+	+	B	Freundt *et al.* (1974)
M. canadense	275C	−	+	(+)	B	Langford *et al.* (1976)
M. canis	PG14	+	−	−	B	
M. capricolum	California Kid	+	+	v	B	Tully *et al.* (1974)
M. caviae	G122	+	+	+	B	Hill (1971)
M. citelli	RG-2C	−	+	−	B	Rose *et al.* (1978)
M. columbinum	MMP1	−	−	−	B	Shimizu *et al.* (1978)
M. columborale	MMP4	+	−	−	B	Shimizu *et al.* (1978)
M. conjunctivae	HRC581	+	−	−	B	Barile *et al.* (1972)
M. cynos	H831	+	−	+	B	
M. dispar	462/2	+	−	−	FF74	
M. edwardii	PG24	+	−	−	N	
M. equigenitalium	T37	+	−	+	B	Kirchhoff (1978b)
M. equirhinis	M432/72	−	+	n.d.	B	Allam and Lemcke (1975)
M. faucium	DC-333	−	+	−	BACY	Freundt *et al.* (1974)
M. feliminutum	Ben	−	−	−	B	
M. felis	CO	+	−	+/−	N	
M. fermentans	PG18	+	+	+/−	B	
M. flocculare	Ms42	−	−	−	FF74	

Species	Strain					Reference
M. gallisepticum	PG31	+	−	−	B	
M. gateae	CS	−	+	−	B	
M. homiais	PG21	−	+	+	B	
M. hyopneumoniae	J	−	−	−	FF74	Subcommittee (1975)
						Rose *et al.* (1979)
M. hyorhinis	BTS-7	+	−	+/−	B	
M. hyosynoviae	S16	−	+	−	B	
M. iners	PG30	−	+/−	−	B	
M. lipophilum	MaBy	−	+/−	n.d.	BACY	Del Giudice *et al.* (1974)
M. maculosum	PG15	−	+	+	N	
M. meleagridis	17529	−	+	+	N	
M. moatsii	MK 405	+	+	n.d.	B	Madden *et al.* (1974)
M. molare	H542	+	−	−	B	Rosendal (1974a)
M. mycoides subsp. *capri*	PG3	+	−	−	B	
M. mycoides subsp. *mycoides*	PG1	+	−	−	B	
M. neurolyticum	Type A	+	−/+	−	B	Rosendal (1975)
M. opalescens	MH5408	−	+	+	B	
M. orale	CH19299	−	+	−	B	† and Freundt *et al.* (1974)
M. ocipneumoniae	Y-98	+	−	−	B	
M. pneumoniae	FH	+	−	+	B	
M. primatum	HRC292	−	+	−	B	
M. pulmonis	Ash (PG34)	+	−	+	B	Tully *et al.* (1974)
M. putrefaciens	KS-1	−	+	+	B	
M. salivarium	PG20	−	−	−	B	
M. spumans	PG13	−	+	+	N	
M. sualvi	Mayfield	+	+	n.d.	B	Gourlay *et al.* (1978)
M. subdolum	TB	−	+	v	B	Lemcke and Kirchhoff (1979)
M. synoviae	WVU 1853	+	n.d.	n.d.	F	
M. verecundum	Strain 107	−	−	+	N	Gourlay *et al.* (1974)

+, positive; (+), weakly positive; +/−, variable, but most strains positive; −, negative; −/+, variable, but most strains negative; v, variable reaction with individual strain; n.d., not determined.
† "Bergey's Manual of Determinative Bacteriology", 8th edn (Freundt, 1974c).
Symbols for growth media: see Appendix A1–A6.

TABLE III
Species of genus *Acholeplasma*

Species	Type strain	Hydrolysis of aesculin	Hydrolysis of arbutin	Phosphatase activity	Growth medium	Reference
A. axanthum	S743	+	+	−	B	†
A. equifetale	C112	n.d.	n.d.	−	B	Kirchhoff (1978a)
A. granularum	BTS-39	−	−	−	B	†
A. hippicon	C1	n.d.	n.d.	−	B	Kirchhoff (1978a)
A. laidlawii	PG8	+/−	−	−/+	B	†
A. modicum	Squire (PG49)	−	−	−	B	†
A. oculi	19-L	+	−	−	B	†

+, positive; −, negative; +/−, variable, but most strains positive; −/+, variable, but most strains negative; n.d., not determined.

† "Bergey's Manual of Determinative Bacteriology", 8th edn (Freundt, 1974c).
Symbols for growth media: see Appendix A1–A6.

examination which proceeds through a series of biochemical tests and ends with serological testing. As a prelude to the more detailed description of these tests in Sections IV and V a brief outline of the general principles is presented here.

A. Biochemical screening

As a first step, the sensitivity to digitonin (1·5%) is determined (Ernø and Stipkovits, 1973). If the strain is resistant or only slightly sensitive, i.e. if the zone of inhibition as determined by a disc method on solid medium is less than 2 mm, the strain can with some assurance be classified as belonging to the genus *Acholeplasma*. In cases of doubt where slight sensitivity to digitonin is observed, a similar test for susceptibility to 5% sodium-polyanethol-sulphonate (Freundt *et al.*, 1973a) may be useful in confirming the genus classification, since all acholeplasmas and also some mycoplasmas are resistant to this compound. The genera *Mycoplasma*, *Ureaplasma* and *Spiroplasma* are all sensitive to digitonin. Members of the genus *Ureaplasma* are distinguished by their urease activity, whereas classification of an organism as belonging to the genus *Spiroplasma* depends primarily on morphological criteria.

Species differentiation within the genus *Mycoplasma* is initiated by four biochemical tests: tests for catabolism of glucose and arginine, phosphatase activity, and digestion of coagulated horse serum. The antisera to be used for the subsequent serological testing are usually selected according to the results of the first two tests which provide a reliable basis for a subdivision of the *Mycoplasma* species into two major biochemically defined groups: (1) the glucose-positive/arginine-negative, and (2) the arginine-positive/ glucose-negative species. So far, the two other possible combinations: (3) glucose-positive/arginine-positive, and (4) glucose-negative/arginine-negative organisms are each represented by only a few species. A further subdivision of these four groups may be attempted on the basis of phosphatase activity but the results should be interpreted with some caution because of the intra-species variability sometimes experienced with respect to this property. The serum digestion test is included as a means of distinguishing strains of *M. mycoides* and *M. capricolum*, the only two species so far known to be capable of liquefying coagulated horse serum. Here again, some strain variability is known to occur within *M. mycoides*.

Biochemical characterisation of members of the genus *Acholeplasma* may be carried out by testing for phosphatase and for the catabolism of aesculin and arbutin. It must be emphasised, however, that our present knowledge about the intraspecific variability of these properties is restricted.

In consequence, it is recommended that only limited importance should be attached to the results of the biochemical tests and that simultaneously strains should be tested serologically with antisera against all of the seven species currently recognised.

There is no basis for a subdivision of the genus *Ureaplasma* on biochemical activities and serological identification essentially employs the same methods that are used for *Mycoplasma* and *Acholeplasma* (Black, 1970, 1973b; Black and Krogsgaard-Jensen, 1974). Until now, only one species has been recognised within the genus *Ureaplasma*, *U. urealyticum*, comprising eight serotypes (Shepard *et al.*, 1974). It should be pointed out here that what are referred to as different "serotypes" of one species of *Ureaplasma* would in fact have been designated as distinct species in the genera *Mycoplasma* and *Acholeplasma* (cf. Subcommittee on the Taxonomy of *Mycoplasmatales*, 1974). The taxonomic status of the ureaplasmas isolated from a variety of different animals, and their relationship to *U. urealyticum*, remains undefined.

B. Serological testing

This is carried out following selection of the antisera which have been suggested as appropriate by the preliminary biochemical screening. As will have appeared from the classification section, there are three methods that should primarily be considered for serological identification, the growth inhibition, metabolism inhibition, and immunofluorescence tests. Up to the present most identifications have probably been performed by means of the growth inhibition and metabolism inhibition tests although unfortunately the former is not sufficiently sensitive and the latter is time-consuming and relatively difficult to perform. The indirect immunofluorescence test approaches the ideal test method, since both specificity and sensitivity are at an adequate level, and it is technically simple when the necessary equipment is at hand.

To summarise, the identification procedure is initiated with the digitonin test, followed by the urease test if the organism proves digitonin-sensitive. Under special circumstances, when the occurrence of a spiroplasma is suspected, a morphological examination should also be carried out. Strains shown to belong to the genus *Mycoplasma* are then tested for glucose fermentation, arginine hydrolysis, phosphatase activity, and possibly serum digestion as a preliminary to the final serological identification. This is most easily performed by the indirect immunofluorescence test, but may be carried out by means of growth inhibition, metabolism inhibition and, in special cases, complement fixation, double immunodiffusion or possibly direct agglutination tests.

IV. BIOCHEMICAL METHODS IN IDENTIFICATION

A. Sensitivity to digitonin

1. Test medium

This is a conventional agar medium containing 10–20% serum and supporting good growth of the test strain. It has been shown that all *Acholeplasma* species are inhibited by 1·5% digitonin when grown on either serum-free medium or on medium containing bovine serum fraction instead of whole serum. For the distinction between *Mycoplasma* and *Acholeplasma* species it is essential, therefore, that the digitonin test is consistently carried out on medium enriched with serum (Freundt *et al.*, 1973a).

2. Reagent

Sterile filter paper discs, about 6·0 mm in diameter, of the type used in antibiotic sensitivity tests, are impregnated with 0·02 ml of a 1·5% (w/v) ethanolic solution of digitonin (E. Merck A.G., Darmstadt, West Germany) dried overnight at 37°C and stored at 4°C until use. To dissolve digitonin in ethanol, heating for 20–30 min in a 50–56°C water bath is required.

3. Test procedure

The test is performed on agar medium which should be free from surface moisture before inoculation. By means of a calibrated platinum loop, a 0·01 ml drop from a liquid stock culture (Section III) containing approximately 10^5 CFU/ml is allowed to run down the surface of a slightly tilted plate (running drop technique). When the fluid of the inoculum has been absorbed, a digitonin disc is placed in the centre of the inoculated area. The agar plates are incubated under optimal growth conditions for the test and a reading is made when growth is visible under a low-power microscope. The width of the zone of inhibition is measured in mm (Ernø and Stipkovits, 1973; Freundt *et al.*, 1973a).

Each batch of digitonin discs should be tested against known positive and negative control strains.

B. Urease activity

1. Reagent

A solution containing 1% urea and 0·8% manganous chloride ($MnCl_2 \cdot 4H_2O$) is the test reagent.

2. Test procedure

An agar block, about 1 cm square, is cut from a plate culture and placed with the colony side upward on a slide. One or two drops of the urea-manganous chloride solution are added to the surface of the agar and the reaction is read immediately under the microscope. *Ureaplasma* colonies turn instantly brownish due to the liberation of ammonia from urea and the subsequent deposition of manganese dioxide on the surface of the colonies according to the reaction:

$$2NH_3 + 2\ H_2O \rightleftarrows 2\ NH_4OH \rightarrow 2\ NH_4^+ + 2OH^-$$

$$MnCl_2 + 2OH^- \rightarrow MnO_2 + 2\ HCl$$

This method, which is the most simple test for demonstrating urease activity is a modification by Shepard (pers. comm.) of the technique originally devised by Shepard and Howard (1970).

It is advisable to include known positive and negative control organisms here and elsewhere in the biochemical test systems.

C. Glucose breakdown

1. *Standard method*

(a) *Test medium.* The test basal medium, b_1, is a modification of the conventional growth medium (B) without yeast extract and with 1% (w/v) PPLO Serum Fraction (Difco) replacing horse serum. It also contains 0·002% (w/v) Phenol red (Appendix, Section B, 1 (a)).

The test medium, bg, is prepared by adding glucose to b_1 to a concentration of 0·6% (w/v) (Appendix, Section B, 1 (b)).

The medium base for b_1 and bg is heart infusion broth (Difco), treated with glucose oxidase, peroxidase and arginine decarboxylase to remove any traces of glucose and arginine. The method used for this, as devised by T. Sander (unpublished), is as follows:

To 1000 ml of heart infusion broth is added 25 mg of glucose oxidase (Type II, Sigma) which also contains traces of peroxidase. The pH is adjusted to 5·4–5·6 with concentrated hydrochloric acid and the enzymatic process is allowed to proceed for 1 h at 35–38°C under constant aeration with compressed air or oxygen. After the addition of 25 mg of arginine decarboxylase (Sigma), the process is continued for another hour. The broth is then cooled to room temperature, and, after adjusting the pH to 7·8, tapped into screw-capped bottles and autoclaved at 121°C for 20 min to destroy the activity of the added enzymes.

Following addition of the other constituents to the medium base and final sterilisation by filtration (Gelman membrane filter, 200 nm pore size), the media are dispensed in 1·7 ml amounts in small test tubes.

(b) *Test procedure.* Media bg (test medium) and b_1 (control without glucose) are inoculated either with single colonies picked from an agar culture or with a drop of liquid culture. The inoculated tubes are incubated aerobically together with controls consisting of uninoculated bg and b_1 tubes. A positive reaction is indicated by a colour change from red to orange or yellow in the inoculated test medium, but not in any of the controls. If the same colour change is observed in either the b_1 medium or in the uninoculated bg medium the result is recorded as equivocal or negative.

(c) *Comments.* The use of media less complex than those normally used for growth is intended to prevent, as far as possible, any decrease of pH resulting from metabolic activities other than catabolism of glucose. For example, the fall in pH that occurs during incubation of M. *bovigenitalium*, which is known to be a non-fermenter of glucose, has been shown to be greater in the presence of yeast extract (Edward and Razin, 1974). The enzymatic removal of pre-existing amounts of glucose and arginine assists further the standardisation of the test, thereby enabling more clear-cut and unequivocal results to be obtained. Since, unfortunately, not all strains will grow in the standard test medium, appropriate enrichments may be required. Whatever medium is used, the importance of including in the test system the two controls, (i) uninoculated medium with glucose, and (ii) inoculated medium without glucose, should be strongly emphasised. Determination of the number of colony forming units occurring during growth may also be required to ensure that there are sufficient organisms to produce a definite change of colour. In our experience about 10^7 CFU/ml are needed to produce such a change, which means that a negative result associated with a peak number of 10^6 CFU/ml or less is unreliable.

2. *Alternative methods*

The relatively simple methods described above for determination of catabolism of glucose may, as indicated already, be unsuitable or give ambiguous results with some strains. More sensitive and specific tests are available, based on (i) determination of glucose disappearance by the glucose oxidase reaction, (ii) determination of acid fermentation products from radioactive glucose, or (iii) determination of hexokinase activity as an indicator of the presence of the glycolytic pathway. Although these methods are probably too laborious as part of the routine identification procedure, the interested reader is referred to a WHO Working Document prepared by Edward and Razin (1974) describing these techniques.

D. Arginine hydrolysis

1. *Test medium*

The basal medium, b_2, is identical with b_1 except that the pH is adjusted to 7·3. The test medium is prepared by supplementing the basal medium with 1% L-arginine (Appendix, Sections B, 2 (a) and (b)).

2. *Test procedure*

This is the same as that described for the glucose test, including similar controls. A distinct alkaline shift in the pH of the test medium indicates a positive reaction.

3. *Comments*

The technical difficulties connected with the glucose test may also be encountered with this test. Enzymatic pretreatment of the heart infusion base is particularly important when testing strains that are both glucose- and arginine-positive, because the acid produced in conventional media tends to mask the alkalisation caused by the hydrolysis of arginine.

The arginine test is based on the demonstration of the presence in some mycoplasma species of the arginine dihydrolase pathway, which involves arginine deiminase, ornithine transcarbamylase, and carbamyl phosphokinase, and results in the release of high-energy phosphate, CO_2 and NH_3 as endproducts (Schimke and Barile, 1963; Barile *et al.*, 1966).

E. Phosphatase activity

1. *Test on solid medium*

(a) *Test medium.* An agar medium, Bph, the specific substrate of which is the sodium salt of phenolphthalein diphosphate, is used. It is important to heat the serum and yeast extract incorporated in the medium at 60°C for 1 h in order to inactivate the enzymes in these components (Appendix, Section B4).

(b) *Test procedure.* A plate is inoculated from the fluid stock culture by the running drop technique. After incubating the inoculated agar plate together with an uninoculated control for 7 days, the plates are tested by flooding the surface with 5 N NaOH. The appearance within half a minute of a red colour in and around the inoculated area indicates a positive reaction.

2. *Test in liquid medium*

(a) *Reagent.* A 0·01% solution of the sodium salt of phenolphthalein diphosphate is used.

(b) *Test procedure.* A log phase culture of the test strain grown in 10–100 ml of appropriate growth medium is centrifuged at about $20\,000 \times g$ for 1 h. The pellet is washed in phosphate-buffered saline and resuspended in 1 ml of the phenolphthalein diphosphate reagent. The pellet from 100 ml of uninoculated growth medium treated in the same way serves as a negative control. After incubation at 37°C for 4 h, two drops of 5 N NaOH are added. A positive reaction is indicated by the appearance of a red colour in the test mixture.

3. Comments

The phosphatase test is based on the liberation of phenolphthalein and its reaction with NaOH. The test on solid medium as recommended by Aluotto *et al.* (1970) is used as the standard method at the FAO/WHO Collaborating Centre for Animal Mycoplasmas. Whereas Aluotto *et al.* perform the test after 3, 7 and 14 days' incubation, using triplicate cultures, one reading made after 7 days' incubation is in our experience sufficient for practical purposes. There is some indication that the plate test is working less satisfactorily in some laboratories, false positive reactions frequently being obtained. This may be explained by the presence in the test medium of uninactivated phosphatase or perhaps by a spontaneous degradation of sodium phenolphthalein diphosphate. Such difficulties may be overcome by using the liquid medium test (2 above) which is slightly modified from that of Bürger *et al.* (1967) and Black (1973a). This method may have the additional advantage of being more sensitive than the standard (plate) method. A fluorogenic test allowing the detection of phosphatase in mycoplasma cultures in 1 h has been described by Bradbury (1977).

F. Liquefaction of coagulated serum

1. Test medium

The substrate is horse serum coagulated as slopes in test tubes (S_d medium, see Appendix, Section B, 5).

2. Test procedure

Two slopes are inoculated with a platinum loop, one from an undiluted stock culture, the other from a dilution of 10^{-3}. Readings are made at frequent intervals during 14 days of incubation at 37°C. Liquefaction becomes visible by the development of a shallow depressed area with a moist base when growth is confluent, and of small pits when growth is more scattered. In the case of heavy liquefaction, fluid accumulates in the angle between the base of the slope and the wall of the tube.

3. Comments

Other methods of demonstrating proteolytic activity, such as hydrolysis of gelatin and digestion of casein have been used to characterise individual strains or species (Freundt, 1958; Aluotto *et al.*, 1970; Czekalowski *et al.*,

14

1973; Woolcock *et al.*, 1973; Watanabe, 1975). The potential value of these methods in the routine identification of mycoplasmas cannot be fully assessed on the basis of currently available data.

G. Hydrolysis of aesculin and arbutin

1. *Media and reagents*

Cultivation is on conventional solid media. Filter paper discs, about 6 mm in diameter, are impregnated first with $0 \cdot 02$ ml of 5% (w/v) ferric citrate; after 24 h $0 \cdot 02$ ml of a 10% (w/v) solution of either aesculin or arbutin is added to each disc.

2. *Test procedure*

The agar plate is inoculated with $0 \cdot 01$ ml of the stock culture by the running-drop technique. Following absorption of the inoculum, a disc containing ferric citrate plus either aesculin or arbutin is placed in the centre of the inoculated area. Incubation is at 37°C, readings being performed daily for up to seven days. The appearance of a brownish colouration of the culture streak, starting at the edge of the disc, indicates a positive reaction.

3. *Comments*

The disc technique developed by G. Askaa (unpublished) for demonstrating hydrolysis of aesculin and arbutin has the practical advantage that the test can be performed on a conventional growth medium without incorporating test substrate and reagent (Williams and Wittler, 1971). The fluoregenic test referred to in Section E, 3 has also been utilised for demonstration of aesculin hydrolysis (Bradbury, 1977).

V. SEROLOGICAL METHODS IN IDENTIFICATION

A. Comments on antisera

No attempt will be made to discuss the methods that are used to prepare reference antisera for classification or identification of mycoplasmas, since these procedures do not differ essentially from those used for other microorganisms. The methods generally used in our respective laboratories for production of immunising antigen and schemes of immunisation are described by Hollingdale and Lemcke (1969) and Ernø *et al.* (1973). However, a problem pertinent to mycoplasmas must be mentioned: the effect of serum proteins adsorbed by mycoplasma antigens from the serum-supplemented media in which they are grown. If the immunising antigen is grown in medium containing serum foreign to the animal which is to be immunised, the resulting antiserum will contain antibodies to the foreign serum proteins. These in turn will react with serum proteins adsorbed to the test antigen, if the latter has been grown in the presence of the same serum. For growth inhibition and metabolism inhibition tests this is unimportant, since antibodies to medium components in the immunising antigen do not affect the growth of the test organisms. For immunofluorescence, complement fixation, double immunodiffusion or direct agglutination, it will result in spurious cross-reactions between species.

The most effective method of preventing such reactions is to grow the immunising antigen in medium containing serum from the animal species to be immunised. Some workers recommend that this should apply to all the medium constituents and that, for example, rabbits should be immunised with organisms grown in a rabbit meat infusion broth supplemented with rabbit serum. However, it is only components of serum which have been shown unequivocally to adsorb to mycoplasmas during growth (Bradbury and Jordan, 1972), and many species grow very poorly in rabbit infusion medium. An alternative method suggested by Kenny (1969) is to grow the immunising antigen in medium supplemented with agamma calf serum and the test organism in medium containing agamma horse serum.

Reference antisera for a total of 29 *Mycoplasma* or *Acholeplasma* species or subspecies, as well as seed reference reagents of the corresponding type strains, are available from the Research Resources Branch, Collaborative Research, National Institute of Allergy and Infectious Diseases, National Institutes of Health, Bethesda, Maryland 20014, U.S.A., and from the FAO/WHO Collaborating Centre for Animal Mycoplasmas, Institute of Medical Microbiology, Bartholin Building, University of Aarhus, DK-8000 Aarhus C, Denmark. These antisera were produced by immunisation of horses or mules with antigens prepared from organisms normally grown in PPLO Broth (Difco) with 10% added autologous preimmune horse or mule serum and 2% yeast extract dialysate (Freundt *et al.*, 1973b; Cunningham, 1978–1980). Antiserum and seed reference reagents for the vast majority of the remaining currently recognised *Mycoplasma* and *Acholeplasma* series are available, in limited quantities, from the FAO/WHO Collaborating Centre for Animal Mycoplasmas. These last-mentioned antisera are produced in rabbits using antigens prepared from organisms grown, whenever possible, in rabbit meat infusion broth supplemented with rabbit serum (or PPLO Serum Fraction, Difco) and yeast extract.

B. Growth inhibition

1. *Disc method*

(a) *Equipment and reagents.* Sterile 6 mm filter paper discs are impregnated with 0·02 ml of antiserum using either a calibrated loop or a micropipette, air-dried at room temperature and stored at 4°C. Discs can be stored at this temperature for several months and possibly for several years. The test is performed on a solid agar medium, approximately 4 mm deep and usually of the composition used for isolating the test organism. The plates should be free from surface moisture before use, for example, by leaving them in the incubator at 37°C for 30 min. Mycoplasma cultures in broth medium or suspensions of organisms containing about 10^5 colony forming units (CFU)/ml are used as antigens. The importance of avoiding too large an inoculum cannot be overemphasised. In order to ensure the optimal density of colonies (well separated, though not too scattered) it is advisable to use, in each test, three culture dilutions containing approximately 10^4, 10^5 and 10^6 CFU/ml, as determined by preliminary titrations.

(b) *Test procedure.* The plates are inoculated by the running-drop method. One drop (0·01 ml) of the antigen is allowed to run down the surface of the agar. When the fluid of the inoculum has been absorbed by the growth medium, the disc is placed in the centre of the inoculated area. Incubation is performed in a moist atmosphere at 37°C until the appearance of microscopically visible colonies. With some organisms this may be achieved in about 48 h. If no growth has appeared by that time, incubation is continued for a further 24–48 h. Readings are made under a conventional low-power microscope or a dissecting microscope. The zone of inhibition is measured from the edge of the disc to the edge of the area of colonies.

Since the borderline between growth and absence of growth is unfortunately not always clear-cut, the extent of inhibition may be recorded in terms of total, nearly total, or relative inhibition. A nearly total inhibition is defined as a clear inhibition with less than ten "break through" colonies appearing within the inhibition zone, and relative inhibition as a less pronounced, although significant reduction in the size and/or number of colonies round the disc. If identification is based solely upon the growth inhibition test, a total or nearly total inhibition of at least 2 mm is required.

2. *Agar well method*

The test is performed as described above, except that wells filled with antiserum are used instead of serum-impregnated discs (Black, 1973b). Following inoculation of the plates by the running-drop method, a well with a diameter of 4 mm is punched out in the centre of the inoculated area, and filled with about 0·05 ml of antiserum. The well is produced by means of a stainless steel cylinder, the agar block being removed by gentle suction.

3. Comments and suggestions for modifications

As mentioned already, the growth inhibition test is used very extensively for identification of mycoplasmas. The test has the advantage of being easy to perform and of exhibiting a high degree of specificity. As it has the disadvantage, on the other hand, of being rather insensitive a variety of modifications have been developed in attempts to increase the sensitivity. In this respect the agar well method is definitely superior to the disc technique. Inhibition zones may be increased still further by refilling the wells at appropriate intervals during incubation (Black, 1973b). A further improvement in sensitivity may be obtained by using suboptimal growth conditions for the test organism, i.e. by lowering the incubation temperature from 37°C to about 27–30°C and/or by reducing the concentration of horse serum in the growth medium from 15–20% to 2% and omitting the yeast extract (Freundt et al., 1973b). All these manoeuvres have the effect of altering the ratio of the reactants, so that, within a given period of time, increased amounts of antiserum react against a smaller number of mycoplasma cells proliferating at a reduced rate. The sensitivity of the test may also be increased by the use of discs impregnated with antisera concentrated with a dry polyacrylamide gel (Windsor and Trigwell, 1976).

It may also be mentioned at this point that the difficulties encountered with the growth inhibition test because of its low sensitivity are more apt to occur with some species than with others, and also that strains within one species may vary. Such variations are not easily understood, but they may be attributed to antigenic differences. This appears to be the case with, for example, M. hominis. The rather frequent failure, experienced by several workers, of antisera produced against the type strain (PG21) of this species to inhibit the growth of heterotypic strains may be compensated for by the use of combined type-specific antisera (Lin et al., 1975). Antisera against three strains of M. hominis that differed serologically by agglutination and a complement-dependent mycoplasmacidal test, although cross-reacting by growth inhibition, proved in the study of Lin and collaborators to inhibit a much wider spectrum of field strains than did each of the individual sera. Apart from antigenic properties, the result of the GI test may also depend on other factors such as membrane stability and the possible existence of surface structures external to the cell membrane.

The choice of the GI procedure to be used for routine identification depends on the given circumstances, such as the species involved and the potency of the antisera available. In general the best results are obtained with the agar well method and the use of suboptimal growth conditions. The agar well method is somewhat more laborious than the disc test, which has the additional advantage of allowing long-term storage of the serum-impregnated discs at 4°C.

4. Alternative methods

The original version of the GI test with antiserum incorporated in the growth medium never obtained any wide use because of the relatively large amounts of antiserum required. A serum-drop method was first described by Herderscheê (1963). The principle of this technique, that may deserve more attention than has hitherto been paid to it, is to allow a drop of antiserum to penetrate into a dried agar medium following inoculation of part of the surface.† Attention should also be drawn to the development of micromethods which utilise the cups of microtitre plastic plates as reservoirs for the growth medium (Jensen, 1963; Campello et al., 1972).

C. Metabolism inhibition

1. Equipment and cleaning procedures

The metabolism inhibition (MI) test is performed in a 0·025 ml microtitre system manufactured by the Cooke Engineering Co., Alexandria, Va., U.S.A., using Disposo Trays (IS-MRC-96) from Linbro Chemical Co., Inc., New Haven, Connecticut. Equipment required includes micropipettes, microdiluters, absorbent "GO NO-GO" test pads, sealing tape and a test reading mirror. Only trays with U-shaped wells, arranged in eight horizontal rows, each with 12 wells, should be used.

The titrations should of course be performed in such a manner that bacterial and fungal contamination is avoided. To that end, the trays are rinsed immediately before use in 70% ethanol for 5 min, and in deionised water for 30 min, and air-dried. Reusable micropipettes are rinsed in saline and distilled water, boiled for 5–10 min in distilled water and air-dried.

Stainless steel microdiluters are rinsed in saline, dried, rinsed again in distilled water, dried and finally lightly flamed. Before reuse, microdiluters should be checked for accuracy of delivery by means of the "GO NO-GO" test pads. Diluters that deliver exactly 0·025 ml, i.e. whose liquid content just covers the circle on the test pad, are ready for use after being gently flamed again. Those that do not deliver accurately should be discarded unless thorough flaming, followed by immediate quenching, results in readjustment.

2. Reagents

These include antigens, i.e. mycoplasma cultures, antisera of reasonably high titre (at least 64) and test media.

---

† A modification of the serum-drop method that appears to be more sensitive than other GI tests used today, and equally specific, was introduced very recently (Møller, B. R. J. appl. Microbiol. (1979). In press).

(a) *Antigens.* As mentioned above, the antigen is a live mycoplasma culture in liquid medium. For production of the antigen, the organism is preferably grown in the test medium. Adaptation to this medium may be achieved by subculturing two or three times. If the strain to be tested is not growing well in medium B or SU (Appendix, Sections A, 1 and 3), another medium must be selected on the basis of growth studies. For most strains, media can be selected that will provide suitable conditions for performance of the test. Difficulties are more likely to occur with glucose- and arginine-negative strains, as the reduction of tetrazolium may be weak and inconsistent with some strains of species that are generally described as being tetrazolium positive. In such cases the problem may be overcome by supplementing the test medium with sodium thioglycollate (0·1%) which reduces the redox potential and thereby potentiates the reduction caused by the mycoplasmas.

A batch of the appropriate medium is inoculated with mycoplasmas, incubated to the end of the log phase of growth and frozen as 1 ml aliquots at $-70°C$.

(b) *Titration of antigen.* Titration of the antigen is carried out to determine the number of colour changing units (CCU) as defined below.

(1) 1 ml volumes of ten-fold serial dilutions (10^{-1}–10^{-8}) are prepared in tubes using test medium as diluent.

(2) Two drops (0·05 ml) of each antigen dilution, and six drops (0·15 ml) of test medium are mixed in the wells of a microtitre tray. Eight drops (0·2 ml) of test medium are placed in each of four wells and serve as medium controls. Endpoint controls consist of four wells each containing eight drops of test medium with the pH adjusted to the endpoint required in the test. This control is not included in tetrazolium reduction tests.

(3) The plate is sealed with tape to avoid evaporation and incubated at 37°C or as otherwise required.

(4) The test is read daily over the reading mirror, and colour changes noted. When the number of wells showing a change is stable, the endpoint is determined. The highest dilution of antigen that causes a colour change is regarded as containing 1 CCU/0·05 ml. From this is calculated the dilution required to make an antigen preparation containing 100–1000 CCU/0·05 ml, the dose to be used in the test proper. At the same time one obtains an idea of the length of the incubation period required for the test proper. The stored antigen will maintain its titre for a considerable period of time, and hence it is usually unnecessary to repeat the antigen titration when performing the test proper. The frozen antigen should not, however, be thawed more than once.

(c) *Test medium.* This consists of a suitable growth medium supplemented with either glucose (1%), arginine (1%), urea (0·45%), or 2,3,5-triphenyl tetrazolium chloride (0·04%). In the first three media Phenol red (0·002%) is used as an indicator of the pH change resulting from growth and metabolism, but in the last, reduction of tetrazolium itself produces a colour change. Since there is no single medium that will support the growth of all mycoplasmas, a formula for a standard growth medium cannot be given. It should be remembered, moreover, that the requirements for a medium to be used for the MI test may be even more exacting than for a mere growth medium. The outcome of the MI test is not simply a matter of growth; the enzymes required for the specific metabolic activity on which the test is based also need to be produced in sufficient quantity to allow a clearly visible colour change to take place. In the Appendix is given the composition of four media referred to as BG, BA, BT and SU. The base of the first three is a modified Hayflick medium (B) that is suitable for the growth of a great many *Mycoplasma* or *Acholeplasma* species. Medium SU is used for urea-plasmas. The initial pH of BG and BT is 7·8, of BA and SU 7·3 and 6·8, respectively.

In order to provide a convenient method of judging the pH changes that occur during incubation of the test system, controls are included which consist of the test medium, BG, BA or SU, the pH of which is adjusted to the desired endpoint, viz. 7·3, 7·8 and 7·3, respectively.

3. *Test proper*

Serum is inactivated (30 min at 56°C) and a primary dilution of 1:10 is made.

(1) Add 0·025 ml (one drop) of test medium to 12 wells, using one horizontal row for the titration of each serum.

(2) Add 0·025 ml of the primary dilution of serum (1:10) by microdiluter, to the first well.

(3) With the microdiluter in the first well prepare two-fold serial dilutions of serum, rotating the diluter about 25 times in each well and transferring 0·025 ml to the next well. From the last well the microdiluter is discharged on to a test pad. This also serves as a check on the delivery volume.

(4) Add 0·05 ml (two drops) of the appropriate dilution of antigen (100–1000 CCU) to each well.

(5) Add 0·125 ml (five drops) of test medium to each well.

(6) One horizontal row is reserved for controls:

(i) Medium control: add 0·2 ml (eight drops) of test medium to four wells.

(ii) Antigen control: add 0·15 ml (six drops) of test medium + 0·05 ml (two drops) of diluted antigen to four wells.

(iii) Endpoint control: add 0·2 ml (eight drops) of test medium with the pH of the desired endpoint to four wells. (Not included for tetrazolium reduction tests.)

(7) Seal the plate with tape and incubate at an appropriate temperature.

(8) Read the test when the antigen control has changed approximately 0·5 pH units, i.e. when it shows the same colour as the "endpoint controls". Tetrazolium reduction readings are made when the antigen control wells show a red precipitate. The endpoint is recorded as the highest serum dilution that prevents a colour change from occurring.

D. Immunofluorescence

1. Equipment

At the FAO/WHO Collaborating Centre for Animal Mycoplasmas, immunofluorescence tests are carried out by the plate epi-immuno-fluorescence (epi-IMF) procedure of Del Giudice et al. (1967). Microscopy is performed with a Zeiss Standard (RA) microscope equipped with an Osram HBO mercury lamp and an incident illumination attachment, allowing illumination of the specimen from above and thereby avoiding the transmission of exciting ultraviolet light through the agar and the plastic plate or glass slide. Two exciting filters, BG 12/4 and BG 3/4 (Zeiss), and a barrier filter No. 47 (Zeiss) are employed. An aperture diaphragm is inserted between the source of light and the exciting filters.

2. Reagents

The test is performed with either unfixed mycoplasma colonies or hot water-fixed impressions of colonies as antigen. Following inoculation of an appropriate agar medium with approximately 10^5 CFU/ml the plates are incubated until the colonies are clearly visible under the microscope but preferably before they are fully grown.

When using unfixed agar colonies, agar blocks about 0·25 cm² in size are cut in different geometric patterns to facilitate their identification and placed on slides with the colonies uppermost. For the hot water-fixation technique, agar blocks are transferred to slides with the colony side downward. The slides are placed at an angle of 45° in a bath with distilled water at 80°C. The temperature of the water is maintained at 80°C until the blocks drop off, which usually occurs after 30 to 60 s. The slides are then immediately washed in distilled water and air-dried.

For the direct immunofluorescence technique, the antisera are conjugated according to Nairn (1968). The globulin is precipitated with 40% saturated ammonium sulphate. Conjugation with fluorescein isothiocyanate (FITC) I (Sigma) is performed at 25°C and pH 9·5 in phosphate buffer, adding one part of FITC to 80 parts of protein. Non-conjugated fluorescein is removed by filtration through a Sephadex G-25 column equilibrated with 0·01 M phosphate buffered saline at pH 7·2. The molecular relationship between FITC and protein (F/P ratio) is calculated by estimating the protein concentration (C_p) of the conjugate according to the method of Lowry *et al.* (1951), and measuring the optical density of the conjugate, diluted to contain 0·5 mg protein per ml ($O.D._{0·5}$), in a Zeiss photometer at a wavelength of 495 nm. The molecular relationship is calculated according to the following formula (Nairn, 1968):

$$F/P = 2·8 \frac{O.D._{0·5}}{C_p}$$

For the indirect immunofluorescence test, two serum reagents are required: an antimycoplasma immune serum produced in rabbits,† and a commercial fluorescein-conjugated antirabbit immunoglobulin (HAR-C, Centraal Laboratorium van de Bloedtransfusiedienst van het Nederlandsche Roode Kruis).

For further details about the use of fluorescent-labelled antibody techniques the reader is referred to Volume 5A of this series (Walker *et al.*, 1971).

3. *Test procedures*

Testing by immunofluorescence may be carried out either by the direct or the indirect method, and each of these techniques may be applied both to unfixed agar colonies and to hot water-fixed colonies.

For the examination of unfixed agar colonies by the direct method, a drop of suitably diluted fluorescein-conjugated antiserum is placed on each agar block with the colonies *in situ*. The blocks are then incubated for 30 min in a moist chamber at room temperature. Unbound conjugate is removed by washing the blocks twice with phosphate buffered saline (PBS), pH 7·2, for 10 min in test tubes placed in an automatic test tube rotator (HETO, Birkerød, Denmark). The blocks are then placed on slides again and are now ready for microscopy by the epi-immunofluorescence technique.

Application of the direct immunofluorescence method to hot water-fixed colonies involves essentially the same procedure as that used for unfixed colonies except that slides carrying the hot water-fixed colony impressions are washed in slide carriers. After the second wash with PBS the slides are rinsed for half a minute in distilled water. The hot water-fixed specimens

---

† Polyvalent antisera may also be used. They have the advantage of permitting final identification by means of a limited number of monospecific antisera and without previous biochemical testing (Ernø, 1977).

may be read by means of an ordinary fluorescence microscope, using transmitted ultraviolet light, as well as by the epi-immunofluorescence method.

When applying the indirect immunofluorescence method to unfixed agar colonies, the blocks with colonies are incubated with a drop of appropriately diluted antiserum for 30 min at 20°C in a moist chamber. The blocks are washed twice for 10 min with PBS, pH 7·2, in test-tubes on the test-tube rotator. A drop of appropriately diluted fluorescein-conjugated antirabbit globulin is placed on each block, and the blocks are incubated for another 30 min at 20°C. After washing twice in PBS the blocks are mounted on slides and examined by the epi-immunofluorescence method. A technical modification of the epi-IMF test was recently described by Møller (1979). It has the advantage, *inter alia*, that the agar blocks are held in the same position during the whole procedure and do not have to be moved from slides to test tubes and back again.

Indirect immunofluorescence of hot water-fixed colonies is carried out in essentially the same way, but here again washing is performed in a slide carrier, the final wash being in distilled water.

(a) *Working dilution of conjugates and antisera.* The determination of appropriate working dilutions for the conjugates and antisera to be used in the test is a prerequisite for the correct procedure.

For the direct immunofluorescence test, fluorescein-conjugated antiserum is tested as described above in two-fold serial dilutions between, for example, 1:10 and 1:320 against the homologous strain. Thereby the optimal dilution, as defined by the highest dilution giving maximal fluorescence, is determined. It is advisable, however, to choose a somewhat lower working dilution, since, to obtain a positive reaction, freshly isolated strains of a species quite often require a higher concentration of antibody than that needed for the type strain used in the production of the antiserum.

For the indirect immunofluorescence test, the working dilution of the antirabbit globulin is determined on the basis of a chessboard titration against a potent rabbit antiserum. Thereafter, the titration of other rabbit antisera may be carried out solely against this dilution of conjugate to determine the optimal dilution of each serum. Here again, in order to avoid false negative results, it is advisable to select a working dilution lower than the optimal for the identification of unknown strains.

The diluted conjugated antirabbit globulin should not be stored for more than 1 week before use, whereas the diluted rabbit antisera are usable for at least 1 month, even if subjected to repeated thawing and freezing.

(b) *Controls.* For each strain being tested, it is necessary to compare the test slide with a negative control in order to detect non-specific binding.

Conjugated normal rabbit serum is used in the direct, and non-conjugated normal rabbit serum in the indirect method. The less experienced investigator is also advised to include positive controls in each test, i.e. to test each antiserum against a homologous reference strain.

4. *Comments*

The epi-IMF test, especially when using unfixed agar-grown colonies as antigen, has several obvious advantages and has in fact contributed very significantly to making IMF a practical and easy tool for rapid identification of mycoplasmas. In the hands of an experienced investigator the epi-IMF test is as easy to perform as the GI test, highly reliable, and may lead to a specific diagnosis within a few hours. In contrast to the GI test the reading is not affected by the inoculum size, as long as the colonies are small and well separated. The use of colonies grown on less complex media (serum-free or reduced concentration of serum) is advantageous. Moreover, the test can be carried out on a mixed culture and may even help to detect the occurrence of a mixture of organisms in a culture or a single colony (Del Giudice *et al.*, 1967; Al-Aubaidi and Fabricant, 1971).

Although both the direct and the indirect versions of the IMF test have been described above, the indirect method would seem for several reasons to be superior to the direct method: (1) only one conjugate is needed, (2) the non-specific background fluorescence that may interfere with reading and interpretation of the test is less pronounced, and (3) it has been shown to be more sensitive, yielding two to eight-fold higher titres than the direct method (Rosendal and Black, 1972; Black and Krogsgaard-Jensen, 1974). Recently, a promising new method, the indirect immunoperoxidase test, has been introduced and compared with the indirect IMF test (Polak-Vogelzang *et al.*, 1978).

E. Direct agglutination

Although the authors have very little personal experience with direct agglutination as applied to mycoplasmas it is felt that the basic techniques of this very simple method should be briefly described. The description that follows of the tube agglutination test is based essentially on the procedure used by Edward (1950).

1. *Tube agglutination*

(a) *Preparation of antigen.* A culture in 200–1000 ml of liquid medium is harvested at the end of the logarithmic growth phase in a refrigerated angle centrifuge for 1 h at approximately 10 000 rev/min ($12\,000 \times g$). The deposit is washed once by suspending in saline and recentrifuging. For agglutination the deposit is finally suspended in formol saline of the following composition (Klieneberger, 1938):

0·2 M citric acid	0·15 ml
0·2 M Na$_2$HPO$_4$·12 H$_2$O	1·85 ml
Formalin	0·25 ml
NaCl, 0·9%	to 100·00 ml

The opacity of the suspension to be used for agglutination is suggested by Edward (1950) as half of that of a Brown's Opacity tube No. 1 (Wellcome Reagents Ltd, Beckenham, Kent, England). Alternatively, suspensions can be adjusted to give a 50% light transmittance at 650 nm in a spectrophotometer (Tully, 1963).

(b) *Test procedure.* The test is made in small test-tubes containing 0·1 ml of serial two-fold dilutions of antiserum and 0·1 ml of antigen suspension. Controls with saline and normal rabbit serum should be included. Tubes are incubated in a water bath at 52–56°C. Agglutination is read after 4 and 6 h by the naked eye under a bright light, preferably in a dark room. The endpoint of the titration is confirmed using a hand lens.

2. *Slide agglutination*

Several modifications of this alternative technique for direct agglutination have been described. The workability and relative merits of each of these modifications are difficult to assess from the literature and the authors must therefore withhold recommendation of any single procedure. According to Hromatka and Adler (1969) the sensitivity of the antigen is maximal when suspended in citrate–phosphate buffer of pH 6–7, an atmospheric pressure of 4·1–4·9, and a molarity of 0·06–0·1. Reference may be further made to Adler (1954, 1958), Adler and DaMassa (1968) and Roberts (1970).

3. *Comments*

A major technical obstacle to the agglutination test is the tendency of some antigen suspensions to autoagglutinate. A mere mechanical breaking up of clumps and aggregates of particles is said to be more effective than other methods tried for the preparation of homogeneous and stable suspensions (Klieneberger, 1938). This was accomplished by Hromatka and Adler (1969) by stirring the antigen suspension for 6 h at 4°C with a magnetic stirrer and glass beads.

F. Complement fixation test

The test to be described is a two-dimensional one based on that designed by Bradstreet and Taylor (1962) and Taylor (1967) for detecting virus antibodies, but adapted to a microtitre system using 0·025 ml as unit volume. In the test proper, the primary reaction mixture comprises one volume each of antiserum, antigen and complement, fixation of the latter being allowed to proceed overnight at 4°C. The indicator system consists of two volumes of sheep erythrocytes optimally sensitised with haemolysin. The endpoint of all titrations is recorded as that dilution which shows 50% haemolysis of the indicator system. The procedures to be described minimise cross-reactions between different species.

1. *Equipment and cleaning procedures*

The microtitre system used is essentially the same as that described for the MI test. Permanent lucite plates with U-shaped wells are satisfactory if washed by the procedure of Sever (1962). Plates are soaked in detergent (Haemosol, Mainecke and Co., Inc., Baltimore, U.S.A.: available from Alfred Cox, Surgical Ltd, Coulsdon, Surrey, England) at 40–50°C for 30 min, rinsed in warm tap water, then in deionised water and finally dried at 37–50°C. During each wash or rinse operation, the plates are scrubbed with a hand brush, filled and shaken empty several times.

2. *Reagents*

These include diluent, sheep erythrocytes, complement, haemolysin, antigen and antiserum.

(a) *Diluent.* A barbitone buffer containing Ca^{2+} and Mg^{2+} is used through-out. This is most conveniently prepared from Oxoid Barbitone CFT diluent tablets (Code BR16, Oxoid Ltd, London S.E.1). Water for pre-paring the diluent should be freshly glass-distilled or, if more crudely distilled, also deionised. Alternatively, barbitone buffer may be prepared from its constituent chemicals as described by Cruickshank *et al.* (1975).

(b) *Sheep erythrocytes preserved in Alsever's solution.* Satisfactory products are available from commercial sources (Wellcome Reagents Ltd) and can be stored at 4°C. Alsever's solution and buffy coat are aspirated after an initial centrifugation, and the red cells washed three times in 0·85% saline. The packed erythrocytes are then resuspended in diluent to a concentration of 2%. The addition of an equal volume of the appropriate dilution of haemolysin to this suspension means that the final concentration of red cells is 1%. This is suitable for the microtitre system because it gives cell buttons and/or coloured supernatants which are easily interpreted in terms of percentage haemolysis.

(c) *Complement.* Pooled guinea pig serum preserved by Richardson's method (Cruickshank *et al.*, 1975) and stored at 4°C, or frozen without preservative and stored at −70°C, is the source of complement. Preserved complement is also available commercially (Wellcome Reagents Ltd, Beckenham, Kent, England). In our experience, guinea pig serum which is separated rapidly in the cold and stored frozen in small amounts (0·2 and 0·5 ml are convenient) gives superior results to preserved complement. The former invariably has a higher titre and annoying "trace" reactions which sometimes occur even with excess of preserved complement are not observed. Providing storage is below −50°C, complement maintains its titre for many months. Storage at −25 to −30°C is unsatisfactory. During a test, complement is removed from the freezing cabinet only when both serum and antigen are in the wells, and tubes for diluting the complement are ready on ice. It is then rapidly thawed in lukewarm water, appropriately diluted (*vide infra*) and kept on ice. Plates are transferred to the refrigerator as soon as complement has been added.

(d) *Haemolytic serum (haemolysin).* Antiserum to sheep erythrocytes prepared in rabbits is available commercially (e.g. Wellcome Reagents Ltd). It is stable for long periods at 4°C, but should not be diluted to the required concentration until immediately before use.

(e) *Titration of haemolysin and complement.* The optimal concentration of haemolysin and the titre of complement are determined in a single chessboard titration (Bradstreet and Taylor, 1962; Taylor 1967).

Using a micropipette, two drops of buffered diluent are placed in each of 70 wells (7 rows of 10 columns) in a microtitre plate (Table V). Wells in the last (tenth) column receive a third drop of diluent in place of complement, and serve as controls to establish that the haemolysin does not lyse red cells in the absence of complement. A series of complement dilutions with a 20% difference in concentration is prepared in suitable tubes, starting with a dilution of 1:30. If frozen complement is used, it is thawed and 3 ml of 1:30 dilution prepared in diluent. For preserved complement, which is hypertonic and already diluted by the preservatives, a 1:10

TABLE IV

Scheme for the dilution of complement for haemolysin-complement titration

Reciprocal of complement dilution	38	47	59	73	92	114	143	179
Diluent (ml)	0·5	0·5	0·5	0·5	0·5	0·5	0·5	0·5
Diluted complement (ml)	2·0 (1:30)	2·0 (1:38)	2·0 (1:47)	2·0 (1:59)	2·0 (1:73)	2·0 (1:92)	2·0 (1:114)	2·0 (1:143)

TABLE V
Titration of haemolysin and complement

Row	Reciprocal dilutions of haemolysin	Reciprocal dilution of complement									Control (no C′)
		30	38	47	59	73	92	114	143	179	
1	25	0	0	0	0	1	3	4	4	4	4
2	50	0	0	0	0	0	2	4	4	4	4
3	100	0	0	0	0	0	1	3	4	4	4
4	200	0	0	0	0	0	0	2	4	4	4
5	400	0	0	0	0	1	2	3	4	4	4
6	800	1	1	2	3	3	4	4	4	4	4
7	Control (no haemolysin)	4	4	4	4	4	4	4	4	4	4

dilution is prepared by diluting one volume (0·2 ml) with seven volumes (1·4 ml) of distilled water. This is then diluted to 1:30 with 3·2 ml of diluent. The other dilutions are prepared according to the scheme set out in Table IV. One drop of each complement dilution is then added to each well of the appropriate column as shown in Table V. The plate is covered with another plate, or if disposable plates are used, with one of the plastic lids manufactured for this purpose, placed in a plastic box containing damp cotton wool to prevent evaporation, and left at 4°C overnight.

Next morning, a standard 2% suspension of washed sheep cells is prepared, as described in Section (b) and a series of six doubling dilutions of haemolysin between 1:25 and 1:800 is made. One millilitre of each dilution of haemolysin is added to an equal volume of red cells. A seventh tube comprises 1 ml each of diluent and red cells.

The tubes are incubated in a 37°C waterbath for 60 min and shaken gently at intervals. Ten minutes before sensitisation is complete, the plate containing the complement dilutions is transferred to room temperature. The red cell suspensions are shaken gently and two drops are added to each well in the appropriate row, starting at row 7. This control row without haemolysin is included to check that unsensitised red cells do not lyse in the presence of complement. The plates are tapped gently to mix the contents of the wells, sealed with tape and incubated at 37°C for 60 min. After removing the tape, the red cells are allowed to settle at room temperature for at least 1 h. The amount of haemolysis is assessed visually according to the size of the red cell button and the colour of the supernatant, and scored 0, 1, 2, 3 and 4, where 0 indicates 100% haemolysis and 1–4 indicate, respectively, 75, 50, 25 and 0% haemolysis.

The optimal sensitising concentration (OSC) of the haemolytic serum is in that dilution which gives most lysis with the highest dilution of complement. In the example shown in Table V, it is 1:200 and the haemolytic serum should be used at that dilution in subsequent titrations. One unit (HC50) of complement is contained in that dilution which gives 50% lysis (reading 2) at the OSC of the haemolytic serum. In the titration shown in Table V, the HC50 is in the 1:114 dilution. In the test proper 3 HC50 are used, so the working dilution of complement is 1:38.

Since haemolytic serum is stable for long periods, a full chessboard titration need not be done frequently. Once the OSC is established, a simple "line" titration of complement can be carried out with that dilution of haemolysin.

(f) *Production of antigen.* Mycoplasma antigens for CF tests usually consist of washed suspensions of whole organisms. The practice at the Lister Institute has been to grow the organisms in 500 ml of medium containing 10–20% serum. Cultures are usually incubated for 6–7 days, because the anticomplementary activity of many antigens is reduced by this extended incubation period. This may be due to the loss of complement-inactivating enzymes as the organisms age. After harvesting, the mycoplasmas are washed twice in 0·85% saline, resuspended in 5–10 ml saline containing 1:10 000 merthiolate (Eli Lilly and Co. Ltd, Basingstoke, England) and stored at 4°C. Antigens thus prepared retain their activity for many months.

The problem of anticomplementary activity is a real one. Most preparations are anticomplementary at low dilutions, but the optimal antigen dilution is usually found to be in dilutions higher than those which are anticomplementary. If an antigen proves to be too anticomplementary to use, it must be discarded and a fresh culture initiated as there is little one can do to "improve" an antigen. Absorption with complement has, for example, never been effective in our hands. Extending the incubation period still further (15–24 days) has effectively overcome the anticomplementary activity of *A. laidlawii* antigens. The use of 1–2% PPLO serum fraction (Difco) instead of whole serum may also reduce the anticomplementary activity of the resulting antigen. For species which possess lipid haptens, for example, *M. pneumoniae* and *M. fermentans*, the lipid extracted with chloroform–methanol (2:1) is usually less anticomplementary than whole cell suspensions. Heating may also reduce the anticomplementary activity. For example, the CF antigen used for the diagnosis of contagious bovine pleuropneumonia is an aqueous suspension of *M. mycoides* subsp. *mycoides* boiled for 10 min in a brine bath and then made isotonic with sodium chloride (Campbell, 1938). It is noteworthy that

TABLE VI

(a) **Titration of antigen for anticomplementary activity**

Dilution of antigen	HC50 units of complement					
	3	2	1	$\frac{1}{2}$	0	
1:2	1	2	3	4	4	Anticomplementary
1:4	0	1	3	4	4	Anticomplementary
1:8	0	0	2	4	4	Satisfactory
1:16	0	0	2	4	4	Satisfactory
1:32	0	0	2	4	4	Satisfactory
1:64	0	0	2	4	4	Satisfactory
1:128	0	0	2	4	4	Satisfactory
1:256	0	0	2	4	4	Satisfactory
0 (control)	0	0	2	4	4	
0 (control)	0	0	2	4	4	

(b) **Titration of serum for anticomplementary or procomplementary activity**

	Reciprocal of serum dilution				
	10	20	40	80	
Serum + 3 HC50 complement	0	0	0	0	Not anticomplementary
Serum without complement	4	4	4	4	Not procomplementary

suspensions boiled in saline instead of water were highly anticomplementary. The anticomplementary activity of suspensions of *M. pneumoniae* has also been eliminated by heating at 56°C for 2 h or at 100°C for 5 min (Kenny and Grayston, 1965). Unfortunately not all species retain their serological activity after such treatment, and boiled aqueous suspensions of many species give much lower titres with homologous antisera than unheated antigens (Lemcke, 1961).

(g) *Tests for anticomplementary activity.* It is essential that neither the antigen nor the antiserum used in a test should, when tested separately, be capable of fixing complement. It is often convenient to run tests of antigens and antisera simultaneously as shown in Table VI(a) and (b).

An initial antigen dilution of 1:2 is prepared in a tube: from this, two-fold dilutions up to 1:256 are made in the plates with microdiluters. Five wells of each dilution are prepared so that each can be tested against

3, 2, 1, $\frac{1}{2}$ and 0 HC50 units of complement. One drop each of diluent and the appropriate complement dilution completes the primary reaction mixture. Two rows in which diluent is substituted for antigen are included as controls.

Serum to be tested is first diluted 1:10 in diluent and inactivated at 56°C for 30 min. From this, two-fold dilutions up to 1:80 are made in the plate as described for the MI test. One drop each of diluent and 3 HC50 complement is added to each well. To establish that there is no pro-complementary activity, a second row of serum dilutions in which complement is replaced by diluent may be set up.

Next morning, a 2% suspension of washed sheep cells is sensitised for 60 min at 37°C with an equal volume of haemolysin at the OSC. Two drops are added to each well and the tests completed as described for the haemolysin-complement titration.

Antigen is only acceptable at a given dilution if it gives the same reaction as the control rows without antigen. Since by definition 1 HC50 of complement lyses 50% of the red cells, the control rows should give the reactions shown in Table VI(a). In this example, the antigen is not anti-complementary at 1:8 and above. Dilutions which are not anticomplementary are then titrated to determine the optimal antigen concentration.

Antisera or preimmunisation sera which are anticomplementary can be diluted to 1:10 with a 1:10 dilution of complement and reacted for 24–48 h at 4°C. After inactivation at 56°C for 30 min, sera are retested, but the second row from which complement has been omitted must be included to establish that the complement used for absorption has been inactivated. Absorbed inactivated sera can be stored as dilutions for short periods at or below −30°C. The guinea pig serum used as the source of complement could contain antibodies which react with mycoplasma antigens, although such antibodies have not been detected in the batches of pooled serum which we have used.

(h) *Titration of antigen.* It has been shown that there is an equivalence zone with respect to concentrations of antigen and antibody, and that fixation of complement is optimal at or near this zone. Inhibition of complement fixation occurs as the quantity of antigen is increased beyond that optimum (Mayer, 1961). To determine the optimal antigen dilution, therefore, a series of antigen dilutions is titrated against dilutions of the homologous antiserum or an antiserum against a strain of the same species. For this test, antiserum is diluted in the plates, antigen dilutions are prepared in tubes and one drop added to the wells. One drop of 3 HC50 complement completes the primary reaction mixture. The test is completed as described in the preceding Section.

The optimal antigen dilution is that which gives a 50% haemolysis endpoint with the highest dilution of antiserum. To avoid or at least minimise false positive reactions due to anticomplementary activity, Mayer (1961) suggests that antigen when tested without antiserum should not be anticomplementary at twice the concentration used in the test proper. Quite frequently, two consecutive dilutions of antigen give the same optimal endpoint. In such a case, the rule quoted above may suggest the use of the higher antigen dilution.

To select the working concentration of an antigen whose relationships are unknown and for which no homologous antiserum is available, a titration for anticomplementary activity should be carried out. The two lowest dilutions that are not anticomplementary should then be used in the test proper.

3. *The test proper*

The OSC of haemolysin, the dilution of complement containing 3 HC50 units and the optimal dilutions of the antigens have already been determined.

The test is set up as described for the antigen titration in Section 2 (h): antiserum is diluted in the plates from a primary dilution of 1:10 or 1:20; one volume each of diluted antigen and complement are added to each well and fixation is allowed to proceed at 4°C overnight; two volumes of freshly washed and sensitised sheep erythrocytes are added to each well and the test incubated for 60 min at 37°C.

A properly controlled test to identify an unknown mycoplasma should be designed so that (i) each reference antiserum is titrated against the unknown antigen, its homologous antigen (positive control) and diluent without antigen; (ii) where the corresponding preimmunization serum is available, a similar titration is set up with serum dilutions between 1:10 and 1:80; (iii) each antigen is tested against 3, 1, $\frac{1}{2}$ and 0 HC50 units of complement (for anticomplementary activity).

In our experience, a strain which reacts with a reference antiserum to within 50 or 25% (i.e. one or two dilutions) of the homologous titre, can be considered to belong to the same species as the homologous strain.

G. Double immunodiffusion

Two types of double immunodiffusion (DID) tests are described. The first is a micro-test, which is carried out on microscope slides and is used to screen sera and antigens for specific and non-specific activity, and for determining the optimal antigen concentration. The second is on a slightly larger scale and requires more materials, but has the advantage that individual precipitation lines in the often complex patterns are better resolved. The macro-test is therefore preferred for the final tests in any comparative investigation.

1. *Equipment*

For the micro-test, a length of 4·5 cm is marked off with a diamond in the centre of a 7·5 × 2·5 cm microscope slide. Slides are soaked in chromic acid for 48 h, rinsed in distilled water, air-dried (at 37–50°C), and stored in acetone.

For the macro-test, 8 × 8 cm slide cover glasses are similarly treated except for diamond marking. On these plates, agar is confined to a circular area approximately 6·4 cm in diameter by means of a metal ring.

For the micro-test, the template is drilled to give wells 3 mm in diameter with centres 7 mm apart, and for the macro-test 6 mm in diameter and 12 mm apart.

2. *Reagents*

(a) *Gelling agent.* The following formula for an agar gel has proved satisfactory for most purposes:

Sodium chloride (Analar)	4·5 g
Difco Noble Agar	5·0 g
Sodium azide	0·325 g
Deionised water buffered to pH 7·0 with 0·5 M Na$_2$HPO$_4$	500 ml

The sodium chloride is dissolved in the buffered water in a 1-litre conical flask and heated to 98°C on a magnetic stirrer. The agar is added and allowed to dissolve at this temperature, which is not exceeded. Sodium azide is added just before the flask is removed from the heat. Universal containers are filled with agar, screwed down tightly and stored at 4°C. If the agar is not calcium-free, phosphate ions from the buffer may precipitate and produce turbid gels. This can be overcome by adding diaminoethane tetra-acetic acid (disodium salt) at 30 mg/litre to chelate Ca^{2+} ions (Feinberg, 1957).

If a very clear gel is required, specially purified agar can be prepared by the method of Feinberg (1956) or that of Crowle (1961). Alternatively, agarose (Miles Laboratories Ltd, Stoke Poges, England) may be substituted for agar (Lemcke *et al.*, 1965; Kenny 1969).

(b) *Stains and decolourising solution.* Amidoschwarz or Ponceau fuchsin are prepared as follows:

Amidoschwarz (Naphthalene black)	10B
or	
Ponceau fuchsin	0·1 g
6% acetic acid	45 ml
0·1 M sodium acetate	45 ml
Glycerol	10 ml

The decolourising solution consists of:

Glycerol	15 ml
Glacial acetic acid	2 ml
Distilled or deionised water	83 ml

(c) *Antisera*. For DID tests, rabbit antisera have advantages over horse antisera, because the antigen–antibody precipitates obtained with the former are much less soluble in the presence of excess antigen or antibody. Also, rabbit serum is less likely to give rise to artifacts due to changes in temperature (Crowle, 1961). Antisera for DID tests should be free from antibodies to foreign serum proteins (Section V.A). Moreover, satisfactory precipitating sera usually require the use of Freund's complete or incomplete adjuvant in their production. With mycoplasma antigens, more lines are given by adjuvant-produced antisera than by antisera produced by intravenous inoculation. However, the additional lines given by adjuvant-produced sera are mainly due to cytoplasmic antigens (Lemcke, 1973).

Tests should be set up to make sure that neither preimmunisation sera nor antisera react with any of the medium constituents used to grow the antigens. The sediment obtained after centrifugation of the medium has often been used, but media supplemented with bovine serum fraction (Difco) are virtually clear and give very little deposit. It may therefore be more convenient to test the undiluted sera against the complete medium and those medium constituents which are most likely to be antigenic. The test may be set up with serum in the central well and medium constituents or complete medium alternating with the homologous antigen in the peripheral wells. Such a test will also indicate whether any of the lines given by the antigen are due to adsorbed medium constituents, since the reaction of an adsorbed component will be enhanced by the greater concentration of that component in the adjacent well. In our experience, cross-reactions between antisera from rabbits immunised with mycoplasmas grown in rabbit serum broth and components adsorbed to antigens grown in serum fraction media have not been a problem. If such reactions are detected, the reacting antibody can be absorbed from the antiserum by including the medium component or the complete medium in the gel–diffusion agar at a concentration of 10% (v/v) (Hollingdale and Lemcke,

1972). Alternatively, antisera can be absorbed with lyophilised medium at a rate of 10 mg ml^{-1} (Ross and Karmon, 1970).

(d) *Production of antigen.* Cultures in the late log-phase are harvested by centrifugation at 4°C and washed three times in PBS. If the last wash is carried out in a known volume and samples are taken before centrifugation for protein estimation (Lowry *et al.*, 1951), the protein content of the final pellets can be determined. Pellets are drained well and stored frozen at or below −30°C.

For DID tests, the mycoplasmas need to be disrupted or lysed. The methods usually employed are: alternate cycles of freezing and thawing, ultrasonic treatment, or lysis with detergents. For freezing and thawing. pellets are resuspended in deionised water to an appropriate protein concentration (5–10 mg protein ml^{-1}), frozen rapidly in a mixture of dry ice and acetone, and thawed in luke-warm water. This is carried out 10–20 times. The details of ultrasonic treatment depend on the instrument available. With a Branson Sonifier S-75 operating at 20 kc s^{-1}, mycoplasmas suspended in PBS are treated for five periods of 1 min, each period being separated by a cooling interval of 1 min. Tubes are kept in ice and water throughout.

Most of the precipitin bands produced by frozen–thawed or ultrasonically treated suspensions are due to cytoplasmic antigens (Hollingdale and Lemcke, 1970; Lemcke, 1973). To release additional antigens which are bound to the membrane and thus to obtain a more complete precipitin pattern, it is necessary to use a detergent. We have found that the non-ionic detergent Triton X-100 is best for this purpose because it does not give non-specific precipitates with rabbit serum as do sodium dodecyl sulphate and sodium deoxycholate. Triton X-100 is added to suspensions in PBS at the rate of 5 mg mg^{-1} of mycoplasma protein and incubated for 15 min at 37°C.

Antigens prepared by freezing and thawing, ultrasonic treatment or lysis with Triton X-100 are usually satisfactory at 5–10 mg protein ml^{-1}, but it is advisable to first test an antigen against its homologous antiserum over a range of concentrations (e.g. 1·25, 2·5, 5·0 and 10 mg ml^{-1}) to see at which concentration the lines are sharpest and best resolved. The choice of concentration is inevitably somewhat arbitrary since each "antigen" is, in fact, a complex mixture of antigens and the concentration selected cannot be optimal for all the components.

3. Test procedure

The slides or cover glasses are wiped free of acetone, set out on a levelling table (Shandon Southern Instruments Ltd) and the metal rings centred on the cover glasses. Agar is melted rapidly and allowed to stand for a few minutes so that air bubbles can disperse. One millilitre is pipetted over the 4.5×2.5 cm area marked on each microscope slide to give a layer of agar 1 mm deep. For the macro-test, 7 ml of agar is pipetted into the ring, giving a depth of 1.8 mm. The agar sets in a few minutes and the rings can be removed. Before the wells are punched, slides should be transferred to 4°C in a suitable container with moist filter paper to avoid drying out. The wells should not be punched until the slides or plates are required for use and must be sealed by rapidly pipetting hot agar in and out. Wells are filled with antigen or antiserum so that the fluid is just level with the agar. Recharging the wells should be avoided since this practice may result in displacement and multiplication of precipitin bands, especially if concentrations equal to or higher than those originally used are added. Reactions are allowed to develop at room temperature and are usually optimal between 24 and 48 h. Thereafter, the reactions often become too diffuse to record. If the reactions develop too rapidly, it may be advantageous to conduct the test at 4°C. Plates and slides should not be subjected to changes of temperature, since this tends to produce artifacts. Gels should be inspected at regular intervals and the results recorded, preferably photographically, at different stages.

For storage, the gels can be dried and stained. Slides or plates are washed for 2 days in saline containing the concentration of NaCl used in the gel with two changes each day, and finally in distilled or deionised water for a further day. The agar is covered with filter paper and dried on to the glass in a warm place over a period of several days. Staining is by immersion for 20–30 min in Amidoschwarz (Naphthalene black) 10B, or Ponceau fuchsin (Section 2). Gels are decolourised in at least two changes of decolourising solution until the background is clear.

Preliminary micro-tests are set up to select suitable reference antisera and to establish that these sera do not react with constituents of the medium (Section 2). The concentration of antigen which gives optimal precipitation with the homologous antiserum is also determined. The actual test, in which homologous and heterologous antigens (peripheral wells) are compared against a reference antiserum (central well), is then carried out. Although reference sera are pretested for the presence of antibodies against constituents of the medium, it is desirable that one of the peripheral wells should contain the growth medium as a control. As a further control, another slide is set up with antigens in the same positions, but with pre-immunisation serum in the central well. In the first series of tests, the arrangement of the heterologous antigens in relation to the homologous antigen is necessarily arbitrary, but in subsequent tests cross-reacting and homologous antigens can be placed in adjacent wells.

4. Interpretation and comments

Results are usually recorded as the number of bands which develop between each antigen and the antiserum well. If the precipitin lines are well resolved it may be possible to assign a letter or number to each line and to construct an antigenic formula for an organism.

With strains belonging to the same species, most if not all of the lines produced by the homologous strain will, under optimal conditions, also be given by the heterologous strains, the common lines fusing in reactions of identity.

A general discussion of immunodiffusion is presented in Volume 5A of this Series (Oakely, 1971).

REFERENCES

Adler, H. E. (1954). *In* "Proceedings Book", Am. Vet. Med. Ass. Ninety-first Annual Meeting, pp. 346–349.
Adler, H. E. (1958). *Poult. Sci.* **37**, 1116–1123.
Adler, H. E. and DaMassa, A. J. (1964). *Proc. Soc. exp. Biol. Med.* **116**, 608–610.
Adler, H. E. and DaMassa, A. J. (1967). *Proc. Soc. exp. Biol. Med.* **124**, 1064–1067.
Adler, H. E. and DaMassa, A. J. (1968). *Appl. Microbiol.* **16**, 558–562.
Adler, H. E. and Yamamoto, R. (1956). *Am. J. vet. Res.* **17**, 290–293.
Al-Aubaidi, J. M. and Fabricant, J. (1971). *Cornell Vet.* **61**, 519–542.
Allam, N. M. and Lemcke, R. M. (1975). *J. Hyg., Camb.* **74**, 385–408.
Aluotto, B. B., Wittler, R. G., Williams, C. O. and Faber, J. E. (1970). *Int. J. syst. Bact.* **20**, 35–58.
Bak, A. L., Black, F. T., Christiansen, C. and Freundt, E. A. (1969). *Nature, Lond.* **224**, 1209–1210.
Barile, M. F., Schimke, R. T. and Riggs, D. B. (1966). *J. Bact.* **91**, 189–192.
Barile, M. F., Del Giudice, R. A. and Tully, J. G. (1972). *Infect. Immun.* **5**, 70–76.
Barile, M. F., Razin, S., Tully, J. G. and Whitcomb, R. F. (Eds) (1979). "The Mycoplasmas", Vol. I–III. Academic Press, New York and London.
Belly, R. T., Bohlool, B. B. and Brock, T. D. (1973). *Ann. N.Y. Acad. Sci.* **225**, 94–107.
Black, F. T. (1970). *In* "Proceedings of Vth International Congress of Infectious Diseases", Vienna, pp. 407–411.
Black, F. T. (1973a). *Int. J. syst. Bact.* **23**, 65–66.
Black, F. T. (1973b). *Appl. Microbiol.* **25**, 528–533.
Black, F. T. and Krogsgaard-Jensen, A. (1974). *Acta path. microbiol. scand. Sect. B* **82**, 345–353.
Bové, J. M. and Duplan, J. F. (Eds) (1974). "Les Mycoplasmes de l'Homme, des Animaux, des Végétaux et des Insectes", Colloq. INSERM, Paris, Vol. 33.
Bradbury, J. M. (1977). *J. clin. Microbiol.* **5**, 531–539.
Bradbury, J. M. and Jordan, F. T. W. (1972). *J. Hyg., Camb.* **70**, 267–278.
Bradstreet, C. M. P. and Taylor, C. E. D. (1962). *Mon. Bull. Minist. Hlth, Lond.* **21**, 96–104.
Bürger, H., Doss, M., Mannheim, W. and Schüler, A. (1967). *Z. med. Mikrobiol. Immunol.* **153**, 138–148.

426   E. A. FREUNDT *ET AL.*

Campbell, A. D. (1938). *J. Coun. scient. ind. Res. Aust.* **11**, 112–118.
Campello, C., Ionadi, F. and Majori, L. (1972). *Boll. Ist. Sieroter. Milanese* **51**, 354–363.
Card, D. H. (1959). *Br. J. vener. Dis.* **35**, 27–34.
Chanock, R. M., Hayflick, L. and Barile, M. F. (1962). *Proc. natn. Acad. Sci., U.S.A.* **48**, 41–49.
Cho, H. J., Ruhnke, H. L. and Langford, E. V. (1976). *Can. J. comp. Med.* **40**, 20–29.
Christiansen, C., Freundt, E. A. and Black, F. T. (1975). *Int. J. syst. Bact.* **25**, 99–101.
Clyde, W. A. (1961). *Proc. Soc. exp. Biol. Med.* **107**, 716–718.
Clyde, W. A. (1964). *J. Immunol.* **92**, 958–965.
Crowle, A. J. (1961). "Immunodiffusion". Academic Press, New York and London.
Cruickshank, R., Duguid, J. P., Marmion, B. P. and Swain, R. H. A. (1975). *In* "Medical Microbiology", 12th edn, Vol. 2, p. 258. Churchill Livingstone, Edinburgh, London and New York.
Cunningham, S. (Ed.) (1978–1980). "NIAID Catalog of Research Reagents", pp. 859–959. DHEW Publication No. (NIH) 78–899.
Czekalowski, J. W., Hall, D. A. and Woolcock, P. R. (1973). *J. gen. Microbiol.* **75**, 125–133.
Darland, G., Brock, T. D., Samsonoff, W. and Conti, S. F. (1970). *Science, N.Y.* **170**, 1416–1418.
Del Giudice, R. A., Robillard, N. F. and Carski, T. R. (1967). *J. Bact.* **93**, 1205–1209.
Del Giudice, R. A., Purcell, R. H., Carski, T. D. and Chanock, R. M. (1974). *Int. J. syst. Bact.* **24**, 147–153.
Dowdle, W. R. and Robinson, R. Q. (1964). *Proc. Soc. exp. Biol. Med.* **116**, 947–950.
Edward, D. G. ff. (1950). *J. gen. Microbiol.* **4**, 4–15.
Edward, D. G. ff. (1967). *Ann. N.Y. Acad. Sci.* **143**, 7–8.
Edward, D. G. ff. (1974). *In* "Les Mycoplasmes de l'Homme, des Animaux, des Végétaux et des Insectes" (J. M. Bové and J. F. Duplan, Eds), Colloq. INSERM, Paris, **33**, 13–18.
Edward, D. G. ff. and Fitzgerald, W. A. (1951). *J. gen. Microbiol.* **5**, 566–575.
Edward, D. G. ff. and Fitzgerald, W. A. (1954). *J. Path. Bact.* **68**, 23–30.
Edward, D. G. ff. and Freundt, E. A. (1967). *Int. J. syst. Bact.* **17**, 267–268.
Edward, D. G. ff. and Kanarek, A. D. (1960). *Ann. N.Y. Acad. Sci.* **79**, 696–702.
Edward, D. G. ff. and Razin, S. (1974). World Health Organisation, VPH/MIC/ 74–2, pp. 1–3.
Elliott, K. and Birch, J. (Eds) (1972). "Pathogenic Mycoplasmas", Ciba Foundation Symposium. Elsevier, Excerpta Medica, North-Holland, Amsterdam, London and New York.
Ernø, H. (1977). *Acta vet. Scand.* **18**, 176–186.
Ernø, H. and Jurmanová, K. (1973). *Acta vet. scand.* **14**, 524–537.
Ernø, H., Jurmanová, K. and Leach, R. H. (1973). *Acta vet. scand.* **14**, 511–523.
Ernø, H. and Stipkovits, L. (1973). *Acta vet. scand.* **14**, 436–449.
Fallon, R. J. and Whittlestone, P. (1969). *In* "Methods in Microbiology" (J. R. Norris and D. W. Ribbons, Eds), Vol. 3B, pp. 211–267. Academic Press, London and New York.
Feinberg, J. G. (1956). *Nature, Lond.* **178**, 1406.
Feinberg, J. G. (1957). *Int. Arch. Allerg. appl. Immunol.* **11**, 129–152.
Forshaw, K. A. and Fallon, R. J. (1972). *J. gen. Microbiol.* **72**, 501–510.

Fox, H., Purcell, R. H. and Chanock, R. M. (1969). *J. Bact.* **98**, 36–43.

Freundt, E. A. (1958). "The Mycoplasmataceae (The Pleuropneumonia Group of Organisms)", pp. 55–56. Munksgaard, Copenhagen.

Freundt, E. A. (1973). *Ann. N.Y. Acad. Sci.* **225**, 7–13.

Freundt, E. A. (1974a). *Path. Microbiol.* **40**, 155–187.

Freundt, E. A. (1974b). In "Les Mycoplasmes de l'Homme, des Animaux, des Végétaux et des Insectes" (J. M. Bové, J. F. Duplan, Eds), Colloq. INSERM, Paris, **33**, 19–25.

Freundt, E. A. (1974c). In "Bergey's Manual of Determinative Bacteriology", 8th edn (R. E. Buchanan and N. E. Gibbons, Eds), pp. 929–954. Williams & Wilkins Company, Baltimore.

Freundt, E. A., Andrews, B. E., Ernø, H., Kunze, M. and Black, F. T. (1973a). *Zbl. Bakt. Abt. I Orig. A* **225**, 104–112.

Freundt, E. A., Ernø, H., Black, F. T., Krogsgaard-Jensen, A. and Rosendal, S. (1973b). *Ann. N.Y. Acad. Sci.* **225**, 161–171.

Freundt, E. A., Taylor-Robinson, D., Purcell, R. H., Chanock, R. M. and Black F. T. (1974). *Int. J. syst. Bact.* **24**, 252–255.

Goodburn, G. M. and Marmion, B. P. (1962). *J. gen. Microbiol.* **29**, 271–290.

Gourlay, R. N., Leach, R. H. and Howard, C. J. (1974). *J. gen. Microbiol.* **81**, 475–484.

Gourlay, R. N., Wyld, S. G. and Leach, R. H. (1977). *Int. J. syst. Bact.* **27**, 86–96.

Gourlay, R. N., Wyld, S. G. and Leach, R. H. (1978). *Int. J. syst. Bact.* **28**, 289–292.

Hayflick, L. (Ed.) (1969). "The Mycoplasmatales and the L-phase of Bacteria". Appleton–Century–Crofts, New York.

Herderscheê, D. (1963). *Antonie van Leeuwenhoek* **29**, 154–156.

Hers, J. F. Ph. (1963). *Am. Rev. resp. Dis.* **88**, 316–333.

Hill, A. (1971). *J. gen. Microbiol.* **65**, 109–113.

Hollingdale, M. R. and Lemcke, R. M. (1969). *J. Hyg., Camb.* **67**, 585–602.

Hollingdale, M. R. and Lemcke, R. M. (1970). *J. Hyg., Camb.* **68**, 469–477.

Hollingdale, M. R. and Lemcke, R. M. (1972). *J. Hyg., Camb.* **70**, 85-98.

Holmgren, N. (1973). *Acta vet. scand.* **14**, 353–355.

Hromatka, L. and Adler, H. E. (1969). *Avian Dis.* **13**, 452–461.

Huijsmans–Evers, A. G. M. and Ruys, C. (1956). *Antonie van Leeuwenhoek* **22**, 377–384.

Jensen, K. E. (1963). *J. Bact.* **86**, 1349–1350.

Johnson, J. L. (1973). *Int. J. syst. Bact.* **23**, 308–315.

Kende, M. (1969). *Appl. Microbiol.* **17**, 275–279.

Kenny, G. E. (1969). *J. Bact.* **98**, 1044–1055.

Kenny, G. E. (1972). *Med. Microbiol. Immunol.* **157**, 174.

Kenny, G. E. (1973). *J. infect. Dis. (Suppl.)* **127**, 52–55.

Kenny, G. E. and Grayston, J. T. (1965). *J. Immunol.* **95**, 19–25.

Kirchhoff, H. (1978a). *Int. J. syst. Bact.* **28**, 76–81.

Kirchhoff, H. (1978b). *Int. J. syst. Bact.* **28**, 496–502.

Klieneberger, E. (1938). *J. Hyg., Camb.* **38**, 458–476.

Krogsgaard-Jensen, A. (1971). *Appl. Microbiol.* **22**, 756–759.

Krogsgaard-Jensen, A. (1972). *Appl. Microbiol.* **23**, 553–558.

Laidlaw, P. B. and Elford, W. J. (1936). *Proc. R. Soc. Lond. Ser. B* **120**, 292–303.

Lam, G. T. and Morton, H. E. (1974). *Appl. Microbiol.* **27**, 356–359.

Langford, E. V. and Leach, R. H. (1973). *Can. J. Microbiol.* **19**, 1435–1444.

Langford, E. V., Ruhnke, H. L. and Onoviran, O. (1976). *Int. J. syst. Bact.* **26**, 212–219.

Lapage, S. P., Clark, W. A., Lessel, E. F., Seeliger, H. P. R. and Sneath, P. H. A. (1973). *Int. J. syst. Bact.* **23**, 83–108.

Leach, R. H. (1967). *Ann. N.Y. Acad. Sci.* **143**, 305–316.

Lemcke, R. M. (1961). *J. Hyg., Camb.* **59**, 401–412.

Lemcke, R. M. (1964). *J. Hyg., Camb.* **62**, 199–219.

Lemcke, R. M. (1965). *J. gen. Microbiol.* **38**, 91–100.

Lemcke, R. M. (1973). *Ann. N.Y. Acad. Sci.* **225**, 46–53.

Lemcke, R. M. and Allam, N. M. (1974). *In* "Les Mycoplasmes de l'Homme, des Animaux, des Végétaux et des Insectes" (J. M. Bové and J. F. Duplan, Eds), Colloq. INSERM, Paris, **33**, 153–159.

Lemcke, R. M. and Kirchhoff, H. (1979). *Int. J. syst. Bact.* **29**, 42–50.

Lemcke, R. M., Shaw, E. J. and Marmion, B. P. (1965). *Aust. J. exp. Biol. med. Sci.* **43**, 761–770.

Lin, J.-S. L., Alpert, S. and Radnay, K. M. (1975). *J. infect. Dis.* **131**, 727–730.

Lind, K. (1968). *Acta path. microbiol. scand.* **73**, 459–472.

Lind, K. (1970). *Acta path. microbiol. scand. Section B* **78**, 149–152.

Liu, C. (1957). *J. exp. Med.* **106**, 455–466.

Lowry, O. H., Rosebrough, N. J., Farr, A. L. and Randall, R. J. (1951). *J. biol. Chem.* **193**, 262–275.

Madden, D. L., Moats, K. E., London, W. T., Matthew, E. B. and Sever, J. L. (1974). *Int. J. syst. Bact.* **24**, 459–464.

Maramorosch, K. (Ed.) (1973). "Mycoplasma and mycoplasma-like agents of human, animal, and plant diseases" *Ann. N.Y. Acad. Sci.* Vol. 225.

Mayer, M. M. (1961). *In* "Experimental Immunochemistry", 2nd edn (E. A. Kabat and M. M. Mayer, Eds), pp. 133–240. Charles C. Thomas, Springfield, Illinois.

Møller, B. R. (1976). *J. appl. Bact.* **46**, 185–188.

Morton, H. E. (1966). *J. Bact.* **92**, 1196–1205.

Nairn, R. C. (1968). *Clin. exp. Immunol.* **3**, 465–476.

Nicol, C. S. and Edward, D. G. ff. (1953). *Br. J. vener. Dis.* **29**, 141–150.

Oakley, C. L. (1971). *In* "Methods in Microbiology." (J. R. Norris and D. W. Ribbons, Eds), Vol. 5A, pp. 173–218. Academic Press, London and New York.

Polak-Vogelzang, A. A., Hagenaarts, R. and Nagel, J. (1978). *J. gen. Microbiol.* **106**, 241–249.

Pollack, J. D., Somerson, N. L. and Senterfit, L. B. (1970). *Infect. Immun.* **2**, 326–339.

Purcell, R. H., Taylor-Robinson, D., Wong, D. C. and Chanock, R. M. (1966a). *Am. J. Epidem.* **84**, 51–66.

Purcell, R. H., Taylor-Robinson, D., Wong, D. C. and Chanock, R. M. (1966b). *J. Bact.* **92**, 6–12.

Purcell, R. H., Wong, D., Chanock, R. M., Taylor-Robinson, D., Canchola, J. and Valdesuso, J. (1967). *Ann. N.Y. Acad. Sci.* **143**, 665–675.

Razin, S. (1969). *Ann. Rev. Microbiol.* **23**, 317–356.

Razin, S. (1978). *Microbiol. Rev.* **42**, 414–470.

Roberts, D. H. (1970). *Vet. Rec.* **87**, 125–126.

Roberts, D. H. (1971). *J. Hyg., Camb.* **69**, 361–368.

Roberts, D. H. and Olesiuk, O. M. (1967). *Avian Dis.* **11**, 104–119.

Roberts, D. H. and Pijoan, C. (1971). *Br. vet. J.* **127**, 582–586.

Robinson, I. M. and Allison, M. J. (1975). *Int. J. syst. Bact.* **25**, 182–186.

Robinson, I. M., Allison, M. J. and Hartman, P. A. (1975). *Int. J. syst. Bact.* **25**, 173–181.

Robinson, J. P. and Hungate, R. E. (1973). *Int. J. syst. Bact.* **23**, 171–181.
Rose, D. L., Tully, J. G. and Langford, E. V. (1978). *Int. J. syst. Bact.* **28**, 567–572.
Rose, D. L., Tully, J. G. and Wittler, R. G. (1979). *Int. J. syst. Bact.* **29**, 83–91.
Rosendal, S. (1974a). *Int. J. syst. Bact.* **24**, 125–130.
Rosendal, S. (1974b). *Acta path. microbiol. scand.*, Section B **82**, 25–32.
Rosendal, S. (1975). *Acta path. microbiol. scand.*, Section B **83**, 463–470.
Rosendal, S. and Black, F. T. (1972). *Acta path. microbiol. scand.* Section B **80**, 615–622.
Ross, R. F. and Karmon, J. A. (1970). *J. Bact.* **103**, 707–713.
Saglio, P. Lhospital, M., Laflèche, D., Dupont, G., Bové, J. M., Tully, J. G. and Freundt, E. A. (1973). *Int. J. syst. Bact.* **23**, 191–204.
Schimke, R. T. and Barile, M. F. (1963). *J. Bact.* **86**, 195–206.
Senterfit, L. B. and Jensen, K. E. (1966). *Proc. Soc. exp. Biol. Med.* **122**, 786–790.
Sever, J. L. (1962). *J. Immunol.* **88**, 320–329.
Sharp, J. T. (Ed.) (1970). "The Role of Mycoplasmas and L Forms of Bacteria in Disease". Charles C. Thomas, Springfield, Illinois.
Shepard, M. C. and Howard, D. R. (1970). *Ann. N.Y. Acad. Sci.* **174**, 809–819.
Shepard, M. C., Lunceford, C. D., Ford, D. K., Purcell, R. H., Taylor-Robinson, D., Razin, S. and Black, F. T. (1974). *Int. J. syst. Bact.* **24**, 160–171.
Shimizu, T., Ernø, H. and Nagatomo, H. (1978). *Int. J. syst. Bact.* **28**, 538–546.
Skripal, I. G. (1974). *Mikrobiol. Zh.* **36**, 462–467.
Subcommittee on the Taxonomy of *Mycoplasmatales* (1967). *Science, N.Y.* **155**, 1694–1696.
Subcommittee on the Taxonomy of *Mycoplasmatales* (1972). *Int. J. syst. Bact.* **22**, 184–188.
Subcommittee on the Taxonomy of *Mycoplasmatales* (1974). *Int. J. syst. Bact.* **24**, 390–392.
Subcommittee on the Taxonomy of *Mycoplasmatales* (1975). *Int. J. syst. Bact.* **25**, 237–239.
Subcommittee on the Taxonomy of *Mycoplasmatales* (1977). *Int. J. syst. Bact.* **27**, 392–394.
Taylor, C. D. E. (1967). *In* "Progress in Microbiological Techniques" (C. H. Collins, Ed.), pp. 1–14. Butterworths, London.
Taylor-Robinson, D. and Berry, D. M. (1969). *J. gen. Microbiol.* **55**, 127–137.
Taylor-Robinson, D., Canchola, J., Fox, H. and Chanock, R. M. (1964). *Am. J. Hyg.* **80**, 135–148.
Taylor-Robinson, D., Fox, H. and Chanock, R. M. (1965a). *Am. J. Epidemiol.* **81**, 180–191.
Taylor-Robinson, D., Ludwig, W. M., Purcell, R. H., Mufson, M. A. and Chanock, R. M. (1965b). *Proc. Soc. exp. Biol. Med.* **118**, 1073–1083.
Taylor-Robinson, D., Somerson, N. L., Turner, H. C. and Chanock, R. M. (1963). *J. Bact.* **85**, 1261–1273.
Taylor-Robinson, D., Purcell, R. H., Wong, D. C. and Chanock, R. M. (1966). *J. Hyg., Camb.* **64**, 91–104.
Therkelsen, A. J. (1961). *Biochem. Pharmac.* **8**, 269–279.
Thirkill, C. E. and Kenny, G. E. (1974). *Infect. Immun.* **10**, 624–632.
Tully, J. G. (1963). *Proc. Soc. exp. Biol. Med.* **114**, 704–709.
Tully, J. G., Barile, M. F., Edward, D. G. ff., Theodore, T. S. and Ernø, H. (1974). *J. gen. Microbiol.* **85**, 102–120.

Walker, P. D., Batty, I. and Thomson, R. O. (1971). *In* "Methods in Microbiology" (J. R. Norris and D. W. Ribbons, Eds), Vol, 5A, pp. 219–254. Academic Press, London and New York.
Watanabe, T. (1975). *Med. Microbiol. Immunol.* **161**, 127–132.
Williams, C. O. and Wittler, R. G. (1971). *Int. J. syst. Bact.* **21**, 73–77.
Windsor, G. D. and Trigwell, J. A. (1976). *Res. vet. Sci.* **20**, 221–222.
Woode, G. N. and McMartin, D. A. (1973). *J. gen. Microbiol.* **75**, 43–50.
Woolcock, P. R., Czekalowski, J. W. and Hall, D. A. (1973). *J. gen. Microbiol.* **78**, 23–32.

APPENDIX

FORMULAE OF GROWTH AND TEST MEDIA

A. Growth media

1. *Medium B: modified Hayflick medium*

(a) *Liquid medium*

Heart Infusion Broth (Difco)	90·0 ml
(Sterilise by autoclaving)	
Horse serum (unheated)	20·0 ml
Yeast extract (25%) (Taylor-Robinson *et al.*, 1963)	10·0 ml
Thallium acetate (1% (w/v) solution)	1·0 ml
Penicillin (20 000 IU ml^{-1})	0·25 ml
Deoxyribonucleic acid† (0·2% (w/v) solution)	1·2 ml
Adjust pH to 7·8	

(b) *Solid medium*

This is prepared by replacing heart infusion broth with heart infusion agar (Difco)

2. *Medium N*

(a) *Liquid medium*

Bacto brain heart infusion (Difco)	3·7 g
Yeast extract (Difco)	0·5 g
Distilled water	100·0 ml
Sterilise by autoclaving	
Horse serum (unheated)	20·0 ml
Yeast extract (25%) (Taylor-Robinson *et al.*, 1963)	10·0 ml
Thallium acetate (1% (w/v) solution)	1·0 ml
Penicillin (20 000 IU ml^{-1})	0·25 ml
Deoxyribonucleic acid (Sigma, 0·2% (w/v) solution)	1·3 ml
Glucose (50% (w/v) solution)	2·0 ml
Adjust pH to 7·8	

(b) *Solid medium*

This is prepared by adding 1·4 g Ionagar No. 2 (Code L 12, Oxoid) before autoclaving.

† From calf thymus, Sigma Chemical Company, St Louis, Miss., U.S.A.; Type 1, Catalogue No. 1501.

3. *Medium SU: Shepard medium* (for *Ureaplasma* spp.)

(a) *Liquid medium*

Trypticase soy broth powder (Baltimore Biological Laboratories) (3% (w/v) solution)	100·0 ml
Horse serum (unheated)	22·0 ml
Yeast extract (25%) (Taylor-Robinson *et al.*, 1963)	10·0 ml
Urea (40% (w/v) solution)	1·6 ml
Phenol red (0·06% (w/v) solution)	5·0 ml
Penicillin (20 000 IU ml⁻¹)	1·0 ml

Adjust pH to 6·0.

(b) *Solid medium*

This is prepared by adding 1·4 g Purified Agar (Code L 28 Oxoid).

4. *Medium BACY* (for *M. faucium* and *M. lipophilum*)

(a) *Liquid medium*

Heart infusion broth (Difco)	90·0 ml
(Sterilise by autoclaving)	
Horse serum (unheated)	20·0 ml
Yeast extract (25%) (Taylor-Robinson *et al.*, 1963)	10·0 ml
Deoxyribonucleic acid (Sigma, 0·2% (w/v) solution)	1·3 ml
Cystein hydrochloride (10% (w/v) solution)	1·24 ml
L-arginine (30% (w/v) solution)	4·25 ml
Phenol red (0·06% (w/v) solution)	5·0 ml
Thallium acetate (1·0% (w/v) solution)	1·0 ml
Penicillin (20 000 IU ml⁻¹)	0·25 ml

Adjust pH to 7·3

(b) *Solid medium*

Replace heart infusion broth with heart infusion agar (Difco).

5. *Medium F: Frey medium* (for *M. synoviae*)

(a) *Liquid medium*

Bacto PPLO broth w/o CV (Difco)	2·25 g
Distilled water	90·0 ml
Eagle's essential vitamins (× 100)†	0·025 ml
Glucose (50% (w/v) solution)	2·0 ml
Swine serum (inactivated 56°C 30 min⁻¹)	12·0 ml
β-NADH‡ (1% (w/v) solution)	1·0 ml
Cystein hydrochloride (1% (w/v) solution)	1·0 ml
Phenol red (0·06% (w/v) solution)	5·0 ml
Penicillin (20 000 IU ml⁻¹)	0·25 ml

The solutions of β-NADH and cystein hydrochloride are mixed, and after 10 min added to the other ingredients.
Adjust pH to 7·8.

† Therkelsen (1961).
‡ β-Nicotinamide dinucleotide, reduced form, Sigma, Grade III.

(b) *Solid medium*

Bacto PPLO broth w/o CV (Difco)	2·25 g
Ionagar No. 2 (Code L 12, Oxoid)	1·4 g
Distilled water	90·0 ml
Adjust pH to 7·8	
Sterilise by autoclaving (121 °C 20 min⁻¹)	

Sterilise by autoclaving (121 °C 20 min^{-1})

Eagle's essential vitamins (× 100)	0·025 ml
Glucose (50% (w/v) solution)	2·0 ml
Swine serum (unheated)	12·0 ml
β-NADH (1% (w/v) solution)	1·0 ml
Cystein hydrochloride (1% (w/v) solution)	1·0 ml
Phenol red (0·06 (w/v) solution)	5·0 ml
Penicillin (20 000 IU ml⁻¹)	0·25 ml

6. *Medium FF74: Friis medium* (for *M. flocculare* and *M. hyopneumoniae*)

(a) *Liquid medium*

Brain heart infusion broth (Difco)	0·82 g
Double distilled water	75·0 ml
Bacto PPLO broth w/o CV (Difco)	0·87 g
Hanks's balanced salt solution (modified, see (c) below)	50·0 ml
(Sterilise by autoclaving 121 °C 2–5 min⁻¹)	
Fresh yeast extract (FG, see (d) below)	6·0 ml
Horse serum (inactivated)	15·0 ml
Swine serum (from SPF pigs) inactivated	15·0 ml
Phenol red (0·06% (w/v) solution)	5·0 ml
Bacitracin (2·5% (w/v) solution)	1·0 ml
Meticillin (2·5% (w/v) solution)	1·0 ml

(b) *Solid medium*

Agar-Agar (Merck: art. 1613)	1·3 g
DEAE-dextran	0·017 g
Hanks's balanced salt solution (modified, see (c) below)	16·0 ml
(Sterilise by autoclaving 121 °C 2–5 min⁻¹)	

After cooling below 100 °C this solution is added to one portion of preheated FF74 broth.

(c) *Hanks's balanced salt solution (HBSS), modified*

The solution is prepared from two stock solutions, A and B.
(A) NaCl, 80·0 g; KCl, 4·0 g; $MgSO_4$, $7H_2O$, 1·0 g; $MgCl_2$, $6H_2O$, 1·0 g; dissolve in 400 ml water, add $CaCl_2$ (anhydrous) 1·4 g, and water to 500 ml.
(B) $Na_2HPO_4 \cdot 12H_2O$, 1·5 g in 400 ml water; add KH_2PO_4, 0·6 g and water to 500 ml.
HBSS: 25 ml of A is added to 400 ml of water, mixed with 25 ml of B and 50 ml of water.

(d) *Yeast extract FG*

Fleischmann pure dry yeast, type 2040	125 g
Double distilled water	750 ml

The suspension is heated to 37°C for 20 min, thereafter heated to 90–100°C for 5 min. After cooling, centrifuge at 100 × g for 30 min. The supernatant is dispensed in appropriate volumes and autoclaved at 115°C for 2–5 min. Stored at −20°C.

B. Media for biochemical tests

1. *Glucose fermentation*

   (a) *Basal medium: medium b₁*

Heart infusion broth (Difco) enzyme-treated see page 398	120·0 ml
Sterilise by autoclaving	
PPLO serum fraction (Difco)	1·0 ml
Deoxyribonucleic acid (Sigma, 0·2% (w/v) solution)	1·2 ml
Phenol red (0·06% (w/v) solution)	5·0 ml
Thallium acetate (1% (w/v) solution)	1·0 ml
Penicillin (20 000 IU ml⁻¹)	0·25 ml
Adjust pH to 7·8	

   (b) *Test medium: medium bg*

Medium b₁	128·0 ml
Glucose (50% (w/v) solution)	1·6 ml
Adjust pH to 7·8	

2. *Arginine hydrolysis*

   (a) *Basal medium: medium b₂*

   As b₁, except that pH is adjusted to 7·3

   (b) *Test medium: medium ba*

Medium b₂	128·0 ml
L-arginine (30% (w/v) solution)	4·25 ml
Adjust pH to 7·3	

3. *Urea hydrolysis*

   Test medium: medium SU (as A.3) with pH adjusted to 6·8

4. *Phosphatase activity*

Test medium: medium Bph	
Heart infusion agar (Difco)	74·0 ml
(Sterilise by autoclaving)	
Horse serum (heated at 60°C for 60 min)	20·0 ml
Yeast extract (25%) (Taylor-Robinson *et al.*, 1963)	5·0 ml
Sodium phenolphthalein diphosphate (1% (w/v) solution)	1·0 ml
Penicillin (20 000 IU ml⁻¹)	0·2 ml
Thallium acetate (1% (w/v) solution)	1·0 ml
Adjust pH to 7·8	

15

5. *Proteolytic activity (digestion of coagulated serum)*

Test medium: medium S$_d$

Heart infusion broth (Difco)	8·0 ml
(Sterilise by autoclaving)	
Horse serum	30·0 ml
Yeast extract (25%) (Taylor-Robinson *et al.*, 1963)	0·8 ml
Sterile water	1·2 ml

Adjust pH to 7·8

The medium is dispensed in 2 ml volumes into screw cap tubes and heated in a slanted position in flowing steam for 45 min.

C. Media for metabolism inhibition tests

1. *Glucose: medium BG*

Heart infusion broth (Difco)	90·0 ml
Horse serum (unheated)	20·0 ml
Yeast extract (25%) (Taylor-Robinson *et al.*, 1963)	10·0 ml
Deoxyribonucleic acid (Sigma 0·2% (w/v) solution)	1·3 ml
Glucose (50% (w/v) solution)	2·6 ml
Phenol red (60 mg/100 ml)	5·0 ml
Thallium acetate (1% (w/v) solution)	1·0 ml
Penicillin (20 000 IU ml^{-1})	0·25 ml

Adjust pH to 7·8

2. *Arginine: medium BA*

Heart Infusion Broth (Difco)	90·0 ml
Horse serum (unheated)	20·0 ml
Yeast extract (25%) (Taylor-Robinson *et al.*, 1963)	10·0 ml
Deoxyribonucleic acid (Sigma, 0·2% (w/v) solution)	1·3 ml
L-arginine (30% (w/v) solution)	4·25 ml
Phenol red (60 mg/100 ml)	5·0 ml
Thallium acetate (1% (w/v) solution)	1·0 ml
Penicillin (20 000 IU ml^{-1})	0·25 ml

Adjust pH to 7·3

3. *Urea: medium SU* (see Section A.3) with pH adjusted to 6·8

4. *Tetrazolium: medium BT*

Heart infusion broth (Difco)	90·0 ml
Horse serum (unheated)	20·0 ml
Yeast extract	10·0 ml
DNA (Sigma, 0·2% (w/v) solution)	1·2 ml
2, 3, 5-triphenyl tetrazolium chloride (1% (w/v) solution)	5·0 ml
Thallium acetate (1% (w/v) solution)	1·0 ml
Penicillin (20 000 IU ml^{-1})	0·25 ml

Adjust pH to 7·8

CHAPTER X

Methods in *Campylobacter*

UWE ULLMANN

Department of Microbiology, Institute of Hygiene, University of Tübingen,
Silcherstr. 7, 74 Tübingen, West Germany

I. INTRODUCTION

Infection with *Campylobacter fetus* has been familiar to veterinarians for more than 60 years. The organism was first described as a cause of infectious abortion in sheep and cattle by McFadyean and Stockman in 1913. The pathogenesis of campylobacteriosis in animals is largely known.

C. fetus subspecies *fetus* causes a veneral disease in cattle characterised by transient infertility, endometritis and abortion of infected cows. The organisms are carried on the prepuce of the bull, which does not show any symptom of disease. They are introduced into the cervicovaginal area at oestrus when cows are served by infected bulls. The bacteria are also pathogenic for guinea pigs, hamsters and embryonated chicken eggs. Not susceptible are rabbits, mice and rats. The organisms do not multiply in the intestinal tract of animals and man, but do cause human infections.

Campylobacter fetus subspecies *intestinalis* causes infectious abortion during late pregnancy and perinatal lamb abortion as well as sporadic abortion in cattle; it is transmitted orally. This subspecies can be isolated from faeces, bile, and blood of sheep, cattle, rabbit, guinea pig and mouse, as well as from the intestinal contents of sparrows, starlings, magpies, pigeons, blackbirds, chickens, and turkeys. The micro-organism also grows in the intestinal tract and gallbladder of man.

Campylobacter fetus subspecies *jejuni* can be isolated from the stomach contents and placentae of fetuses of aborted sheep. It is also detected in faeces of normal sheep, cattle, swine, goats and various birds. The organism produces gastroenteritis in man.

A review of literature and veterinary importance of *Campylobacter* sp. was published by Winkler and Ullmann (1973).

II. TAXONOMY

Up to a few years ago *C. fetus* was assigned to the genus *Vibrio* because of its morphological characteristics. The new taxonomic position is justified in consequence of the microaerophilic strictly aerobic metabolism, their failure to produce acid in media containing carbohydrates and the composition of their DNA with a $(G+C)$ content between 29 and 36 mol% (Sebald and Veron, 1963; Veron and Chatelain, 1973). *Campylobacter* constitutes the genus II of the family *Spirillaceae*, comprising two species, *C. fetus* and *C. sputorum* (Smibert, 1974). *C. fetus* is subdivided into three subspecies: *C. fetus* subspecies *fetus* (*Vibrio fetus* var. *venerealis*), *C. fetus* subsp. *intestinalis* (*Vibrio fetus* var. *intestinalis*) and *C. fetus* subsp. *jejuni* (*V. jejuni*, related *Vibrio* spp.).

III. MORPHOLOGY

C. fetus is a Gram-negative, non-sporogenous rod, characterised by short, comma-shaped bacteria in 48-h cultures (Fig. 1). In old cultures (96 h) long spiral filaments appear (Fig. 2). The curved rods are 0·2–0·5 μm wide and 1·5–5 μm long. They have a single polar flagellum (Fig. 3) and are very actively motile.

IV. BIOCHEMICAL CHARACTERISTICS

The metabolic activity of *C. fetus* is comparatively slight: there is no production of acid from carbohydrates, negative reactions for Methyl red, indole and Voges-Proskauer reaction; lysine, ornithine, and arginine are not decarboxylated, and malonate is not utilised. The most important morphological and biochemical properties are summarised in Table I.

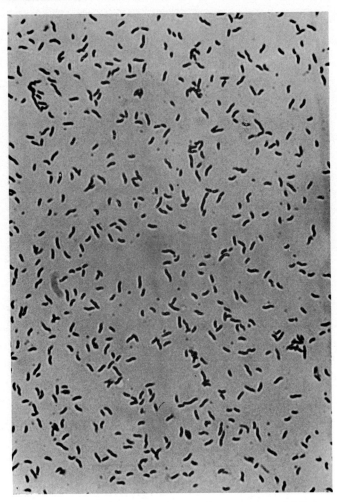

Fig. 1. *Campylobacter fetus* from 48-h liquid culture, comma and S-shaped forms (magnification × 1000).

V. ISOLATION OF *C. FETUS*

A. Material useful for isolation in animal and man

For diagnosis of infected bulls, samples of preputial secretions collected by pipette are helpful. Also suitable are specimens obtained by washing the preputial mucosa with 10 ml of thioglycollate medium (Oxoid, London), or washing out samples from the artificial vagina after removing sperm (Bisping, 1974). Sperm itself is less useful; only a small number of *Campylo-*

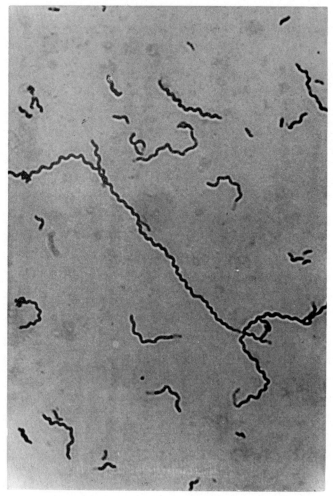

FIG. 2. *Campylobacter fetus* from 96-h liquid culture, long spiral filaments (magnification × 1000).

bacter can be expected because the testes are not infected. In bovine campylobacteriosis samples of faeces are the most frequently investigated, and after abortion specimens from placenta, chorion, or fetal stomach contents.

Samples for diagnosis of *Campylobacter* infection in man are listed according to frequency of hitherto existing isolation (Ullmann, 1976): blood cultures, cerebrospinal fluid, pus from abscess, synovial fluid, placental tissue, bile, pleural fluid, swabs from pericardium, content of ovarian cysts, fetus, swabs from cervix, urine, and faeces in cases of gastroenteritis.

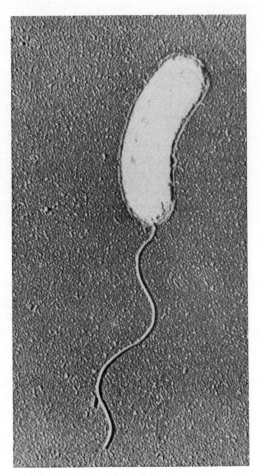

Fɪɢ. 3. Electron micrograph of *Campylobacter fetus* showing single polar flagellation (magnification × 15,000). (The author is grateful to Professor Dr R.-E. Bader (Tübingen) for permission to reproduce this figure.)

TABLE I

Some differential characteristics of subspecies of C. fetus

	C. fetus subsp. fetus	C. fetus subsp. intestinalis	C. fetus subsp. jejuni
Motility	+	+	+
Growth at 25°C	+	+	∅
Growth at 37°C	+	+	+
Growth at 42–45°C	∅	∅	+
Growth in 1% glycine	∅	+	+
Growth in 3·5% NaCl	∅	∅	∅
Sodium selenite-reduction	∅	+	+
Nitrate-reduction	+	+	+
H₂S lead acetate strips	∅	+	+
Hydrolysis of palmitic acid	+	+	?
Oxidase	+	+	+
Catalase	+	+	+
Carbohydrate fermentation	∅	∅	∅

B. Transport of specimen

In principle all samples should be investigated in the bacteriological laboratory within 6–8 h because of the oxygen susceptibility and the danger of overgrowing aerobic bacteria. The transport time can be prolonged by use of Stuart's medium or other substrates useful for anaerobic isolation such as Port-A-Cul tube for specimens on swabs, or Port-A-Cul vial for fluid specimens (BBL, Cockeysville, U.S.A.).

C. Treatment of mixed specimens

The isolation of C. fetus from samples that contain mixed flora requires application of selection methods: (1) the thioglycollate medium is diluted with the specimen 1:8 (v/v), homogenised with a vortex vibrator, the suspension is left to settle for 1 h to eliminate the large particles, the supernatant is centrifuged at $1500 \times g$ for 2–3 min; 4 ml of the surface liquid are sucked into a syringe for subsequent filtration through a Millipore filter with a mean pore size of 0·65 μm (Dekeyser et al., 1972; Bisping, 1974); the first 3 ml are discarded and some drops of the remaining suspension are inoculated on liquid and solid media. This technique is based on the fact that cells of C. fetus are sufficiently small to pass through filters that hold back other bacteria. (2) The mixed specimens are inoculated on media containing antibiotics ineffective against C. fetus, but killing concomitant micro-organisms.

The following antibiotic mixtures can be used. (a) Bacitracin 15 i.u./ml, Novobiocin 10 mg/litre, and Polymyxin 2 i.u./ml (Bisping et al., 1964).

(b) Bacitracin 25 i.u./ml, Polymyxin B 10 i.u./ml, Novobiocin 5 mg/litre, and Actidione 50 mg/litre (Dekeyser *et al.*, 1972). (c) Vancomycin 10 mg/ litre, Polymyxin B 2·5 i.u./ml, and Trimethoprim 5 mg/litre (Skirrow, 1977).

The antibiotics are added to liquid or solid media. Cefalotin discs (30 μg) may also be placed on solid media. Colonies growing within the inhibition zone suggest *C. fetus* (Ullmann, 1976).

(3) The Brilliant-green medium of Florent (1959), is used for mixed cultures also: fluid thioglycollate medium is supplemented with 10% defibrinated sheep blood plus 11·4 ml of 0·22% aqueous Brilliant-green solution/litre (Bisping, 1974).

D. Liquid and solid media

Useful for isolation are liquid media designed for isolation of fastidious bacteria, such as Tarozzi's liver broth, liver broth (CM78 Oxoid, London). Brewer's thioglycollate medium (BBL, Difco, Oxoid), brain heart infusion (BBL, Difco, Oxoid), tryptone soya broth (Oxoid) with addition of 0·05 g *p*-amino-benzoic acid per litre of medium (Ruckdeschel, 1973). The growth of *C. fetus* will be accelerated by addition of defibrinated cooked blood from sheep or ox to the following solid media: Columbia agar (BBL, Oxoid), brain heart infusion agar (BBL, Difco, Oxoid), D.S.T. agar (Oxoid,) *Brucella* agar (Albimi). Some investigators supplement these media with 5 or 10% defibrinated fresh blood. My own experience shows that some strains do not grow on fresh blood media but grow spontaneously on chocolate blood agar. Nutrient agar, Endo-agar, media for *Salmonella* and *Shigella* isolation, C.L.E.D. medium or other simple nutrient substrates are useless.

E. Incubation of cultures

The inoculated plates and tubes are incubated at 37°C in an anaerobic incubator in which 70% of the volume of air has been replaced by a mixture of 10% carbon dioxide and 90% nitrogen. Solid media can also be put into the GasPak system using the carbon dioxide generator; incubation following the Fortner method is very useful for isolation. Thereby one half of the solid medium is inoculated with an oxygen consuming bacterial strain (for example *Serratia marcescens*): on the other is spread the sample required for anaerobes. The Petri dishes are closed with paraffin or an airtight plastic tape (Leukoflex, Beiersdorf, Hamburg), and incubated upside down. The aerobic bacteria grow, consuming the oxygen and giving rise to a CO_2-enriched atmosphere for anaerobes. Sometimes the growth fails to appear in subcultures on solid media especially if the GasPak system has been used. In such cases, the subculture must be restarted from liquid culture perhaps also changing the anaerobic incubation.

Solid and liquid cultures are first incubated for three to five days. The liquid media show a slightly diffuse turbidity with a whitish streaky sediment. If the thioglycollate medium contains 0·2% agar, the micro-organisms grow in the upper third of the medium.

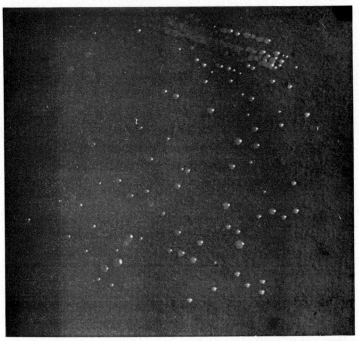

Fig. 4. Colonies of *Campylobacter fetus* subspecies *intestinalis* grown for 72 h on 5% cooked sheep blood Columbia agar (magnification × 1·5).

After three to five days on solid media the colonies are smooth, low convex, entire spherical and opalescent. Colony size varies between strains, ranging from a diameter less than 0·5 up to 2 mm (Fig. 4). Sometimes a slightly α-haemolytic zone appears around the colonies.

F. Immunofluorescent technique

For diagnosis of the *Campylobacter* carrier state in bulls examination of preputial washing by the fluorescent antibody technique is useful in addition to cultural methods (Philpott, 1966; Winter *et al.*, 1967). The immunofluorescent method may distinguish between *C. fetus* and *C. bubulus* (Bingöl and Blobel, 1970). Following the results of Winter *et al.* (1967), the fluorescent antibody technique or the cultural technique is satisfactory for routine detection of carrier bulls. But the presence of *C. fetus* cannot be ruled out by either test alone on the basis of a single negative result.

1. *Preparation of specimen*

The preputial washing is filtered through a coarse paper (Whatman No. 1) under slight negative pressure and the filtrate is centrifuged at $1000 \times g$ for 15 min. Then the supernatant fluid is centrifuged for 30 min at $12,000 \times g$ at 4°C. Smears are made from sediments, fixed, and stained with a fluorescent conjugate (Winter *et al.* 1967).

2. *The fluorescent conjugate*

Immunoglobulins against *C. fetus* subsp. *fetus* and subsp. *intestinalis* are obtained by immunisation of rabbits (Bingöl and Blobel, 1970). Sera possessing an agglutination titre of more than 1:1280 are pooled and the immunoglobulins are separated using a DEAE chromatography column (Bingöl *et al.*, 1969). Then 0·04 mg of fluorescein isothiocyanate per milligram protein are used for labelling. The free dye is removed by filtration on G25 Sephadex coarse (Philpott, 1966; Bingöl *et al.*, 1969); the conjugate is kept at -80°C.

G. Conservation of *C. fetus*

Tarrozi's liver broth is an exceptionally good substrate for short-term preservation of micro-organisms in the laboratory (Ruckdeschel, 1973). The cultures are incubated over four days at 37°C, closed with a rubber stopper and stored at room temperature. In this way the micro-organisms remain viable up to half a year. In the case of reisolation, 0·5 ml from the shaken-up cultures is added to liquid culture and incubated as described. For long-term storage, freeze-drying of cultures is the method of choice. After growth for 36 h on liquid media, the culture is collected with a pipette and suspended in sterile whole milk at a ratio of 1:2. This suspension is freeze-dried in 0·2 ml quantities and kept at 4°C. According to the investigations of Jakovljevic (1972), cultures have been viable after eight years. 0·5 ml of normal saline are necessary for rehydration of cultures; they are then inoculated into liquid media. Mixture of cultures with sterile horse serum 1:2 is also suitable for freeze-drying. Before dispatch of viable strains, they are inoculated on chocolate media and incubated for two days using the Fortner method.

H. Biochemical identification

The substrate for performance of biochemical tests is the thioglycollate medium without agar and dextrose or material of colonies. The following methods can be used.

1. *Catalase production* is tested on specimen holders by application of three to four drops of 3% solution of hydrogen peroxide into which material is rubbed from a single colony. Bubbling occurs immediately if the isolate is catalase positive.

2. *Oxidase* is proved by scraping some colonies from a pure culture and rubbing them on a filter paper; this is soaked with two or three drops of 1% aqueous tetramethyl-*p*-phenylene diamine dihydrochloride; blue colour within 10–20 sec indicates a positive reaction.

3. *Nitrate reduction* is investigated by adding 0·1% potassium nitrate to thioglycollate medium. After four days incubation dimethyl-α-naphthyl-amine reagent is dropped into the culture; a red colour complex indicates reduction to nitrite.

4. *Sodium selenite reduction* is examined on *Brucella* agar (Albimi) which contains 0·1% sodium selenite. The test medium is inoculated with a streak and incubated four days using the Fortner method or in an anaerobic incubator. A positive reaction is indicated by red pigmented colonies.

5. *Tween hydrolysis* is tested on D.S.T. agar: 40 g D.S.T. agar (Oxoid) + 0·1 g $CaCl_2 \cdot H_2O$ + 2·0 g $MgSO_4 \cdot 7H_2O$ is dissolved in 990 ml of distilled water and sterilised by autoclaving. Tween-40 (palmitic acid ester) or one of the other Tweens are autoclaved separately and then added to the base medium which has been cooled to 50–60°C and poured into plates. The test medium is inoculated with a streak and incubated for four days as described above. Hydrolysis of Tweens is revealed by speckled and cloudy zones around the colonies (Ullmann and Blasius, 1974).

6. *Hydrogen sulphide production* is demonstrated on thioglycollate medium with 0·02% cysteine; on the top of the cultures is placed a strip of filter paper impregnated with a solution of 10% basic lead acetate; blackening of the paper after four days indicates that the isolate produces H_2S.

7. *For the tolerance test* 1% glycine or 3·5% NaCl are added to thiogly-collate medium and multiplication is examined after four days incubation.

8. *Motility* is checked by the hanging drop method or in the S.I.M. medium after four days incubation.

9. *Temperature tolerance* is tested by simultaneous inoculation of three thioglycollate media; all tubes are placed in a water bath, one at 25°C, the other at 37°C and the third at 42°C. Growth is estimated after four days incubation.

VI. SEROLOGICAL IDENTIFICATION OF *CAMPYLOBACTER*

A. O Antigens

According to the investigations of Smibert (1970), the hydrolysates of whole cells of 50 strains of the three *C. fetus* subspecies contain small amounts of *meso*-diaminopimelic acid (DAP). The *meso*-DAP cannot be detected in purified cell walls. The carbohydrate composition of the cell walls of *C. fetus* subsp. *fetus* shows galactose and mannose, apart from one strain with galactose and glucose. Most strains of *C. fetus* subsp. *intestinalis* have galactose and mannose, some strains possess galactose and rhamnose; however, one strain had galactose alone and another galactose, glucose and mannose. Most of the *C. fetus* subsp. *jejuni* strains had galactose and glucose; two strains contained galactose, glucose and mannose; one strain galactose alone.

The species *C. fetus* has three thermostable antigens (O antigens) and a lot of different thermolabile antigens (K and H antigens) (Berg *et al.*, 1971; Mitscherlich and Liess, 1958a, b). The three O antigen serotypes correspond to the three subspecies. The nomenclature of the O antigens is confusing and inconsistent. Table II gives a synopsis of the most usual designations. The somatic antigens can be typed by slide agglutination, tube agglutination, indirect bacterial haemagglutination, and complement fixation tests. The same methods are used for detection of antibodies in animals and man.

1. *O antigen production*

For optimal production it is necessary to regenerate the strains by double passage on thioglycollate medium. During the logarithmic growth phase of the third subculture, chocolate media are inoculated with this population

TABLE II

Nomenclature used for O antigens of *C. fetus*

			Campylobacter fetus
Serotype II, III, V[a]	O-AG[b]	O-AG A[d]	subspecies *venerealis*
Serotype II, III, V[a]	O-AG 2[b]	O-AG B[d]	subspecies *intestinalis*
Serotype I	O-AG 13[c]	O-AG C[d]	subspecies *jejuni*

[a] Marsh and Firehammer (1953).
[b] Mitscherlich and Liess (1958a, b).
[c] Winkenwerder (1967).
[d] Berg *et al.* (1971).

and incubated as described. The bacterial growth from each Petri dish is harvested in 5 ml of 0·8% NaCl (pH 7·0), checked for purity and pooled (Bokkenheuser, 1972). For production of O antigens the samples are centrifuged at 4080 × g for 20 min, then the cells are resuspended in 0·85% saline solution and autoclaved for 2 h at 121°C. The cells are then removed from saline solution by pelleting (Berg et al., 1971). For use in slide agglutination tests, the antigen is diluted with saline solution to an optical density of 0·8 at 700 nm. For application in the tube agglutination test, the antigen suspension is adjusted to an optical density of 0·15 (Berg et al., 1971).

Extraction of antigen for the indirect haemagglutination test after Bokkenheuser (1972): the bacterial suspension is heated to 100°C for 2 h, cooled to room temperature, and centrifuged at 12,000 × g for 45 min. Two volumes of the supernatant fraction are treated with one volume of 0·25 N NaOH at 56°C for 30 min. The mixture is cooled to room temperature, adjusted to pH 7·2 with 0·25 N HCl, and centrifuged at 20,000 × g for 45 min. This antigenic extract occasionally caused clumping of the erythrocytes. Dialysis in Visking tubes, pore size 0·0024 μm, against 0·15 M NaCl at 4°C for 48 h removed the erythrocyte-damaging factor. Antigenetically weak solutions are concentrated in the same type of tubes by dialysis against a saturated solution of sucrose at room temperature. The purified product is stored at −12°C.

Also useful is phenol extraction after O'Neill and Todd (1961): acetone-dried cells are extracted for 3 h successively with 0·25 N trichloroacetic acid (10 ml/g dried bacteria) at 4°C. The suspension is then centrifuged, the sediment washed, and extracted with 45% phenol at 65°C for 1 h. The aqueous phase is separated and dialysed against distilled water, centrifuged and the supernatant freeze dried in 3 ml portions. For use the extract is dissolved in 1 ml 0·85 saline solution.

The antigen prepared for haemagglutination can also be employed for the complement fixation test in the Kolmer technique.

2. Alteration of O antigens and R forms

The somatic antigen can change its specific character. According to the results of Mitscherlich and Heider (1968) C. fetus is able to change its O antigen from type 1 to type 2 but not from type 2 to type 1. This phenomenon is observed in strains that are subcultured in live broth for months or years.

Rough strains are also found in viable strain collections. Then the growth is granulated in liquid media. They will become smooth after one or two subcultures in liver broth or on chocolate solid media containing 10–20% serum from rabbit or guinea pig without complement.

B. H antigens

1. *Preparation of H antigen*

The *C. fetus* strains are cultured as described for production of O antigens. The pooled saline suspensions are centrifuged for 20 min at 3000 rev/min and then for 30 min at 15,000 rev/min. The supernatant is discarded, the sediment is carefully resuspended and washed twice in phosphate buffered 0·8% NaCl (pH 7·2). Finally the suspension is placed into an ultramixer (Bühler, Tübingen) for 2 min to shear off the flagella from cell bodies. For sedimentation of intact cells and broken cell debris, the suspension must be centrifuged at 21,000 rev/min for 1 h. The flagella are found in the supernatant fluid (Keeler *et al.*, 1966; Ullmann, 1973). The purity of the material is investigated by electron microscopy. The flagella can be enriched by deep freezing 100 ml of the suspension at −40°C for 24 h. After thawing out the suspension, the flagella are clumped in the upper third of the liquid column. After centrifugation at 3000 rev/min, for 10 min, they can easily be separated (Ullmann, 1973, 1976). For agglutination tests, the flagella are resuspended homogeneously using a Vortex mixer.

Another possibility for purification of flagella is described by McCoy *et al.* (1975). Flagella are separated from cells by five cycles of low and high speed centrifugation, essentially as performed by Miwatani *et al.* (1970). The final product is devoid of cells but contains globular bodies. It is stored at 5°C with sodium azide (0·02%) as preservative. Flagella are further purified by isopycnic gradient centrifugation in CsCl by the method of Shapiro and Maizel (1973) at 25,000 rev/min, for two days at 22°C. Fractions are obtained by collecting drops either through a puncture in the bottom of the tube, or by upward displacement using an automatic system (ISCO Gradient Fractionator). CsCl is removed by dialysis.

2. *Ultrastructure of flagella*

Ultrastructural examination of the flagellum of *C. fetus* (McCoy *et al.*, 1975) shows a filament diameter exceeding 17·6 nm during the exponential growth phase. Filament diameters increase during the course of growth, reaching a mean width of 21·2 nm about the middle to late stationary phase. No evidence of a flagellar sheath is observed after different treatments (0·01 N HCl, 6 M urea, Tris buffer, warm water).

3. *The amino-acid composition*

The amino-acid composition of flagella from *C. fetus* subsp. *fetus* is different from that of *C. fetus* subsp. *intestinalis* (Table III). Cysteine is absent and proline present in traces only (Ullmann, 1976).

TABLE III

Amino-acid analysis of flagellar protein of *C. fetus*, in g amino-acid per 100 g protein (Ullmann, 1976)

Amino-acid	H antigen subspecies *venerealis*	H antigen subspecies *intestinalis*
Asp.	14·06	13·18
Thr.	6·46	5·22
Ser.	5·10	5·41
Glu.	8·93	10·94
Pro.	+	+
Gly.	4·92	5·05
Ala.	6·76	6·58
Cys.	Ø	Ø
Val.	6·49	7·03
Met.	1·33	1·27
Ileu.	5·67	6·56
Leu.	9·66	9·25
Tyr.	2·87	3·23
Phe.	5·10	6·41
Lys.	9·71	8·55
His.	2·88	2·51
NH$_3$	4·82	2·05
Arg.	4·04	5·51

4. *H antigen factors*

Using specific H sera from rabbits, agglutination occurs only with the homologous antigen if flagellar antisera have been absorbed with heterologous antigens. Investigations of ultrasonic break-down products of flagella with immunoprecipitation and immunoelectrophoresis show two common antigenic components: a and c and a partially common antigenic factor bb (Ullmann, 1973, 1976). In *C. fetus* subsp. *fetus*, McCoy *et al.* (1975) also found three H antigen factors after acid treatment of flagella, designated a, d, and e (Table IV).

TABLE IV

Formula of H antigen factors of *C. fetus*

Subspecies *fetus*	Subspecies *intestinalis*
a, c, bb$_1$	a, c, bb$_2$
a, d, e	

C. K antigen

Wiidik and Hlidar (1955) demonstrated that the capsular antigen may inhibit O agglutination of living or formolised *C. fetus* suspensions in homologous O serum. The presence of K antigen in *C. fetus* in this investigation is detected by (1) K agglutination with diluted K serum performed in the slide agglutination test, (2) the K agglutination with diluted K serum in the tube agglutination test, (3) the swelling reaction which is usual for pneumococci, and (4) by heating the cultures at 120°C for 2 h. It seems that this surface antigen appears as a layer of different thickness in various strains and that different strains vary only quantitatively in their K antigen content.

VII. PHAGES OF *C. FETUS*

C. fetus is lysogenic and carries temperate bacteriophage (Bryner *et al.*, 1968; Firehammer and Border, 1968).

A. Maintenance of phages

The fluid thioglycollate medium is useful for maintenance and enrichment of lysogenic strains of *C. fetus*. The method is described by Bryner *et al.* (1970): Broth for shaken cultures and phage dilution blanks is *Brucella* broth (Albimi) supplemented with 3 g sodium succinate and 0·1 g cysteine/litre (pH 6·9). *Brucella* agar (Albimi) is the nutrient base for double agar layer cultures; the soft top layer contains per litre of water: agar (Colab Ion agar no. 2) 5 g; NH_4Cl 1 g; KH_2PO_4 6·5 g; Na_2HPO_4 3·5 g; sodium succinate 3 g; glucose 1 g. The agar cultures are incubated as described.

B. Phage isolation (Bryner *et al.*, 1970)

For isolation of lysogenic strains of *C. fetus*, each test strain is inoculated in a separate tube containing 9 ml of thioglycollate medium and then 1 ml of the host culture is added to each tube. The mixed cultures are incubated for three days at 36°C and then centrifuged at $2500 \times g$ for 20 min to remove cells from unadsorbed phage in the supernatant fluid. This fluid is transferred to separate 100 ml flasks containing 5×10^9 log phase host cells in 20 ml supplemented Albimi broth (as described above) and incubated for 5 h. One millilitre samples from each flask are plated in overlay agar to test for plaque formation.

Bryner *et al.* (1970) isolated 22 bacteriophages from 38 strains of *C. fetus* and characterised the phage V-45 according to its stability in broth, growth kinetics, morphology, and physicochemical properties. The minimum latent period was 135 min, rise time was 75 min, and average burst size 35 plaque forming units per infected cell.

Chang and Ogg (1970) discovered a temperate bacteriophage designated VPF-11 that was capable of transducing the Streptomycin-resistance characteristic to strains of *C. fetus*. The transduction is possible between subsp. *fetus* and subsp. *intestinalis* strains and has a frequency of 4–5×10^{-6} transduction per infecting phage particle. The authors (Chang and Ogg, 1971) could also demonstrate the transduction of the glycine tolerance from a glycine-resistant donor to the glycine-sensitive recipient strain. Therefore, use of the glycine characteristic seems to be questionable for differentiation of subspecies *fetus* and subspecies *intestinalis*.

VIII. TESTING OF *C. FETUS* AGAINST ANTIBIOTICS

The disc methods are standardised for testing of rapid growing pathogens only. Micro-organisms that are fastidious in their nutritional requirments and need an anaerobic/microaerophilic atmosphere or increased concentrations of CO_2 for growth show a poor growth on Mueller–Hinton agar. Strain to strain variation in the growth rate is often observed. Following the recommendations of the National Committee for Clinical Laboratory Standards (1975), micro-organisms like *C. fetus* should not be tested by disc diffusion methods since the results cannot be properly interpreted. To use diffusion methods in the laboratory for testing of *Campylobacter* the Mueller–Hinton agar (Oxoid, BBL) must be supplemented with 10% defibrinated, cooked sheep or ox blood without antimicrobial activity.

The agar dilution method using the medium described above under appropriate incubation conditions is the method of choice for susceptibility testing of *C. fetus*.

Brown and Sautter (1977) describe detailed susceptibility testing of *C. fetus* subsp. *intestinalis* using the microtube dilution technique and Mueller–Hinton broth. This method can be effective in many isolates, but our experience shows that a number of strains, especially subspecies *fetus* do not grow in Mueller–Hinton broth. Most strains are resistant to Penicillin G, Cefalotin (including Cefoxitin), and susceptible to Clindamycin, Erythromycin, Ticarcillin, and Gentamicin (Butzler *et al.*, 1974; Chow *et al.*, 1977; Ullmann, 1978). Differences in sensitivity between the subspecies have not yet been noted.

IX. EPIDEMIOLOGY OF *C. FETUS*

The epidemiology of *C. fetus* in animals is generally understood. In man, however, there are epidemiological problems. In the literature, about 150 human infections are documented. Many isolations are not published and most infections are not diagnosed. Infections are described all over the world: Africa, America, Australia, Europe, India, Japan, Korea, etc. Seasonal fluctuations of infection cannot be demonstrated, nor differences in infection frequency in country or urban areas, or in various social classes (Ullmann, 1969). The majority of infections have been reported in males (57%) followed by women (32%) and, newborns or children (11%) (Bokkenheuser, 1970; Ullmann, 1975). Most human infections are caused by the subspecies *intestinalis*, several by the subspecies *fetus* (Bokkenheuser, 1970). Recently there have been increasing infections by the subspecies *jejuni*, especially in cases of gastroenteritis (King, 1957, 1962; Butzler *et al.*, 1973; Skirrow, 1977). People who have contact with domestic animals, farmers, veterinarians and slaughterhouse workers are not often infected more than others. Most reported cases have not had animal contact. The common features of all infections published are chronic debilitating disease or immunosuppression in the patients. Three possible modes of acquiring the infection are discussed: (1) direct contact with infected animals; (2) contamination of food, e.g. milk, vegatables, meat, especially liver (Soonattrakul *et al.*, 1971); (3) endogenous infection; there are some reports describing *Campylobacter* as a transient part of the normal flora of the gastrointestinal tract in man (Slee, 1972; Butzler *et al.*, 1977).

REFERENCES

Berg, R. L., Jutila, J. W. and Firehammer, B. D. (1971). *Am. J. vet. Res.* **32**, 11–22.
Bingöl, R. and Blobel, H. (1970). *Zbl. Bakt. Abt. I Orig.* **215**, 316–319.
Bingöl, R., Blobel, H. and Scharmann, W. (1969). *Zbl. Vet. Med. B* **16**, 799–807.
Bisping, W., Langenegger, J. and Winkenwerder, W. (1964). *Dt. tierärztl. Wschr.* **71**, 285–291.
Bisping, W. (1974). *Dt. tierärztl. Wschr.* **87**, 330–333.
Bokkenheuser, V. (1970). *Am. J. Epidemiol.* **91**, 400–409.
Bokkenheuser, V. (1972). *Infect. Immun.* **5**, 222–226.
Brown, W. J. and Sautter, R. (1977). *J. clin. Microbiol.* **6**, 72–75.
Bryner, J. H., Bermann, D. T. and Ritchie, A. E. (1968). *Bact. Proc.* **115**.
Bryner, J. H., Ritchie, A. E., Foley, J. W. and Berman, D. T. (1970). *J. Virol.* **6**, 94–99.
Butzler, J.-P., Dekeyser, P., Detrain, M. and Dehaen, F. (1973). *J. Pediat.* **82**, 493–495.
Butzler, J. P., Dekeyser, P. and Lafontaine, T. (1974). *Antimicrob. Ag. Chemother.* **5**, 86–89.
Butzler, J.-P., Dereume, J.-P., Barbier, P., Smekens, L. and Dekeyser, J. (1977). *Nouv. Presse méd.* **6**, 1033–1035.
Chang, W. and Ogg, E. (1970). *Am. J. vet. Res.* **31**, 919–924.
Chang, W. and Ogg, E. (1971). *Am. J. vet. Res.* **32**, 649–653.
Chow, A. M., Patten, V. and Bednorz, N. (1977). 17th Intersci. Conf. Antimicrob. Ag. Chemother, New York, Abstract No. 59.
Dekeyser, P., Gossuin-Detrain, M., Butzler, J.-P. and Sternon, J. (1972). *J infect. Dis.* **125**, 390–392.

Firehammer, B. D. and Border, M. (1968). *Am. J. vet. Res.* **29**, 2229–2235.

Florent, A. (1959). *Meded. Veeartsenijschool Gent* **3**, 1–60.

Jakovljevic, D. (1972). *Aust. vet. J.* **48**, 421.

Keeler, R. F., Ritchie, A. E., Bryner, J. H. and Elmore, J. (1966). *J. gen. Microbiol.* **43**, 439–454.

King, E. O. (1957) *J. infect. Dis.* **101**, 119–128.

King, E. O. (1962). *Ann. N.Y. Acad. Sci.* **98**, 700–711.

Marsh, H. and Firehammer, B. D. (1953). *Am. J. vet. Res.* **14**, 396–398.

McCoy, E. C., Doyle, D., Wiltberger, H., Burda, K. and Winter, A. J. (1975). *J. Bact.* **122**, 307–315.

McFadyean, F. and Stockman, S. (1913). Report of the Departmental Committee appointed by Board of Agriculture and Fisheries to inquire into epizootic abortion, Vol. 3. His Majesty's Stationery Office, London.

Mitscherlich, E. and Liess, B. (1958a). *Dt. tierärztl. Wschr.* **65**, 2–5.

Mitscherlich, E. and Liess, B. (1958b). *Dt. tierärztl. Wschr.* **65**, 36–39.

Mitscherlich and Heider, R. (1968). *Zbl. Vet. Med.* (*B*) **15**, 486–493.

Miwatani, T., Shinoda, S. and Fujino, T. (1970). *Biken J.* **13**, 149–155.

National Committee for Clinical Laboratory Standards (1975). Performance standards for antimicrobial disc susceptibility tests, pp. 1–11.

O'Neill, G. J. and Todd, J. P. (1971). *Nature, Lond.* **190**, 344–345.

Philpott, M. (1966). *Vet. Rec.* **79**, 811–812.

Ruckdeschel, G. (1973). *Arztl. Lab.* **19**, 23–27.

Sebald, M. and Véron, M. (1963). *Ann. Inst. Pasteur* **105**, 897–910.

Shapiro, L. and Maizel, J. V. (1973). *J. Bact.* **113**, 478–485.

Skirrow, M. B. (1977). *Br. med. J.* **2**, 9–11.

Slee, K. J. (1972). *Aust. J. med. technol.* **3**, 7–12.

Smibert, R. M. (1970). *Int. J. syst. Bact.* **20**, 407–412.

Smibert, R. M. (1974). *In* "Bergey's Manual of Determinative Bacteriology", 8th edn (R. E. Buchanan and N. E. Gibbons, Eds), pp. 207–212. Williams and Wilkins Co., Baltimore.

Soonattrakul, W., Andersen, B. R. and Bryner, J. H. (1971). *Am. J. med. Sci.* **261**, 245–249.

Ullmann, U. (1969). *Dt. med. Wschr.* **94**, 2399–2402.

Ullmann, U. (1973). Bakteriologische, biochemische und serologische Untersuchungen an vier *Vibrio-fetus*-Stämmen. Habilitationsschrift Universität Tübingen, pp. S.1–94.

Ullmann, U. (1975). *Zbl. Bakt. Abt. I Orig.* **230**, 480–491.

Ullmann, U. and Blasius, C. (1974). *Zbl Bakt. Abt. I Orig.* **229**, 264–267.

Ullmann, U. (1976). *Zbl. Bakt. Abt. I Orig.* **234**, 346–361.

Ullmann, U. (1978). *In* "Neonatale Infektionen" (C. Simon and V. von Loewenich, Eds) Klinische Pädiatrie.

Véron, M. and Chatelain, R. (1973). *Int. J. syst. Bact.* **23**, 122–134.

Wiidik, R. W. and Hlidar, G. E. (1955). *Zbl. Vet. Med.* **2**, 238–250.

Winkenwerder, W. (1967). *Zbl. Vet. Med.* **14**, 737–745.

Winkler, H. and Ullmann, U. (1973). *In* "Handbuch der experimentellen Pharmakologie". (Ed. O. Eichler), pp. 55–184. Bd. XVI/11B, Berlin, Heidelberg, New York, Springer.

Winter, A. J., Samuelson, J. D. and Elkana, M. (1967). *J. Am. vet. med. Ass.* **150**, 499–502.

Subject Index

A

Acholeplasma, 379 et seq.
  species, 381, 394
A. laidlawii, 387, 391
Acholeplasmataceae, 379 et seq.
Actinomyces, 287 et seq.
  antigens of, 315
    soluble, 314
  antiserum,
    production of, 293
    for speciating, 308
  biochemical reactions of, 291
  identification of, 290
    cell wall agglutination for, 313
    fluorescent antibody techniques for, 287–319
    in clinical material, 311
    in dental plaque, 312
    pure cultures, 310
  isolation of, 290
  production of fluorescent antibody reagents for, 292
  reactions with fluorescein isothiocyanate conjugated antisera, 298
  serological cross-reactions between *Arachnia* and, 300
  serotype specific fluorescein isothiocyanate conjugates for, 309
  species, fluorescent antigen reactions of, 299
A. baudetii, 288
A. bovis, 288 et seq.
A. eriksonii, 288, 289
A. humiferus, 288, 289
A. israelii, 288 et seq.
  serotypes, 301
A. meyerii, 288
A. naeslundii, 288 et seq.
  serotypes, 302
A. ondontolyticus, 288 et seq.
A. propionicus, 288 et seq.
A. suis, 288 et seq.
A. viscosus, 288 et seq.
  serotypes, 304

Actinomycetaceae, 288
actinomycosis, serological diagnosis of, 315
Anaeroplasma, 379 et seq.
Arachnia, 287 et seq.
  antigens, soluble, 314
  antiserum,
    production of, 293
    for speciating, 308
  biochemical reactions of, 291
  identification of, 290
    cell wall agglutination for, 313
    fluorescent antibody techniques for, 287–319
    in clinical material, 311
    in dental plaque, 312
    pure cultures, 310
  isolation of, 290
  production of fluorescent antibody reagents for, 292
  serological cross-reactions between *Actinomyces* and, 300
  serotype specific fluorescein isothiocyanate conjugates for, 309
A. propionica, 289 et seq.
  fluorescent antigen reactions of, 299

B

Bacillus, 5
Bacterionema, 288 et seq., 310
B. matruchotii, 289 et seq.
Bifidobacterium, 288 et seq., 310
Boticin, 9

C

Campylobacter, 435–451
  antigens,
    H, 447
    K, 449
    O, 445
  biochemical characteristics, 437